Fire Ants
and Leaf-Cutting Ants

About the Book and Editors

The 1985 Research Conference on Fire Ants and Leaf-Cutting Ants covered the most recent developments in research and control of these insect pests of the New World tropical and subtropical zones, the southern United States, South and Central America, and the Caribbean Islands. This volume contains chapters that discuss the history and economics, biology and ecology, behavior, pheromones and other semiochemicals, physiology, and biochemistry of fire ants and leaf-cutting ants, as well as current and future control strategies. The information provided illustrates past and present agricultural and medical problems associated with these pest ants; however, it also brings out the point that they may at times be beneficial. The chapters on basic aspects of the biology and ecology of the ants provide up-to-date information that is useful for a more complete understanding of their social behavior and the unique symbiotic relationship between leaf-cutting ants and their fungi. New approaches to control are illustrated by innovative research on anti-feedant chemicals from plants that prevent feeding by leaf-cutting ants. The present status of chemical baits and biocontrol is addressed, as well as the possibilities of future novel methods based on the use of anti-metabolites, insect hormones, behavior modifying substances, and species-specific toxic bait systems to create integrated pest management systems.

Clifford S. Lofgren is research leader and entomologist and **Robert K. Vander Meer** is research chemist for the Imported Fire Ant Unit, Insects Affecting Man and Animals Research Laboratory, Agricultural Research Service, United States Department of Agriculture.

Fire Ants
and Leaf-Cutting Ants
Biology and Management

edited by Clifford S. Lofgren
and Robert K. Vander Meer

Routledge
Taylor & Francis Group

LONDON AND NEW YORK

First published 1986 by Westview Press

Published 2018 by Routledge
52 Vanderbilt Avenue, New York, NY 10017
2 Park Square, Milton Park, Abingdon, Oxon OX14 4RN

Routledge is an imprint of the Taylor & Francis Group, an informa business

Library of Congress Cataloging-in-Publication Data
Main entry under title:
Fire ants and leaf-cutting ants.
  (Studies in insect biology)
  1. Fire ants. 2. Leaf-cutting ants. I. Lofgren,
Clifford S. II. Vander Meer, Robert K. III. Series.
QL568.F7F45 1986   595.79′6   85-15259

ISBN 13: 978-0-367-00840-6 (hbk)

ISBN 13: 978-0-367-15827-9 (pbk)

Contents

BIOLOGY AND ECOLOGY

BEHAVIOR, PHEROMONES, AND OTHER SEMIOCHEMICALS

PHYSIOLOGY AND BIOCHEMISTRY

CURRENT AND FUTURE CONTROL STRATEGIES

Foreword

Repeated episodes in the history of entomology have taught us that long-term effective control of pest species can be accomplished only through sound research. Often the studies that provide the definitive clue, the one that reveals a weak link in the life cycle and leads to efficient control techniques, are among those that were least expected as the research began. Measures such as male sterilization and juvenile hormone analogs seem in retrospect to have come, as it were, out of left field.

Fire and leaf-cutter ants are among the most intractable of insect pests, in part due to their highly efficient colony organization. Social life permits them to live deep underground and simultaneously to forage over large areas into the open. If you strike at the individuals doing the damage, you merely eliminate a few of the sterile colony members, rather like removing a bit of skin and blood from an offending mouse. The queen, as the highly fertile progenitrix of the colony, remains securely hidden in the soil. To find the magic bullet of safe, cost-efficient control requires some knowledge not only of ants as individual organisms but their colonies as organized societies.

Research on virtually every conceivable aspect of the biology of fire and leaf-cutter ants, entailing every level of organization, has moved far ahead during the past 20 years and now appears to be accelerating. The key species, including Solenopsis invicta, Atta cephalotes, A. sexdens, and Acromyrmex octospinosus, are well on their way to becoming paradigms of biological research, in the same league with malarial mosquitoes and migratory locusts. The present volume is notable for being the first synthesis of existing knowledge. It was contributed by a majority of the active researchers in the world, many of whom met for the first time in Gainesville. The conference organizers and editors, Clifford S. Lofgren and Robert K. Vander Meer, are to be congratulated for having made this event possible.

E. O. Wilson

Preface

The history of the biology and control of fire ants and leaf-cutting ants is long and extensive; however, research on these two ant groups has proceeded independently in the United States, Europe, and Latin America with limited exchange of information. Considering their economic and medical importance and the many common problems associated with developing control strategies, a joint conference to bring researchers of both groups together was desirable.

The 1982 meeting of the International Union for the Study of Social Insects, Boulder, Colorado, provided the forum for a direct exchange of ideas between researchers from both ant groups and brought into greater focus the need for a conference on fire ant and leaf-cutting ant research. From the Boulder, Colorado meeting, Dr. Alain Kermarrec spearheaded a drive to convene a joint colloquium at the I.N.R.A. Station de Zoologie et lutte Biologique, Petit-Bourg Guadeloupe, F.W.I. This effort was unsuccessful because financial support could not be obtained. The need, potential benefits, and momentum for a joint research meeting were great enough that the editors and other staff of the Imported Fire Ant Unit, Insects Affecting Man and Animals Research Laboratory, ARS, USDA, in cooperation with the Department of Entomology and Nematology, University of Florida took up the challenge to convene the conference at Gainesville, Florida, USA. An organizing committee composed of the editors, Dr. D. F. Williams (ARS), and Dr. P. G. Koehler (UF), established guidelines, raised funds, enlisted participants, developed a program, and made local arrangements. These efforts culminated on March 5-7, 1985 in a colloquium entitled, "Research Conference on Fire Ants and Leaf-Cutting Ants: Assessment of Basic and Applied Research for Improving Pest Management Strategies."

The Colloquium was composed of invited speakers grouped according to the broad subject areas indicated in our Table of

Contents.   One sad note was experienced when Dr. J. C. M. Jonkman of The Netherlands, an invited participant, passed away a few months prior to the conference.   In addition to the invited speakers, the meeting was attended by an impressive group of scientists who contributed significantly to formal and informal discussions.   The colloquium was a great success and the dual goals of exchange of research information and personal contacts between the researchers were achieved.

The information in the chapters that follow is of such diverse nature that the book should be useful to persons conducting basic and applied research and/or control programs on social insects, as well as fire ants and leaf-cutting ants. We hope the Colloquium and the published proceedings will form the foundation for a series of meetings between the two research groups.   The royalties from the sale of the book will be used in support of future meetings and/or publication costs of Attini, a newsletter on pest ants.

Clifford S. Lofgren
Robert K. Vander Meer

Acknowledgments

The editors wish to thank all of the persons who contributed to the successful publication of these proceedings. In particular, we thank Dr. Alain Kermarrec and Gerard Febvay, whose fortuitous visit to our laboratory in 1982 provided the initial impetus for the conference. We are indebted also to Drs. Malcolm Cherrett, P. E. Howse, and K. Jaffe for suggestions and encouragement in the development of the conference agenda. We thank Professor E. O. Wilson for his keynote address and the Foreword to this book. Diane Bowers and Renee Lofgren are thanked for the many hours they spent in typing and editing the manuscripts; Mrs. Bowers deserves an additional thanks for her diligence and patience in formatting the typescript of each chapter. The staff of our USDA Fire Ant Research Unit contributed to the editing process by reviewing and proofing papers. Dr. D. P. Wojcik provided invaluable editorial assistance as our consultant on references and scientific names. Dr. P. G. Koehler and Dr. J. L. Stimac of the University of Florida provided assistance in organizing the conference.

Each chapter was peer reviewed by two scientists. These reviewers are listed below and we thank them for their critiques.

C. T. Adams, F. A. Alvarez, C. S. Apperson, D. L. Bailey,
W. A. Banks, M. S. Blum, R. E. Brown, G. R. Buckingham,
D. A. Carlson, J. M. Cherrett, J. A. Coffelt, H. L. Collins,
D. J. C. Fletcher, D. A. Focks, B. M. Glancey, L. Greenberg,
J. J. Howard, P. E. Howse, K. Jaffe, D. P. Jouvenaz,
P. G. Koehler, J. N. McNeil, R. W. Mankin, M. E. Mispagel,
J. C. Moser, J. Nation, M. S. Obin, S. A. Phillips, S. D. Porter,
F. Punzo, T. E. Reagan, D. L. Silhacek, A. A. Sorensen,
J. C. Trager, J. H. Tumlinson, A. H. Undeen, T. J. Walker,
D. A. Waller, N. A. Weber, D. F. Wiemer, D. F. Williams, and
D. P. Wojcik.

Finally, we thank the congenial staff of Westview Press for making our sometimes frustrating job much easier every step of the way, including the final review of all manuscripts by their scientific advisor, Mike Breed.

EDITORIAL NOTE

The inclusion of accent marks in French, German, Portuguese, and Spanish words has not been possible due to the limitations set by the word processing system used in the preparation of camera-ready copy.

C.S.L.
R.K.V.

1

The Defining Traits of Fire Ants and Leaf-Cutting Ants

E. O. Wilson

Why do some insect species become major insect pests, while others teeter on the brink of extinction? Experience has shown that certain traits predispose species for life in man-made habitats. For example, the dominant cosmopolitan household ants, including the Pharaoh ant, Monomorium pharaonis, the dimorphic myrmicine Pheidole megacephala, and the dolichoderines Iridomyrmex humilis and Tapinoma melanocephalum, are distinguished by their resistance to drying, possession of multiple fertile queens, and swift, efficient colony emigration. Other pest ants have entirely different life cycles and physiological responses. Consider the carpenter ants of the Camponotus herculeanus group, the leading timber pests of the north temperate zones, and the giant Paraponera clavata of tropical American forests, most notable for its paralyzing sting, neither of which bears much resemblance to the cosmopolitan house pests. Obviously, there is no single syndrome that allows a confident prediction of which species will attain economic or medical significance.

Nevertheless, in order to cope with pests as serious as fire ants and leaf-cutting ants it is important to know them thoroughly—not just the details of their life cycle that might render them vulnerable to control measures but also their evolutionary history, world distribution, and major adaptive traits. In short, their total biology must be known, including the qualities that distinguish them from otherwise similar species. It is with this need in mind that I will offer a few observations on how these insects fit in among the ants as a whole.

At a very general level, I believe it important that fire ants and leaf-cutter ants are among the most phylogenetically advanced, highly organized of all ant taxa. Their colonies are notable for exceptionally large populations, deep earthen retreats that make the queens all but invulnerable to enemies, and trail systems along which food can be quickly retrieved over relatively long distances. Both kinds of ants originated in the New World tropics, probably South America, during the Tertiary period. Perhaps because they are bound so closely to the earth, neither group of species has been successful as tramps. With the exception of the fire ant Solenopsis geminata (which is still not to be ranked among the great tropicopolitan pests) they have limited dispersal ability, and their influence is limited to their native ranges and relatively few localities in the New World where they have attained a foothold with the help of human commerce. For the same reason—that is, they have yet to be tested—they must be regarded as a threat to substantial parts of the tropics and warm temperate zones. A colonization of any part of the Old World tropics by Acromyrmex or Atta might easily develop into an ecological disaster.

FIRE ANTS

Fire ants are members of Solenopsis, which is one of the largest ant genera in the world, comprising approximately 160 currently unchallenged species and a great many more in collections waiting to be described. Fire ants constitute a relatively small, exclusively New World assemblage within Solenopsis. There are at least the 12 species recognized in revisions by Wilson (1952), Snelling (1963), and Buren (1972), and the number could be as much as several times higher if research initiated by the late William F. Buren and being continued by James Trager continues to define sibling species in the manner suggested by Buren (who, in fact, suggested that the number is about 30).

Fire ants (traditionally, the subgenus Solenopsis) are distinguished from the great majority of other Solenopsis by their polymorphism, relatively large size, large eyes (30 ommatidia or more), and possession of second and third funicular joints at least 1.5X as long as broad (Creighton 1930). In a tribal-level revision, Ettershank (1966) synonymized the subgenus but gave no explicit reason. If this conservative move is accepted, it would be appropriate to refer formally to fire ant species as the "S. geminata group." I prefer, in contrast, to keep the subgenus as the more convenient label unless species are discovered that are clearly intermediate to other subgenera. A list of the known species is given in Table 1.

The genus Solenopsis as a whole is more clearly delimited than its constituent subgenera, being characterized by the possession in the worker of 10-segmented antennae (10- or 11-segmented in the queen) with sharply defined clubs; geniculate, two-segmented maxillary palpi; a cleft labrum; paired carinae on the clypeus that terminate in anteriorly projecting teeth; and a median clypeal seta (Ettershank 1966). More than 90% of the species are grouped in the subgenus Diplorhoptrum, loosely referred to as "thief ants": small in size, with reduced eyes, relatively thick funicular segments, usually monomorphic (but a few are weakly polymorphic, such as S. fugax), and usually lightly pigmented and soil-dwelling (but S. picta, at least, is darkly pigmented and arboricolous). A few but far from all species raid the brood of larger ant species (Holldobler 1973).

TABLE 1. Fire ants (Solenopsis sens. str.): Distribution of species.

Solenopsis aurea. Southwestern United States and California to northern Mexico.
S. blumi. Uruguay.
S. bondari. Brazil to Peru and Guianas.
S. gayi. Chile, Peru.
S. geminata. Tropicopolitan. In the New World, southern United States south through West Indies, Mexico, Central America, to Brazil, Bolivia, and the Galapagos.
S. interrupta. Argentina, northern Uruguay.
S. invicta. Rondonia and Mato Grosso, Brazil, to northernmost Argentina; introduced into southern United States.
S. quinquecuspi. Uruguay and adjacent regions of Argentina and Brazil.
S. richteri. Argentina, southernmost Brazil, Uruguay; introduced into southern United States.
S. saevissima. Guianas to Argentina and Paraguay.
S. tridens. Argentina, Brazil.
S. xyloni. California to Florida, United States

Solenopsis almost certainly arose in South America during the early or middle part of the Tertiary period. The maximum number of species groups and individual species now occur on that continent. In fact, a major adaptive radiation has occurred, whose products include granivores (geminata, granivora), a plethora of tiny subterranean forms (many Diplorhoptrum), social parasites ("Labauchena" daguerri), arboreal parabionts (parabiotica), an extreme littoral specialist (globularia), and still others. A second piece of evidence is the fact that the principal sister genus, Oxyepoecus, is entirely limited to South America.

Solenopsis is unknown in the Old World fossil record, including the rich and well documented Baltic amber which is usually regarded to be of early Oligocene age. I have recently been able to examine the first Solenopsis fossils of any kind. I discovered them in amber from the Dominican Republic, which is probably late Oligocene or early Miocene in age. The specimens, which are beautifully preserved, include a typical fire ant worker (Solenopsis subgenus Solenopsis), as well as several small "thief ants" (Solenopsis subgenus Diplorhoptrum). From this material we know that the genus had assumed its diagnostic anatomical traits and also undergone some degree of radiation no later than the early Miocene. Solenopsis in fact may be considerably older, because one contemporary Diplorhoptrum species is known from Tasmania (froggatti) and another from southern Australia (insculpta), but apparently none from the remainder of Australia. This pattern suggests the possibility of a trans-Antarctic invasion from South America and hence a quite early Tertiary origin for the genus as a whole. In any case, Solenopsis evidently reached the southern United States early, where a rich Diplorhoptrum fauna coexists with three native fire ant species. A similarly diverse Diplorhoptrum assemblage exists in Asia, Europe, and North Africa, presumably derived from stocks that spread across the Bering land bridge. Consistent with this interpretation is the steady decline of diversity southward across the Indo-Australian region, with only 3 species as yet having been discovered in the islands of the western Pacific and 8 in sub-Saharan Africa.

Why has Solenopsis been so successful in both geographic spread and local abundance? The truth is that with the exception of two thief ants (fugax, molesta) and two fire ants (geminata, invicta), we know almost nothing about the biology of this large genus. A substantial opportunity exists to address the Neotropical species as products of a major adaptive radiation, analyzing the entire assemblage with the aid of cladistic analysis and ecological theory.

In assessing the causes of the biogeographic advance or retreat of a group such as Solenopsis, it is customary to look for unusual or unique traits possessed by the group, as well as its most effective predators and competitors. The only truly distinctive qualities of which I am aware are those associated with the venom. The poisons of fire ants consist largely of piperidines (Brand et al. 1973; Blum and Hermann 1978), while those of at least one Diplorhoptrum, the South African "fire ant" S. (D.) punctaticeps, contain pyrrolidines and pyrrolines (Pedder et al. 1976). The burning sensation in human beings caused by the piperidines, rather like that induced by a match struck and held close to the skin, is the source of the fire ants' common name. As Holldobler (1973) has further demonstrated, Solenopsis fugax discharges a highly affective and long-lasting auxiliary substance from the poison gland that repels defending workers when the little thief ant is raiding other species. Judging from the behavior of other thief and fire ants I have observed, the same or a closely similar technique is employed to a widespread degree in the genus. However, a similar method is also used by Monomorium pharaonis (Holldobler 1973). The truth is that we know very little about the evolution of defense in myrmicine ants. The venom and defensive techniques of the great majority of genera remain to be studied, including a number of geographically restricted and relatively uncommon assemblages anatomically near Solenopsis, such as Brownidris, Chelaner, Diplomorium, Nothidris, and Tranopelta.

Whatever the particular physiological and behavioral mechanisms involved, I agree with Buren et al. (1974) that competition with other species of ants is a key factor in restricting the abundance of fire ants in South America. It seems likely that Pheidole is particularly important. This genus is enormously diverse, with over 200 described Neotropical species (and probably several times that number in actuality), and it is the most abundant of all ant genera in numbers of colonies and individuals (Wilson 1976a). The soldier caste of at least two species, Pheidole dentata (Wilson 1976b) and P. lamia (Buren et al. 1977) are specialized for defense against Solenopsis. It is entirely possible that Solenopsis invicta is kept within its present relatively modest range and low population densities in Brazil and Argentina by pressure from Pheidole and other ant enemies that have adapted to its presence during thousands of years of coevolution.

LEAF-CUTTING ANTS

Leaf-cutting ants comprise 24 known species of Acromyrmex (Table 2) and 15 of Atta (Table 3). They represent the apex of evolution of the exclusively New World tribe Attini, which in turn is the only ant group to have evolved the ability to cultivate fungus as food (Weber 1972, 1982). Besides this unique adaptation and the many peculiar behavioral and physiological changes associated with it, the Attini are distinguished anatomically from other ants by a combination of traits in antennal segmentation; a less-than-absolute tendency toward hard, spinose or tuberculate bodies; and a proportionately large, casement-like first gastric segment. It is reasonable to suppose that the Attini are monophyletic, having arisen once in South America and spread—with rapid diminution in numbers of genera and species—northward into the West Indies, Central America, Mexico, and the southern United States. Extinct but modern-looking species of Trachymyrmex (Baroni Urbani 1980) and Cyphomyrmex (Wilson, unpublished) have been recorded from the Dominican amber, establishing the origin of the Attini well back into the Tertiary and possibly as early as the Oligocene Period.

Like Solenopsis, the Attini have undergone an impressive adaptive radiation. Approximately 190 species are known, constituting 12 currently recognized genera. They include minute species that form small colonies on the floor of tropical forests (Mycocepurus, Myrmicocrypta), specialists of grasslands and the littoral zone (Mycetophylax), social parasites (Pseudoatta), and others. Most of the smaller attines grow their fungi either on insect excrement and carcasses or pieces of rotting vegetation. Mycetophylax is unusual in utilizing fragments of grass, while Trachymyrmex often collects flower pet ls.

The advance of greatest importance, however, is the very efficient utilization of almost all forms of fresh vegetation, including flowers, fruits, leaves, and stems, by the true "leaf-cutter" genera Acromyrmex and Atta. So unusual is this adaptation and so successful have been the ants possessing it that it can be properly called one of the major breakthroughs in animal evolution. The leaf-cutters consume more vegetation than any other comparable number of herbivorous species, including mammals, birds, and other kinds of insects. One consequence is that the species of Atta are the most important insect pests of the New World tropics.

TABLE 2. Leaf-cutting ants of the genus Acromyrmex: Distribution of species.

Subgenus Acromyrmex

Acromyrmex ambiguus. Argentina, Brazil.
Ac. aspersus. Argentina, Brazil, Peru, Colombia.
Ac. coronatus. Bolivia and Brazil to Costa Rica.
Ac. crassipinus. Argentina, Brazil, Paraguay.
Ac. diasi. Brazil.
Ac. disciger. Brazil.
Ac. gallardoi. Argentina.
Ac. hispidus. Argentina, Bolivia, Brazil.
Ac. hystrix. Guianas, Brazil, Peru.
Ac. laticeps. Bolivia, Uruguay, Brazil.
Ac. lobicornis. Argentina, Bolivia, Brazil.
Ac. lundi. Argentina, Bolivia, Brazil.
Ac. niger. Brazil. (= A. muticonodus of other authors in the
   proceedings)
Ac. nobilis. Brazil.
Ac. octospinosus. Mexico to northern South America,
   Guadeloupe, Cuba.
Ac. rugosus. Colombia to Argentina.
Ac. subterraneus. Brazil and Peru to Argentina.

Subgenus Moellerius

Acromyrmex heyeri. Argentina, Brazil, Paraguay, Uruguay.
Ac. landolti. Northern South America to Argentina.
Ac. mesopotamicus. Argentina.
Ac. pulvereus. Argentina.
Ac. silvestrii. Argentina, Uruguay.
Ac. striatus. Argentina, Bolivia, Brazil.
Ac. versicolor. Arizona, Texas (United States), northern
   Mexico.

TABLE 3. Leaf-cutting ants of the genus <u>Atta</u>: Distribution of species.

<u>Atta</u> bisphaerica. Brazil.
<u>A.</u> capiguara. Brazil, Paraguay.
<u>A.</u> cephalotes. Southernmost Mexico to Ecuador and Brazil;
    Lesser Antilles as far north as Barbados.
<u>A.</u> colombica. Guatemala to Colombia.
<u>A.</u> goiana. Brazil.
<u>A.</u> insularis. Cuba.
<u>A.</u> laevigata. Colombia to Guianas to Paraguay.
<u>A.</u> mexicana. Arizona (United States) to El Salvador.
<u>A.</u> opaciceps. Brazil.
<u>A.</u> robusta. Brazil.
<u>A.</u> saltensis. Argentina, Bolivia, Paraguay.
<u>A.</u> sexdens. Costa Rica to Argentina and Paraguay.
<u>A.</u> silvai. Brazil.
<u>A.</u> texana. Louisiana, Texas (United States).
<u>A.</u> vollenweideri. Argentina, Brazil, Bolivia.

A close examination of the processing of the vegetation by <u>Atta</u> cephalotes and <u>A.</u> sexdens reveals at least some of the reasons that fungus culturing of this particular kind has not occurred more frequently in evolution (Wilson 1980a, b; 1983a, b). First of all, the ants must be relatively large. Workers of the two <u>Atta</u> species with head widths below 1.4 mm have difficulty cutting even the softest leaves and petals (the energetically most efficient size is 2.2 mm). Second, the workers have to be polymorphic. Individuals larger than about 1.2 mm are evidently unable to care for the minute fungi within the nest. Consequently, the cultivation of the fungus entails a remarkable assembly-line operation, as follows. The medias (head width mode 2.2 mm) cut and retrieve the vegetation, smaller medias (1.6 mm) slice it into smaller pieces, still smaller ones (1.4 mm) degrade the pieces into small lumps, and then successively smaller minor workers (1.2 mm to 0.8 mm) place the lumps in the substrate, implant strands of fungus on fresh substrate, and care for the fungus as it proliferates. In addition, the ants use special procedures such as the recycling of cellulases and proteinases from the fungi (Boyd and Martin 1975a, b). Overall, <u>Acromyrmex</u> and <u>Atta</u> have travelled a long path in evolution by mastering the technique of gardening and then shifting to a substitute of fresh vegetation. Because they depend on a fungus to accomplish much of their initial digestion and by this means are able to by-pass the formidable array of terpenoids, alkaloids, and other defensive chemicals that deter most insect herbivores, the leaf-cutters have been able to exploit a very wide range of food plants, including most of the crop species grown in tropical regions.

8

ONE KIND OF SOCIETY MEETS ANOTHER

Fire ants and leaf-cutting ants are thus seen to be among the most advanced social insects that arose in the New World, most likely in South America when that continent was isolated from the remainder of the Western Hemisphere. During the past 10,000 years, a mere eyeblink in geological time, these insects have encountered the most advanced product of mammalian evolution from the Old World, Homo sapiens. Certain difficulties have arisen from this contact, with virtually all the losses occurring on the human side. In order to redress the balance, we need to learn a great deal more about the biology of our opponents, paying particular attention to the weak points that undoubtedly occur in their complicated social systems. Our great advantage in the years to come will be that we can think about these matters, while they cannot.

REFERENCES CITED

Baroni Urbani, C. 1980. First description of fossil gardening ants. Stuttgarter Beitr. Naturk. (B) 54: 1-13.

Blum, M. S., and H. R. Hermann. 1978. Venoms and venom apparatuses of the Formicidae: Myrmeciinae, Ponerinae, Dorylinae, Pseudomyrmecinae, Myrmicinae, and Formicinae. Handbook Exp. Pharmakol. 48: 801-869.

Boyd, N. D., and M. M. Martin. 1975a. Faecal proteinases of the fungus-growing ant, Atta texana. Properties, significance and possible origin. Insect Biochem. 5: 619-635.

Boyd, N. D., and M. M. Martin. 1975b. Faecal proteinases of the fungus-growing ant, Atta texana: Their fungal origin and ecological significance. J. Insect Physiol. 21: 1815-1820.

Brand, J. M., M. S. Blum, and H. H. Ross. 1973. Biochemical evolution in fire ant venoms. Insect Biochem. 3: 45-51.

Buren, W. F. 1972. Revisionary studies on the taxonomy of the imported fire ants. J. Ga. Entomol. Soc. 7: 1-26.

Buren, W. F., G. E. Allen, W. H. Whitcomb, F. E. Lennartz, and R. N. Williams. 1974. Zoogeography of the imported fire ant. J. N. Y. Entomol. Soc. 82: 113-124.

Buren, W. F., M. A. Naves, and T. C. Carlysle. 1977. False phragmosis and apparent specialization for subterranean warfare in Pheidole lamia Wheeler (Hymenoptera: Formicidae) J. Ga. Entomol. Soc. 12: 100-108.

Creighton, W. S. 1930. The New World species of the genus Solenopsis (Hymenop. Formicidae). Proc. Am. Acad. Arts Sci. 66: 39-151, 8 pls.

Ettershank, G. 1966. A generic revision of the world Myrmicinae related to Solenopsis and Pheidologeton (Hymenoptera: Formicidae). Australian J. Zool. 14: 73-171.

Holldobler, B. 1973. Chemische Strategie beim Nahrungserwerb der Diebameise (Solenopsis fugax Latr.) und der Pharaoameise (Monomorium pharaonis L.). Oecologia 11: 371-380.

Pedder, D. J., H. M. Fales, T. Jaouni, M. S. Blum, J. MacConnell, and R. M. Crewe. 1976. Constituents of the venom of a South African fire ant (Solenopsis punctaticeps). Tetrahedron 32: 2275-2279.

Snelling, R. 1963. The United States species of fire ants of the genus Solenopsis, subgenus Solenopsis Westwood with synonymy of Solenopsis aurea Wheeler. Occas. Papers, Bur. Entomol. Calif. Dept. Agric. 3: 1-11.

Weber, N. A. 1972. Gardening ants: The attines. Mem. Am. Phil. Soc. Vol. 92, 146 pp.

Weber, N. A. 1982. Fungus ants, pp. 255-363. In H. Hermann (ed.), Social insects, Vol. IV. Academic Press, New York. 385 pp.

Wilson, E. O. 1952. The Solenopsis saevissima complex in South America. Mem. Inst. Oswaldo Cruz 50: 60-68.

Wilson, E. O. 1976a. What are the most prevalent ant genera? Studia Entomol. 19: 187-200.

Wilson, E. O. 1976b. The organization of colony defense in the ant Pheidole dentata Mayr (Hymenoptera: Formicidae). Behav. Ecol. Sociobiol. 1: 63-81.

Wilson, E. O. 1980a. Caste and division of labor in leaf-cutter ants (Hymenoptera: Formicidae: Atta). I. The overall pattern in A. sexdens. Behav. Ecol. Sociobiol. 7: 143-156.

Wilson, E. O. 1980b. Caste and division of labor in leaf-cutter ants (Hymenoptera: Formicidae: Atta). II. The ergonomic optimization of leaf cutting. Behav. Ecol. Sociobiol. 7: 157-165.

Wilson, E. O. 1983a. Caste and division of labor in leaf-cutter ants (Hymenoptera: Formicidae: Atta). III. Ergonomic resiliency in foraging by A. cephalotes. Behav. Ecol. Sociobiol. 14: 47-54.

Wilson, E. O. 1983b. Caste and division of labor in leaf-cutter ants (Hymenoptera: Formicidae: Atta). IV. Colony ontogeny of A. cephalotes. Behav. Ecol. Sociobiol. 14: 55-60.

2
History of the Leaf-Cutting Ant Problem

J. M. Cherrett

Pest species of leaf-cutting ants are restricted to the two genera Acromyrmex and Atta and, like the whole tribe Attini (the fungus-growing ants) to which they belong, they are confined to the New World approximately between the latitudes 33°N and 44°S. There are 24 species of Acromyrmex and 14 species of Atta; the most recent comprehensive keys to their identification are those of Goncalves (1961) for the genus Acromyrmex (supplemented by Santschi (1925) for the few species not found in Brazil) and Borgmeier (1959) for the genus Atta. With some minor exceptions, all leaf-cutting ants are indigenous to the areas where they cause damage.

Their place of origin is a matter for debate; Weber (1972, 1982) has suggested the Amazon basin as the location, and there leaf-cutters would have used dicotyledonous plants. However, the southern tropics (20-35°S) are particularly species-rich in both genera, and Fowler (1982) has suggested that they may have originated here as grass-cutters in the savannas. Certainly pest problems in the species-poor areas of Central and North America are much less acute than they are in the south.

Leaf-cutting ants are selective in the plant species they exploit; 3 species of Acromyrmex and 3 of Atta have specialized in cutting monocotyledonous plants, whilst 14 Acromyrmex and 9 Atta are restricted to dicotyledons. Even within the plant group on which they specialize, a range of defence mechanisms have been recognized, which prevent exploitation of certain species. However, even in species-rich tropical rain forests, many plants are cut at some time, ranging from 50% (Cherrett 1968a) to 77% (Rockwood 1976) of the species growing in the area. This impressive degree of polyphagy, a prime feature of these insects, has been attributed to the evolution of the fungus-growing habit (Cherrett 1980). The amount of vegetation cut from tropical forests by Atta has been calculated from twelve studies to lie between 12 and 17% of leaf

10

production (Cherrett, in press, a) so that Wheeler (1907) described them as 'dominant invertebrates,' and Wilson (1982) as 'prevalent herbivores.'

PEST STATUS

The 38 Atta and Acromyrmex species are found over a large geographical area with some species specialized to exploit each of the angiosperm classes. The fact that all of these individual species have reputations as pests is attributable to their striking degree of polyphagy, and that they are capable of becoming dominant herbivores. The types and the cost of leaf-cutting ant damage have recently been reviewed by Pollard (1982) and Cherrett (in press, b). Agricultural and horticultural crops thought to be particularly vulnerable include citrus, cocoa, manioc, coffee, maize, and cotton. The importance of grass-cutters as competitors for grass with cattle in natural and semi-natural grasslands is still a matter for debate (Jonkman 1980, Robinson and Fowler 1982), but the spoil heaps of large Atta vollenweideri nests can occupy 3 to 4% of the pasture area in heavy infestations (Jonkman 1977). Old collapsed nests produce holes into which cattle may fall and drown, and newly planted Guinea and Buffalo grass pastures may produce patchy stands with weed colonization of the bare areas (Cherrett et al. 1974). Plantation forestry, particularly eucalyptus and pines, is badly affected by Atta defoliation, and establishment may depend on a reasonable degree of ant control (see Chapters 33 and 34).

In addition to cutting leaves, leaf-cutting ants will take a wide range of dried food stuffs such as cereal grains, flour, dried beans, and cattle meal. Consequently, whenever stored foodstuffs are accessible they are liable to be taken, since the ants readily cut through sacks, paper, and polythene bags (Cherrett, in press, b). Finally, when large Atta nests of 5 m^3 or more (Jonkman 1977) are built under the foundations of buildings, subsidence may result, and Bondar (1927) claimed that, in Bahia, repairs to between 300 and 500 buildings were attributable each year to nest collapses. Nogueira and Martinho (1983) report that nest collapse is often a cause of road subsidence, and in Brazil, they found between 19 and 63 nests per km along roadside verges.

As leaf-cutting ants cause so many types of damage, it is very difficult to calculate a global monetary figure for their economic impact, and as Cherrett (in press, b) has pointed out, the actual losses caused at the present time using current control techniques bear little relation to the potential losses that could result in the absence of control. An oft-quoted figure for the latter is 1000 million U.S. dollars per annum, an amount suggested by Townsend (1923) and never subsequently corrected for inflation. There is little doubt, however, that these ants are prime general pests of crops in tropical and subtropical America.

As they are indigenous pests, the history of leaf-cutting ant problems reflects the history of man's changing ecology, and in particular, the way he obtains food.

IMPACT ON HUNTER GATHERERS

For most of his time in the New World, man lived as a hunter and gatherer, taking whatever food he could find. As the unculti- vated vegetation he used had survived leaf-cutting ant attack, their direct impact on him would not be obvious. However, tropical forest vegetation which has evolved mechanisms to limit herbivore damage provides a scarce food supply for human beings. Leaf-cutting ants would be one among many factors determining the carrying capacity of the land, which supported only low populations of forest-living hunter gatherers. Man probably exploited leaf-cutting ants. Conconi (1982) has shown that in Mexico sexual leaf-cutting ants are eaten, and are very digestible and have a high protein content. Wheeler (1907) quotes Cobo as claiming that the Indians used the locked jaws of Atta soldiers as sutures to hold together the edges of wounds. The presence of the ants must, however, have been a great disincentive to storing surplus food from collecting trips.

IMPACT ON SUBSISTENCE FARMERS

Once man invented agriculture in the New World, some 7000 B.C. (Harris 1967), the pest potential of leaf-cutting ants must soon have become apparent. They were dominant, generalist herbivores which could not be killed with spears or arrows, could not be fenced out, and could not easily be frightened away. The seedlings of few crops survive for any length of time within 150 m of a large Atta nest, and without effective chemicals, such nests would have been virtually impossible to destroy. This situation seemed to breed a sense of hopelessness in the early colonial farmers, which is not fully appreciated today. Mariconi (1970) has collected some of the earliest references to leaf-cutting ants, which illustrate this as follows: "...if it were not for the deplorable plague of ants, Bahia could be called another Land of Promise"; "If there is not much wine in this land it is because of ants which can strip the leaves and fruit..." (de Sousa 1587); "In a word it is the worst scourge that farmers have..." (de Toledo Rondon 1788); "Either Brasil kills the sauva or the sauva will kill Brasil" (Saint Hilaire 1822); "...the sauva is with all certainty the one insect in our climate which causes the most loss to agriculture, and through it, farmers in our country will always be forced to work longer and at a heavy cost..." (Lame 1894). Initially, the only remedies available to farmers were the physical destruction of nests by (1) 'puddling'; i.e., flooding the nest surface with water and then churning the nest up by digging or dancing (Harrison et al. 1916), (2) preventing the ants gaining access to crops by protecting individual plants with circular ceramic troughs (ant pots) filled with water (Sampaio de Azevedo 1894), and (3) tying 'grass skirts' around tree trunks, so 'hiding' their foliage from scouting Atta workers (Belt 1874). The other reaction must have been to concentrate on growing only relatively resistant crops, and not bothering to plant the most susceptible ones (Cherrett, in press, b). Thus, Belt (1874) observed that Nicaraguan farmers did not bother to grow fruit trees in some areas where ant attack was severe, and Cherrett and Jutsum (1983) confirmed this observation from the dearth of citrus trees in parts of Guyana. This is simply an acceptance by farmers of the damaging impact of leaf-cutting ants. Being denied crops which farmers would like to grow is part of the hidden costs of these species.

Clearing forest, and replacing it with agriculture, changes the leaf-cutting ant species present (Cherrett 1981), and Haines (1978) has claimed that ant densities are normally higher in secondary than in primary vegetation, whilst Fowler and Haines (1983) found a reduction in ant frequency in the later stages of succession after disturbance in Paraguayan grasslands. This buildup of ants with agriculture has led to the suggestion that they were one factor favouring a short-term slash and burn system for exploiting neotropical forests, and it has been suggested that they were a contributing factor in the destruction of the Mayan Empire (Fautereau 1952).

IMPACT ON SCIENTIFICALLY-BASED AGRICULTURE

As the results of systematic scientific research were applied to agriculture, man's relations with leaf-cutting ants altered once again. Ways of controlling their depredations have become increasingly available, and chemists have played the major part in this. Mariconi (1970) has given a comprehensive account of the history of control methods, and the importance that was attached to finding suitable ant poisons can be judged by the holding of a prize competition in Brazil in 1920 to find the most effective chemicals and application techniques for killing large Atta sexdens nests. It drew 17 competitors (Anon. 1921). Since then, a wide range of effective techniques for killing large Atta nests have become available, particularly the use of the persistent organochlorine insecticides applied to nests either as liquids, dusts, or fogs, and some of the simpler fumigants such as carbon disulphide, hydrogen cyanide and methyl bromide (Pollard 1982; Cherrett, in press, b). Atta control, however, became a simpler matter once toxic baits were introduced. Initially, the toxicant used was aldrin; but later, it was replaced by the more effective toxicant mirex. For those who can afford these baits, Atta is no longer the limiting factor in agriculture that it once was. However, both mirex and the organochlorines are under attack on environmental grounds, and have been withdrawn from the USA and from some other countries. Unless they are replaced with effective alternatives, the pest potential of Atta could again become apparent.

Modern agriculture has also created problems. As Belt (1874) noted, introduced crop plants are often favoured by leaf-cutting ants and so are vulnerable to damage. The potential impact of Atta attack is made greater by (1) the widespread improvement of indigenous grasslands by reseeding with high-yielding grass species; (2) the replacement of indigenous forests by plantations of eucalyptus, teak, pines, or Gmelina; and (3) the expansion of citrus and cereal production. In some areas, changes in land use have encouraged the spread of species such as Atta capiguara and A. sexdens; both are 'followers of agriculture,' and difficult species to control. In some areas, species of Acromyrmex, such as Ac. landolti, a pest of grassland, and Ac. octospinosus, a general crop pest, have assumed a dominant position. As these species make small nests, they are much more difficult to find and spot-treat than Atta nests, and as nest densities of 6000/ha of Ac. landolti have been recorded in Peru (Cherrett 1981), they present a problem to which we have no immediate solution. To date, there is no evidence of pesticide resistance evolving in leaf-cutting ants, but these species shifts represent the development of an increasingly 'control resistant' ant fauna. The nature of the leaf-cutting ant problem has changed since man first entered their territory, and there is little doubt that this changing relationship will continue—most dramatically if he accidentally introduces them into the Old World in the way that he has into some Caribbean islands (Cherrett 1968b).

Much more research remains to be done because a convincing case could be made for regarding leaf-cutting ants as the most complex and evolutionarily advanced of all the insects, whilst Cramer (1967) considered their pest status, worldwide, comparable with that of locusts. To date, some 1250 articles have been written about Atta and Acromyrmex spp., which is in sharp contrast with the 10,000 locust references listed up to 1978 in Acridological Abstracts. In 1977, an international newsletter, Attini, was started to bring together for the first time people with an interest in leaf-cutting ants, and the present meeting is to be welcomed as evidence that international cooperation has now extended to other economically important groups of ants.

REFERENCES CITED

Anonymous. 1921. Servicio de inspeccao e defesa agricola Resumo do relatorio. Boletim de Agricultura 22: 319-329.

Belt, T. 1874. The naturalist in Nicaragua. E. Bumpus, London, 306 pp.

Bondar, G. 1927. A formiga sauva na Bahia. Correio Agricultura 5: 99-104.

Borgmeier, T. 1959. Revision der Gattung Atta Fabricius (Hymenoptera: Formicidae). Studia Entomol. 2: 321-390.

Cherrett, J. M. 1968a. The foraging behaviour of Atta cephalotes L. (Hymenoptera: Formicidae). 1. Foraging pattern and plant species attacked in tropical rain forest. J. Anim. Ecol. 37: 387-403.

Cherrett, J. M. 1968b. Some aspects of the distribution of pest species of leaf-cutting ants in the Caribbean. Proc. Am. Soc. Hort. Sci., Trop. Reg. 12: 295-310.

Cherrett, J. M. 1980. Possible reasons for the mutualism between leaf-cutting ants (Hymenoptera: Formicidae) and their fungus. Biol. - Ecol. Med. 7: 113-122.

Cherrett, J. M. 1981. The interaction of wild vegetation and crops in leaf-cutting ant attack, pp. 315-325. In J. M. Thresh (ed.), Pests, pathogens and vegetation. Pitman, Boston, London, Melbourne.

Cherrett, J. M. Leaf-cutting ants: Their ecological role, diversity and zoogeography. In H. Lieth and M. J. A. Werger (eds.), Tropical rain forest ecosystems, ecosystems of the world 14B, Elsevier, Amsterdam. (In press, a).

Cherrett, J. M. The economic importance and control of leaf-cutting ants. In S. B. Vinson (ed.), Biology, economic importance and control of social insects. (In press, b).

Cherrett, J. M., and A. R. Jutsum. 1983. The effects of some ant species, especially Atta cephalotes (L.), Acromyrmex octospinosus (Reich) and Azteca sp. (Hym.: Form.) on citrus growing in Trinidad, pp. 155-163. In P. Jaisson (ed.), Social insects in the tropics, Vol. 2. Universite Paris-Nord. 252 pp.

Cherrett, J. M., G. V. Pollard, and J. A. Turner. 1974. Preliminary observations on Acromyrmex landolti (For.) and Atta laevigata (Fr. Smith) as pasture pests in Guyana. Trop. Agric., Trinidad. 51: 69-74.

Conconi, J. R. E. de. 1982. Los insectos como fuente de proteinas en el futuro. Ed. Limusa, Mexico. 144 pp.

Cramer, H. H. 1967. Plant protection and world crop production. "Bayer" Pflanzenschutz, Leverkusen. 524 pp.

Fautereau, E. de. 1952. Etudes d'ecologie humaine dans l'aire Amazonienne. Lussand Freres, Fontenay-le-Comte. 47 pp.

Fowler, H. G. 1982. Evolution of the foraging behaviour of leaf-cutting ants (Atta and Acromyrmex), p. 33. In M. D. Breed, C. D. Michener, and H. E. Evans (eds.), The biology of social insects. Westview Press, Boulder, Colorado. 420 pp.

Fowler, H. G., and B. L. Haines. 1983. Diversidad de especies de hormigas cortadoras y termitas de tumulo en cuanto a la sucesion vegetal en praderas Paraguayas, pp. 187-201. In P. Jaisson (ed.), Social insects in the tropics, Vol. 2. Universite Paris-Nord. 252 pp.

Goncalves, C. R. 1961. O genero Acromyrmex no Brasil (Hym.: Formicidae). Studia Entomol. 4: 113-180.

Haines, B. L. 1978. Element and energy flows through colonies of the leaf-cutting ant Atta colombica, in Panama. Biotropica 10: 270-277.

Harris, D. R. 1967. New light on plant domestication and the origins of agriculture: A review. Geo. Rev. 57: 90-107.

Harrison, J. B., C. K. Bancroft, and G. E. Bodkin. 1916. The cultivation of limes. III. J. Board Agric., British Guiana 9: 122-129.

Jonkman, J. C. M. 1977. Biology and ecology of Atta vollenweideri: Forel 1893 and its impact in Paraguayan pastures. Thesis, Universiteits Bibliotheck, Leiden.

Jonkman, J. C. M. 1980. Average vegetative requirement, colony size and estimated impact of Atta vollenweideri on cattle-raising in Paraguay. Z. Angew. Entomol. 89: 135-143.

Mariconi, F. A. M. 1970. As sauvas. Editora agronomica "Ceres," Sao Paulo. 167 pp.

Nogueira, S. B., and M. R. Martinho. 1983. Leaf-cutting ants (Atta sp.), damage to and distribution along Brazilian roads, pp. 181-186. In P. Jaisson (ed.), Social insects in the tropics, Vol. 2. Universite Paris-Nord. 252 pp.

Pollard, G. V. 1982. A review of the distribution, economic importance and control of leaf-cutting ants in the Caribbean region with an analysis of current control programmes, pp. 43-61. In C. W. D. Brathwaite and G. V. Pollard (eds.), Urgent plant pest and disease problems in the Caribbean. IICA Office in Trinidad and Tobago, Misc. Pub. 378.

Robinson, S. W., and H. G. Fowler. 1982. Foraging and pest potential of Paraguayan grass-cutting ants (Atta and Acromyrmex) to the cattle industry. Z. Angew. Entomol. 93: 42-54.

Rockwood, L. L. 1976. Plant selection and foraging patterns in two species of leaf-cutting ants (Atta). Ecology 57: 48-61.

Sampaio de Azevedo, A. G. 1894. Sauva ou Manhuaara. Monographia como subsidio a historia da fauna paulista. Tipografia Diario Oficial, Sao Paulo. 74 pp.

Santschi, F. 1925. Revision du genre Acromyrmex Mayr. Rev. Suis. Zool. 31: 355-398.

Townsend, C. H. T. 1923. Un inseto de um bilhao de dollares e sua eliminacao. A formiga sauva. Almanaque Agricola Brasileiro, Sao Paulo. 12: 253-254.

Weber, N. A. 1972. Gardening ants: The attines. Mem. Am. Phil. Soc. 92: 1-146.

Weber, N. A. 1982. Fungus ants, pp. 255-363. In H. R. Hermann (ed.), Social insects, Vol. 4. Academic Press, London and New York.

Wheeler, W. M. 1907. The fungus-growing ants of North America. Bull. Am. Mus. Nat. Hist. 23: 669-807.

Wilson, E. O. 1982. Of insects and man, pp. 1-3. In M. D. Breed, C. D. Michener, and H. E. Evans (eds.), The biology of social insects. Westview Press, Boulder, Colorado. 420 pp.

3
Economics of Grass-Cutting Ants

H. G. Fowler, L. C. Forti,
V. Pereira-da-Silva, and N. B. Saes

Leaf-cutting ants of the genera <u>Atta</u> and <u>Acromyrmex</u> are characteristic polyphagous herbivores of the Neotropics, and in the tropical rain-forests of Central and South America, their activities have captured the attention of many researchers and naturalists since the early account of these ants by Belt (1874). However, not all leaf-cutting ants live in the tropical rain-forests or, for that matter, in the tropics. Many species colonize open habitats and rangelands of both the tropical and, much more commonly, the subtropical regions of South America. Twenty-seven of the 38 recognized species of leaf-cutting ants are found in the subtropics. Of these, ten species cut predominantly grasses, while another six cut dicotyledenous (dicot) plants, conifers, and grasses (Table 1).

Those species which harvest grasses for use as fungal substrate differ in behavior and morphology from those that use dicots as a fungal substrate (Fig. 1 and 2). Species which live in forests cut dicots almost exclusively, while species which live in open habitats cut either dicots or grasses, or both. For the species which cut dicots, the harvested vegetation is cut into smaller and smaller pieces until it is in the form of pulp which is then used as the substrate of the fungal garden. Consequently, the conversion of fungal biomass into ant biomass, starting from the plant base, is probably much higher in dicot-cutting than grass-cutting ants, as the latter do not heavily process the grass blades which they bring into the nest (Fowler and Robinson 1979a).

Correspondingly, morphological differences are evident in the two ant types. Species which harvest primarily grasses tend to have massive, short mandibles, while those species which harvest dicots tend to have longer, but less massive mandibles. Grass-cutters also tend to harvest vegetation differently (Fig. 1). In particular, they tend to have shorter metathoracic legs than leaf-cutters and, while dicot cutters use these legs as pivots while cutting, the grass-cutters do not.

TABLE 1. Taxa of leaf-cutting ants which inhabit open habitats or rangeland and type of plants used as fungal substrates.

Species	Type of plants harvested (grasses and/or dicots)
Atta	
capiguara Goncalves	grasses
bisphaerica Forel	grasses
vollenweideri Forel	grasses
opaciceps Borgmeier	both
goiana Goncalves	both
saltensis Forel	dicots
laevigata (F. Smith)	both
silvai Goncalves	both
sexdens piriventris	dicots
mexicana (F. Smith)	dicots
Acromyrmex	
landolti landolti (Forel)	grasses
landolti balzani (Emery)	grasses
landolti fracticornis	grasses
striatus (Roger)	grasses
gallardoi Santschi[a]	both (?)
versicolor Pergande[a]	dicots
pulvereus Santschi[a]	grasses
sylvestrii Emery[a]	grasses
lobicornis (Emery)	both
rugosus (F. Smith)	dicots
muticonodus (Forel)	dicots
lundi (Guerin)	dicots
diasi Goncalves	grasses
mesopotamicus Gallardo[a]	grasses
niger (F. Smith)[a]	dicots
heyeri (Forel)	grasses
disciger (Mayr)	dicots
hispidus Santschi	dicots
laticeps (Emery)	dicots

[a]Taxa of uncertain taxonomic status.

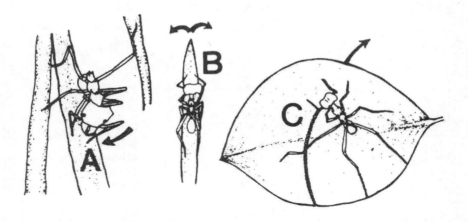

FIGURE 1. Cutting behaviors of leaf-cutting and grass-cutting ants compared. A: A. capiguara orients downward, cutting in one direction. When the grass blade is finally severed, it is transferred to the mandibles and the ant carries it to the nest. B: Cutting behavior of Ac. landolti fracticornis. The worker orients upwards and cuts the grass blade by alternating the force of its mandibles on either side of the blade. C: Cutting behavior of A. sexdens rubropilosa. Note mandibles are used like scissors with the metathoracic legs serving as a pivotal point while cutting.

Obviously, species which have specialized in the harvest of grasses may compete with cattle for forage (Bondar 1925) and even those species which are found in open habitats, but which cut dicots, may have a profound effect on the herbage dynamics of Neotropical rangelands. However, due to space limitations, we will restrict our coverage to those species which cut grasses. Grass-cutting species also harvest monocot crops, in particular, sugarcane.

The geographical distribution of grass-cutting ants is given in Fig. 3. The high concentration of grass-cutting species in the sub-tropics of South America is readily apparent; and not surprisingly, it is here where their current impact is the greatest. In fact, it is surprising to find that probably much more information exists on the impact of grass-cutting ants than dicot-cutting ants. This is not to say that the published information is definitive, as large discrepancies in estimations exist. Most of the research on grass-cutting ants has been performed by a small group of researchers, with each working practically in isolation. Available information, partially reviewed by Cherrett (in press) is limited, but we feel that our observations and synthesis of published data will accurately reflect the current state of knowledge about this group of interesting and important ants. For the sake of brevity, we have chosen to review each aspect of their economic importance separately.

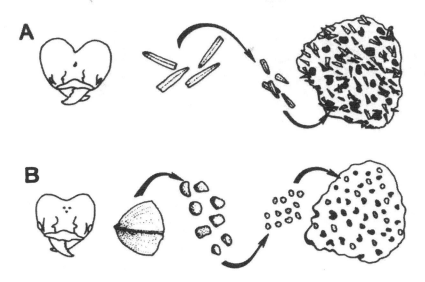

FIGURE 2. Differences in fungal substrate preparation between grass-cutting and leaf-cutting ants. Heads figured are A. robusta (A), a leaf-cutter, and A. bisphaerica (B), a grass-cutter. Note how vegetation is triturated for fungal substrate in leaf-cutters, while only smaller pieces are planted into the fungal garden of grass-cutters.

FORAGE CONSUMPTION AND ITS ESTIMATION

Colony size is a major problem in studying Atta and a few Acromyrmex species because individual colonies send out thousands of foragers, sometimes over an area of up to 1 ha (see Fowler et al., Chapter 11). Even so, since the late 1960s, a number of attempts have been made to estimate the consumption of forage by grass-cutting ants (Table 2). A. capiguara is considered the major pest species of the genus in the state of Sao Paulo in Brazil (Mariconi 1966a), in spite of the fact that it only cuts grasses. In 1972, it was estimated that in this state alone, 130 million dollars (U.S.) in damages were attributable to leaf-cutting ants, with the majority caused by grass-cutting ants (Echandi et al. 1972).

Methods of estimating forage consumption can be divided into three kinds: (1) counts of foraging workers, coupled with weights of the vegetation carried (activity); (2) calculation of the ratio of fresh fungal substrate to spent fungal substrate (conversion); and

22

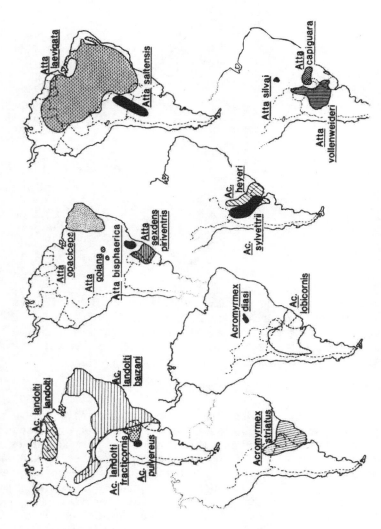

FIGURE 3. The geographical distribution of grass-cutting ants. There is a preponderance of grass-cutting taxa in the southern subtropics.

(3) exclusion plots, in which colonies have been killed for comparison with productivity in plots in which ants are present (exclusion). Each of these methods has been used to estimate forage consumption (Table 2), and each method has its own source of sampling error. These sources of error for the activity method are found in the diel and seasonal estimation of foraging activity, as well as corresponding changes in the types and weights of the vegetation harvested. Foraging intensity is also proportional to colony size. Each of these sources of error becomes magnified if it is extrapolated into yearly estimates. In conversion methods, a nest is generally excavated, and samples of the fungal gardens are taken to estimate the proportion of fresh to spent vegetation. This ratio is taken as fixed and extrapolated for the entire estimated fungal biomass present in the colony. No allowance has been made in published estimates for the decomposition of the spent refuse, and this method probably then tends to underestimate the consumption of forage by grass-cutting ants. The third method, exclusion, is probably the best economic check, yet it requires large sample sizes due to the patchiness of harvest by the ants. Other invertebrates and smaller vertebrates may also respond positively to the absence of ants, and the impact of the ants may be underestimated. These problems, as well as specific methodologies, are discussed in detail in Fowler and Forti (1985). Given these reservations, as well as the calculation of mean colony density levels for regional estimates, these methods, nevertheless, do provide some insight into the economic impact of these ants (Tables 2 and 3).

Amante (1967a, 1967b), using unknown methods, concluded that A. capiguara consumed tremendous amounts of grasses, from 255 to 639 kg of dry matter per colony per year (Table 2), which translates into losses of 512,000 to 870,000 head of cattle per year just in the state of Sao Paulo. Our reanalysis of his data (Fowler and Forti 1985) yielded estimates much lower (Tables 2 and 3), and it seems reasonable that colonies of A. capiguara consume in the neighborhood of 30 to 150 kg of dry matter/colony/year. Colonies of A. vollenweideri consume from 90 to 250 kg/colony/year and colonies of Ac. landolti from 0.4 to 2.2 kg/colony/year. Given these figures, the importance of colony density becomes apparent. Amante (1967a) reported colony densities of A. capiguara in Alta Sorocabana, Sao Paulo, of 18 colonies/ha in pastures five years old or less with a maximum observed density of 28 colonies/ha. Pastures five to six years old had a mean of 10 colonies/ha (Amante 1967a) which probably attests to the natural thinning of colonies through competition. Colony densities of A. vollenweideri probably never surpass five colonies/ha, but densities of Ac. landolti may reach up to 6000 colonies/ha (Table 2). These effects are best observed when regional estimates are performed (Table 3). Although estimates at the regional level are at best sloppy, they are necessary

TABLE 2. Estimated consumption of vegetable matter for various species of grass-cutting ants.

Species and reference[a]	Location	Colonies per ha	Consumption (kg/yr)[b] colony	Consumption (kg/yr)[b] hectare	Estimation methodology[c]
Atta					
capiguara[1]	Sao Paulo	10-18	638.7	6387-11497[d]	unknown
capiguara[2]	Sao Paulo	10-18	255.5	2555-4599[d]	unknown
capiguara[3]	Sao Paulo	10-18	72.5	725-1305	activity[e]
capiguara[4]	Sao Paulo	1-2	33.9	33.9-67.8	conversion
capiguara[4]	Sao Paulo	1-2	110.2	110.2-220.4	activity
capiguara[5]	Paraguay	8-14	55.6	4448-7784	exclusion
vollenweideri[6]	Paraguay	0.4	90.0	36.0	conversion
vollenweideri[5]	Paraguay	1-4	201-217	201-868	activity
vollenweideri[5]	Paraguay	1-4	231.0	231-924	exclusion
Acromyrmex					
landolti[7]	Guyana	94.9	2.2	208	activity
landolti[8]	Paraguay	4400-5850	1.5	6600-8775	activity
landolti[5]	Paraguay	4400-5850	0.82	3595-4779	activity
landolti[9]	Sao Paulo	200-500	0.42	84-210	activity

[a]References according to superscript number as follows: (1) Amante 1967a; (2) Amante 1967b; (3) Amante 1972; (4) Forti 1985; (5) Robinson and Fowler 1982; (6) Jonkman 1980b; (7) Cherrett et al. 1974; (8) Fowler and Robinson 1975; (9) Fowler, unpublished.

[b]Except where indicated, all values are in dry weight.

[c]Activity refers to grass brought into the nest; conversion refers to proportion of fresh to spent substrate; exclusion refers to fencing out cattle and eliminating grass-cutters by toxic baits (see text for details).

[d]Fresh weight.

[e]Based on data from Fowler and Forti (1985).

TABLE 3. Regional estimated damages caused by grass-cutting ants to pastures and sugar cane(*) in Brazil, Paraguay, and Guyana. Surface area has been estimated and calculations based on a consumption of 7 kg of dry matter/day for a 280 kg steer. See Table 2 for estimated nest consumption per day which was used to calculate the remaining estimates.

Species and reference[a]	Area (ha) [x 10^4]	Number of colonies per			Reduction in number of cattle or sacks of sugar(*)	
		ha	total area	animal unit	per ha	per total area
Atta						
capiguara[1]	66.25	13.0	8700000	10.0	1.30	870000
capiguara[2]	51.20	10.0	5120000	13.0	1.00	512000
capiguara[3]	66.25	13.0	8600000	35.0	0.38	251750[b]
capiguara[3]	51.20	10.0	5120000	35.0	0.29	148480[b]
capiguara[4]	1.00	14.0	1400000	46.0	0.30	30000
capiguara[5]	51.20	2.0	1024000	23.3	0.09	46080
capiguara[3]	52.40	2.34	1226160	sugar	6.30(*)	3300000(*)
vollenweideri[6]	800.00	0.40	3200000	28.3	0.01	113074
vollenweideri[4]	800.00	0.40	3200000	11.1	0.04	320000
vollenweideri[4]	800.00	0.40	3200000	12.7	0.03	240000
Acromyrmex						
landolti[7]	3.00	94.9	2847000	1166.7	0.08	2400
landolti[8]	60.00	2000.0	1.2×10^9	1750.0	1.14	684000
landolti[4]	60.00	2000.0	1.2×10^9	3500.0	0.57	342000
landolti[9]	51.20	500.0	2.6×10^9	6363.6	0.08	40960

[a]References according to superscript number are as follows: (1) Amante 1967a; (2) Amante 1967b; (3) Amante 1972; (4) Robinson and Fowler 1982; (5) Forti 1985; (6) Jonkman 1980b; (7) Cherrett et al. 1974; (8) Fowler and Robinson 1975; (9) Fowler, unpublished.
[b]Based on data from Fowler and Forti (1985).

for planning by Ministries of Agriculture and other government agencies as well as for marketing by pesticide companies.

Even though grass–cutting ants may consume a large amount of vegetation, especially when colony density is high, this does not necessarily imply that they are competing with cattle for forage. For example, using this method, Robinson and Fowler (1982) have shown that A. capiguara directly competes with cattle for grasses in eastern Paraguay, but A. vollenweideri does not in the Paraguayan Chaco (Fig. 4) (see also Jonkman 1980b). Competition can be assessed by noting that when both occur together, the total consumption is less than the sum of their individual consumptions when alone (Fig. 4, J. M. Frutos). If total grass consumption is near the sum of the individual consumptions (Fig. 4, Chaco), then competition with cattle is not confirmed, although it may occur.

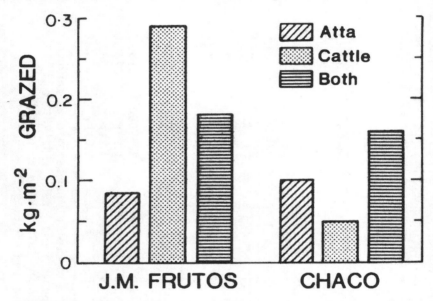

FIGURE 4. The interaction of grass–cutting ants and cattle in two Paraguayan ranching areas, based upon exclusion plots. In J. M. Frutos, competition between cattle and a mixed population of A. capiguara and A. laevigata was intense, while in the Chaco, little competition was evidenced between cattle and A. vollenweideri. Data from Robinson and Fowler (1982).

A. capiguara does not damage newly planted pastures, but serious damage does begin to show 2 to 3 years later (Milan-Neto and Groppo 1980), about the same time that mounds of excavated soil begin to appear in large numbers. Available evidence also suggests that A. capiguara is found only in sun-lit areas in both natural and improved pastures and along roadways but almost always in soils of high fertility (Mariconi 1966b). The same is probably also true of A. bisphaerica. A. vollenweideri is restricted to heavy clay soils (Carvalho 1976) and little, if any improved pastures are currently planted in these areas. In contrast, Ac. landolti does produce serious damage within two years of planting (Everts, unpublished; Fowler, unpublished), a fact that attests to its good colonization ability and high colony densities (see Jaffe, Chapter 34).

Without doubt, the consumption of forage by grass-cutting ants reduces the carrying capacity of natural and improved pastures under high colony densities, although it is hard to say just how much. Many more estimations need to be made at the local level to determine what effect they might have, as well as to determine at what level control becomes economical.

Given these reservations, it is probable under normal colony densities that A. capiguara does not reduce the number of head of cattle by more than 30% and generally not more than 10%/ha (Table 3). Sugar production is also affected proportionally. Other estimates, for example the regional estimates of the impact of A. vollenweideri, may be unrealistic. In the case of A. vollenweideri, the estimated losses are possibly much more than the Paraguyan market could absorb. These estimates, however, may be viewed as potential losses, while the realized losses are probably much less. In the case of A. capiguara, and possibly Ac. landolti, the differential between real and potential losses is probably much less.

SURFACE AREA LOST TO NESTING ACTIVITIES

As colonies of A. capiquara, A. bisphaerica, A. vollenweideri, A. goiana, A. opaciceps, A. sexdens piriventris and other species of Atta mature, the excavated soil rapidly begins to accumulate on the soil surface, resulting in lost area for grass production. Carvalho (1976) reported a mean surface area of 30 m^2/nest for A. vollenweideri in southern Brazil, while Bucher and Zuccardi (1967) reported surface areas of up to 40 m^2/nest in northern Argentina for the same species. Coupled with a zone of erosion around the nest, Jonkman (1980a) estimated that each mature nest of A. vollenweideri occupied up to 100 m^2, with 60 m^2 being due solely to erosion. Estimates of soil surface losses to A. capiguara are 600 m^2/nest in Santo Anastacio, Sao Paulo (Amante 1967a) with a mean for the state of Sao Paulo of 50 m^2/nest (Amante 1967a). In Tres Lagoas, Mato Grosso do Sul, a mean surface area of 71.5 m^2/nest was reported as lost to A. capiguara (Amante 1967a).

In areas of high colony density, it would be expected that mean colony surface area would be less although the summed nest surface areas might be larger. In all cases, grasses do not grow on the mound surface or in the immediate area of erosion. In the case of A. bisphaerica, this phenomenon is so pronounced that the common name of this species is "sauva mata pasto" or the grass-killing ant. As colonies of this species grow, the erosion of the mound surface and the fact that the ants cut the grass at ground level in expanding concentric circles from the mound, causes ever larger areas of soil to be devoid of vegetation.

Without worrying about erosion, and taking into account that 51% of the state of Sao Paulo is covered with pasture, then an estimated 80,000 ha of pasture are lost to the activities of A. bisphaerica and A. capiguara through their nesting activities alone. This assumes an average nest density of 1.2 colonies/ha (Amante 1967a). Coupling this estimate with estimates of forage consumption for the state of Sao Paulo, Gallo et al. (1980) estimated that the carrying capacity of pastures was reduced by 50%, producing an estimated damage of 51 dollars (U.S.)/colony/year (Nakano et al. 1981).

Given the tremendous nest densities of Ac. landolti that have been found in certain parts of their range, the nest surface area of all the colonies would occupy up to 10% of the total surface area. Nests, however, are much smaller and are not as barren. Nevertheless, the estimated forage area lost is probably on the same order of magnitude as that lost to the nesting of species of Atta.

ACCIDENTS WITH ANIMALS AND AGRICULTURAL MACHINERY

Collapsing nest chambers of colonies of A. capiquara, A. laevigata, A. opaciceps, A. saltensis, and A. bisphaerica can form large animal traps. During the year of 1965, Swift Packing Company of Rancharia, Sao Paulo, Brazil, lost 35 head of cattle due to accidents involving collapsed nests of A. capiguara (Amante 1967a). Animals are not the only victims, as costly accidents occur frequently with agricultural machinery (Amante 1967a; Carvalho 1976; Robinson and Fowler 1982).

PROLIFERATION OF WEEDS

The increase in the number of colonies of grass-cutting ants in a pasture is highly correlated with the increase in undesirable weed populations, especially Bidens spp., Sida spp., Tabernaemontana fuchsiaefolia and Solanum spp. (Amante 1967a; Fowler 1977). In the case of A. vollenweideri, nests collapse when they die and are rapidly colonized by woody vegetation, especially Prosopis spp. These form small forest nuclei which expand into the savanna thereby reducing available pasture acreage (Jonkman 1978), and they also probably serve as shelters for many ectoparasitic cattle pests (Fowler, unpublished).

LOSS OF SOIL FERTILITY AND LAND VALUE

Loss of fertility has been documented with respect to A. vollenweideri in Argentina. In these areas, cultivated plants do not develop properly when planted in areas previously occupied by A. vollenweideri (Bucher and Zuccardi 1967). This phenomenon is due to an inversion of the soil profiles caused by the nesting activity of this ant. In contrast, in the state of Sao Paulo, fields previously infested with A. capiguara, when plowed and planted, proved to be highly productive (Amante 1967a). A. opaciceps is also considered beneficial as an agent which promotes soil fertility in the Brazilian northeast (Sales et al. 1980). Nevertheless, large colonies of Atta greatly affect patterns of land usage. Much more work needs to be done to determine where, when and if grass-cutting ants are negative or beneficial soil fertility agents.

The presence of grass-cutting ants greatly diminishes the sale value of pasture lands as potential buyers are well aware of the problems and costs that would be necessary to fully utilize the land (Amante 1967a).

RIGHT-OF-WAY AND STRUCTURAL DAMAGE

Nogueira and Martinho (1983) documented the fact that colonies of Atta located in the right-of-way of highways cause parts of the highway to collapse and recommended systematic control of leaf-cutters in these areas. Sizeable highway damage was also documented in the Jau-Bauru region of Sao Paulo due to the nesting activities of A. capiguara (Fowler, unpublished).

BEHAVIORAL EFFECTS OF ANT FORAGING ON CATTLE

Robinson and Fowler (1982) noted that cattle and grass-cutters often compete for the same grass species in the same stage of development. Preliminary observations by Fowler and Saes (in press) suggest that cattle prefer to avoid areas in which grass-cutting ants are harvesting. In areas of high densities of grass-cutting ants, this behavioral inhibition may lead to less grazing time and a longer time span needed to fatten cattle for slaughter. This effect obviously could influence the grower's monetary returns as well as raise meat costs for the consumer.

PASTURE AREA LOST TO FORAGING TRAILS

Grass-cutters, like many leaf-cutters, maintain a network of foraging trails which are kept clear of vegetation. We may assume that in areas of high colony density, a large portion of the pasture surface area is lost to trail-making as maintenance and trail expansion activities continue. For example, up to 9 m^2 may be lost to each nest of A. vollenweideri and up to 0.7 m^2 to the trails of each nest of Ac. heyeri (Fowler 1978). In certain areas of higher nest densities, up to 50 m^2 may be lost per hectare due to the trails of these ants.

CHEMICAL CONTROL

Control of grass–cutting ants is conducted throughout the year with pesticides (Troppmair 1973). In a survey of 53 municipalities in the state of Sao Paulo, Troppmair (1973) found that land–owners always performed some type of control for leaf–cutting ants, preferentially with baits (67%) or liquid pesticides (21%). Of course, a major problem in conducting surveys was that many land–owners feared legal action by governmental agencies if their control method was in doubt (Troppmair 1973).

Currently, two efficient methods are employed for controlling grass–cutting ants in pastures, in particular A. capiguara and A. bisphaerica in the state of Sao Paulo. The first, and most common method, is the use of toxic baits, principally with mirex as the active ingredient. The second, and apparently most efficient method, is the use of thermal-fogging (Nakano et al. 1978).

Granulated toxic baits, generally based upon an active ingredient of aldrin, heptachlor, or mirex, are impregnated into an attractive matrix. In Paraguay, the Ministry of Agriculture produces and makes available at cost, grass–cutting ant bait using the cheapest active ingredient on the market. The chemical is then formulated into a bait using locally available forage grasses as matrix (Robinson et al. 1980). Even using standard commercial bait, A. capiguara is more costly to control. Mariconi et al. (1970) compared costs of controlling single nests of A. capiguara with those of A. laevigata, A. bisphaerica, and A. sexdens rubropilosa using Mirex 450 bait and found that each colony of A. capiguara cost 2.26 dollars (U.S.) to control, while each of the other species cost only 1.76 dollars (U.S.) per nest for the same level of control.

In the case of Ac. landolti, commerically-produced toxic baits have not provided satisfactory control. These formulations generally use soybean meal or citrus pulp as the bait matrix, products which are not attractive for this species (Labrador et al. 1972). Also, pelleted commercial baits are generally too large to be handled by these ants. Mirex is also rejected by this species (Fowler, unpublished). At present, the only viable control for Ac. landolti is the locally produced grass–cutting ant bait discussed previously (Robinson et al. 1980). Because of the high colony densities found in areas where this species is a problem, individual nest treatments are neither practical nor economical. A bait spreader has been tested with some success in Paraguay (Robinson et al.

1980), and aerial applications may prove to be the most economical (Gonzalez 1976) although the environmental impact of such an application is unacceptable (Lewis 1973).

Thermal-fogging with chlorinated insecticides allows for the injection of the insecticide in the form of minute particles into the nest with a preferred rate of 3.6 ml/m^2 of nest surface (Nakano et al. 1978). However, thermal-fogging is recommended for the hottest part of the year and is not commonly used. Nakano et al. (1978) found that for the control of colonies of A. capiguara using aldrin in a thermal-fogging unit, it took approximately 11.8 minutes to treat each colony using a mean of 202 ml of active ingredient; for the control of colonies of A. laevigata and A. sexdens rubropilosa, only 7.8 minutes were required per colony with a mean of 104 ml of active ingredient used per colony. Additionally, the costs of the equipment and a relatively large amount of equipment maintenance have probably contributed to a reduced usage of thermal-fogging.

Of all species of leaf-cutting ants, A. capiguara is one of the most difficult and expensive to control (Gallo et al. 1980). The fact that the internal nest structure of colonies of A. capiguara, probably the most complex of any species of Atta, remained unknown for many years was the major impediment in formulating adequate control measures against this species as the nest itself is not under the mounds of excavated soil (Mariconi 1966b). This fact demonstrates that much basic research is needed, even for applied objectives.

In Brazil, a rapid growth has been seen in the sale of formicides. In the first trimester of 1983, 950 metric tons with a value of 126,289 dollars (U.S.) were sold, while in the first trimester of 1984, this had increased to 2007 metric tons with a value of 415,929 dollars (U.S.) (ANDEF 1984). Of this, we estimate that at least 50% is used for the control of grass-cutting ants. Research is needed to investigate ways of reducing the amounts of formicides which are used so that the grower can produce his commodity at a lower cost. Nearly all formicides are imported into Latin America which requires much-needed hard currency.

ALTERNATIVE CONTROL STRATEGIES

Mariconi (1970) proposed various alternative methods for the control of A. capiguara. These suggestions were: (1) plowing and grading during the beginning of the annual dry season so pastures would be free of grass for a minimum of 120 days; and (2) changing land usage, especially through the planting of cotton, soybean, or other crops which are not monocots. Mariconi (1970) also suggested planting Eucalyptus in areas in which A. capiguara was a major problem since this species has a high shade tolerance.

Each of the alternatives proposed by Mariconi are marked by their own inconveniences and each necessarily entails high labor,

32

machinery and fuel costs. It does not seem practical to plow under an improved pasture to leave the surface area free of vegetation as the land thus has no return value and erosion is accelerated. Changing the type of land usage implies a radical change in the region's agricultural tradition which would be further aggravated with the cultivation of Eucalyptus, implying a much longer time to realize a return on investment.

Interestingly, pastures planted with Brachiaria decumbens have a lower nest density of Ac. landolti fracticornis than other improved pastures (Fowler and Robinson 1977). The same is true for A. capiguara (Forti, unpublished). The selection of resistant, or low-preference grasses, may be a viable alternative for cultural control of these ants. However, the other potential risks must be evaluated simultaneously. For example, B. decumbens is highly susceptible to spittle-bug attack, and the solution of grass-cutting ant problems may produce other, equally severe, problems for the grower.

IMPACT OF MAN ON GRASS-CUTTING ANTS

As more area is cleared for the production of cattle, primarily on the fringes of the Amazon basin, we anticipate more problems with some grass-cutting ants as has been found with Ac. landolti by Everts (unpublished). This species and A. laevigata and A. capiguara are good colonizers and have been found to expand their range rapidly when they are provided with an opportunity through changes in their habitat. Densities of colonies tend to be much higher in pasture range lands (Fowler 1977; Claver, unpublished), and the burning of certain pasture and savanna habitats may favor the establishment of larger colonies and higher densities of grass-cutting ants.

REFERENCES CITED

Amante, E. 1967a. Sauva tira boi da pastagem. Coopercotia. 23: 38-40.
Amante, E. 1967b. A formiga sauva Atta capiguara, praga das pastagens. O Biologico. 33: 133-120.
Amante, E. 1972. Influencias de alguns fatores microclimaticos sobre a formiga sauva Atta laevigata (F. Smith, 1858), Atta sexdens rubropilosa Forel, 1908, Atta bisphaerica Forel, 1908, e Atta capiguara Goncalves, 1944 (Hymenoptera: Formicidae) em formigueiros localizados no Estado de Sao Paulo. Ph.D. Thesis, Esc. Sup. Agric. "L. Queiroz," Piracicaba, Sao Paulo.
ANDEF, 1984. Aumenta a demanda de defensivos. Defesa Vegetal, Julho-Agosto. 84: 6.
Belt, T. 1874. The Naturalist in Nicaragua. E. Bumpus, London.

Bondar, G. 1925. Formiga "raspa," praga dos pastos. Bol. Path. Veg. Bahia. 2: 45–47.

Bucher, E. H., and R. B. Zuccardi. 1967. Significacion de los hormigueros de Atta vollenweideri Forel como alternadores del suelo en la provincia de Tucuman. Acta Zool. Lill. 23: 83–95.

Carvalho, S. 1976. Atta (Neoatta) vollenweideri Forel, 1908, no Brasil: Ocorrencia, aspectos externos e internos do sauveiro. L.D. Thesis. Univ. Fed., Santa Maria, Rio Grande Sul.

Cherrett, J. M. The economic control of leaf-cutting ants. In S. B. Vinson and J. K. Mauldin (eds.), Biology, economic importance, and control of social insects. (In press).

Cherrett, J. M., G. V. Pollard, and J. A. Turner. 1974. Preliminary observations on Acromyrmex landolti (For.) and Atta laevigata (Fr. Smith) as pasture pests in Guyana. Trop. Agric. 51: 66–74.

Echandi, E., J. K. Knoke, E. L. Nigh, Jr., M. Shenk, and G. T. Weekman. 1972. Crop protection in Brazil, Uruguay, Bolivia, Ecuador and Dominican Republic. USAID Report, Univ. Calif. Berkeley, 65 pp. (mimeo).

Forti, L. C. 1985. Ecologia da atividade forrageira da sauva Atta capiguara Goncalves, 1944 (Hymenoptera: Formicidae) e seu impacto em pastagem. Ph.D. Thesis, Esc. Sup. Agric. "L. Queiroz," Piracicaba, Sao Paulo.

Fowler, H. G. 1977. Some factors influencing colony spacing and survival in the grass-cutting ant, Acromyrmex landolti fracticornis (Forel) (Formicidae: Attini) in Paraguay. Rev. Biol. Trop. 25: 89–99.

Fowler, H. G. 1978. Foraging trails of leaf-cutting ants. J. N. Y. Entomol. Soc. 86: 132–136.

Fowler, H. G., and L. C. Forti. 1985. Methods and sources of error in the estimation of the amount of vegetation harvested by colonies of leaf-cutting ants (Hymenoptera: Formicidae). Rev. Brasil. Entomol.

Fowler, H. G., and S. W. Robinson. 1975. Estimaciones acerca de la accion de Acromyrmex landolti Forel (Hymenoptera: Formicidae) sobre el pastoreo y la ganaderia en el Paraguay. Rev. Soc. Cient. Paraguay. 15: 64–71.

Fowler, H. G., and S. W. Robinson. 1977. Foraging and grass selection by the grass-cutting ant, Acromyrmex landolti fracticornis (Forel) in habitats of introduced forage grasses in Paraguay. Bull. Entomol. Res. 67: 659–666.

Fowler, H. G., and S. W. Robinson. 1979. Foraging ecology of the grass-cutting ant, Acromyrmex landolti fracticornis (Formicidae: Attini) in Paraguay. Int. J. Ecol. Environ. Sci. 5: 29–37.

Fowler, H. G., and N. B. Saes. Behavioral inhibition of grazing by cattle by foraging grass-cutting ants (Atta) in the Southern Neotropics. Z. Angew. Entomol. (In press).

Gallo, D., O. Nakano, S. Silveira-Neto, R. P. L. Carvalho, G. Batista, E. Berti-Filho, J. R. P. Parra, R. A. Zucchi, and S. B. Alves. 1980. Manual de Entomologia Agricola. Ed. Agron. Ceres, Sao Paulo.

Gonzales, R. H. 1976. Plant protection in Latin America. PANS. 22: 26-34.

Jonkman, J. C. M. 1978. Nests of the leaf-cutting ant, Atta vollenweideri, as accelerators of succession in pastures. Z. Angew. Entomol. 86: 25-34.

Jonkman, J. C. M. 1980a. The external and internal structure and growth of nests of the leaf-cutting ant, Atta vollenweideri Forel, (Hymenoptera: Formicidae). Z. Angew. Entomol. 89: 158-173.

Jonkman, J. C. M. 1980b. Average vegetative requirement, colony size and estimated impact of Atta vollenweideri on cattle-raising in Paraguay. Z. Angew. Entomol. 89: 135-143.

Labrador, J. R., and I. J. Martinez. 1972. Acromyrmex landolti Forel plaga del pasto guinea (Panicum maximum) en el estado de Zulia. Rev. Fac. Agron. Univ. Zulia. 2: 27-38.

Lewis, T. 1973. Aerial baiting to control leaf-cutting ants (Formicidae: Attini) in Trinidad. II. Field application, nest mortality and the effect on other animals. Bull. Entomol. Res. 63: 275-287.

Mariconi, F. A. M. 1966a. Nova contribuicao para o conhecimento das sauvas do Estado de Sao Paulo. Bol. An. Esc. Sup. Agric. L. Queiroz. 23: 399-415.

Mariconi, F. A. M. 1966b. Novas informacoes sobre a sauva parda Atta capiguara Goncalves, 1944. Bol. An. Esc. Sup. Agric. L. Queiroz. 23: 1-8.

Mariconi, F. A. M. 1970. As Sauvas. Ed. Agron. Ceres. Sao Paulo.

Milan-Neto, A., and G. A. Groppo. 1980. Sauva Atta capiguara. Bol. Tec. CATI. 154: 1-4.

Nakano, O., J. M. A. Mendes-Filho, and K. A. Harawoto. 1978. Controle de tres especies de sauva atraves da tecnica de termobulizacao de Aldrin. Divulg. Agron. 44: 1-5.

Nakano, O., S. Silveira-Neto, and R. A. Zucchi. 1981. Entomoliogia Economica. Livoceres Ltda. Sao Paulo.

Nogueira, S. B., and M. R. Martinho. 1983. Leaf-cutting ants (Atta spp.) damage to and distribution along Brazilian roads, p. 181-186. In P. Jaisson (ed.), Social insects in the tropics, Vol. II. Universite Paris-Nord. Paris.

Robinson, S. W., A. Aranda, L. Cabello, and H. G. Fowler. 1980. Locally produced toxic baits for leaf-cutting ants for Latin America: Paraguay, a case study. Turrialba. 30: 71-76.

Robinson, S. W., and H. G. Fowler. 1982. Foraging and pest potential of Paraguayan grass-cutting ants (Atta and Acromyrmex) to the cattle industry. Z. Angew. Entomol. 93: 42-54.

Sales, F. M., L. Q. de Oliveira, N. G. Gomes, and V. P. O. Alves. 1980. A sauva do nordeste, Atta opaciceps Borgmeier, 1939, em areas de pastagem do Ceara. II. Acao sobre a fertilidade do soso. Fitossanidade. 4: 41–42.

Troppmair, H. 1973. Estudo zoogeografico e ecologico das formigas do genero Atta (Hymenoptera: Formicidae) com enfase sobre a Atta laevigata (Smith, 1858) no Estado de Sao Paulo. L.D. Thesis, Fac. Filos. Cienc. Letras, Rio Claro, Sao Paulo.

4

History of Imported Fire Ants in the United States

C. S. Lofgren

While hundreds of exotic insect species have found their way into the United States, it is doubtful that any have made their presence any more well-known than the red and black imported fire ants (IFA), Solenopsis invicta and Solenopsis richteri (=Solenopsis saevissima var. richteri). Their mound-building habits, voracious appetite, aggressive stinging behavior, and reproductive capacity have brought them into direct contact and conflict with humans at work and at play, and in urban or rural situations. It is no wonder then that during their relatively short sojourn of about 60 years in this country they have been the center of controversy. While the politics surrounding IFA is an interesting topic in itself, my goal in this paper is to review information on their introduction and spread, and the efforts that have been made to understand and control them.

The earliest records of the IFA are found in reports by Loding (1929), Creighton (1930), Smith (1949), and Wilson and Eads (1949). Their reports allow us to piece together the early history. The first collections of IFA were made in Mobile, Alabama by Loding (1929); they were identified later by Creighton (1930) as S. saevissima var. richteri. At that time the ants were limited to the northwestern part of Mobile and the nearby town of Spring Hill. Creighton (1930) reported that W. P. Loding, who was an amateur entomologist, estimated they were introduced into the Mobile area around 1918, possibly in ballast or dunnage discarded from ships. Loding believed that their early spread was hampered by the Argentine ant, Iridomyrmex humilis, another introduced species that occurred in large numbers in the same area. By 1931 they were found in three other small communities in Mobile County and the city of Fairhope in neighboring Baldwin County (Smith 1949). Six years later they were so abundant in Baldwin County and had caused so much concern that four County, State, and Federal agencies combined in an effort to control them with a calcium cyanide (48%) dust (Eden and Arant 1949). Approximately 2,000 acres of vegetable cropland were involved; over 80% extermination of active mounds was reported.

The onset of World War II apparently caused a temporary cessation of control and research on IFA; but shortly thereafter, a period of intense research and survey began. By this time, the IFA had spread to adjoining counties in Mississippi, and isolated infestations were found over 150 km away in Selma, Alabama, and Meridian and Artesia, Mississippi (Wilson and Eads 1949). This heralded the fact that the IFA had successfully established their foothold in Mobile and were being transported long distances by some unknown means. Research on their biology and control was initiated at Mississippi State University (Lyle and Fortune 1948) and Auburn University (Eden and Arant 1949). In 1948, the state of Mississippi made a $15,000 appropriation to begin a control and eradication program.

In 1949, the U.S. Department of Agriculture (USDA) made a "hurried" survey to determine the IFA distribution and established a research station at Spring Hill, Alabama (Smith 1949). Also in 1949, the Alabama Department of Conservation employed two scientists, E. O. Wilson, Jr. and J. H. Eads, to study the distribution, biology, and economic importance of IFA. Their report contains the first study of the classification, distribution, biology, and economics of IFA and documents their potential for economic damage to crops and wildlife (Wilson and Eads 1949). The first USDA survey report was also released in 1949 and revealed light to heavy IFA infestations in 14 counties in Mississippi, 12 in Alabama, and 2 in Florida (see Fig. 1). They suggested that individual queens or small colonies were artificially spread by car, rail, or air transportation or in commercial or other products. They recognized, but did not realize, the importance of IFA dispersal with nursery plants (Bruce et al. 1949).

A scientific problem also arose at this time concerning the occurrence of black and red forms of IFA. The original collections by Creighton (1930) were a deep brownish-black to blackish color with a reddish-yellow band at the base of the gaster. In the 1940s, however, another atypical form with a reddish color and a blackish gaster without the band was described by Smith (1949). Wilson and Eads (1949) also reported the two color phases and the apparent domination of the red form. In general, these scientists concluded that the black form was introduced first but did not become well-established in Mobile. The red form entered the picture, probably in the 1930s, and slowly began to dominate because of its greater adaptability to the prevailing environmental conditions (Wilson 1959). Wilson (1951) considered the two forms as races of a highly variable South American species. While there were continuing discussions about the significance of the two forms and their biological significance, no firm conclusions were drawn until the revisionary work of Buren (1972), which will be discussed later.

38

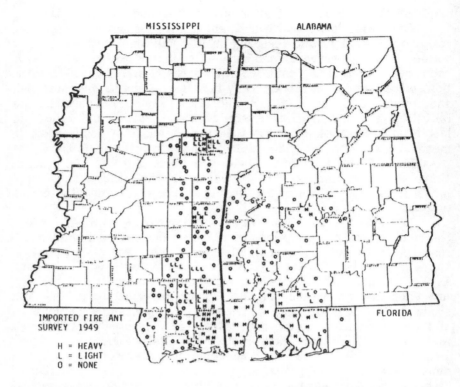

FIGURE 1. Map from Bruce et al. (1949) showing counties surveyed in Alabama and Mississippi and found non-infested (O) or heavily (H) or lightly (L) infested with imported fire ants.

With the continuing expansion of IFA infestations, research was accelerated at Auburn University, Mississippi State University, and a newly organized USDA laboratory at Spring Hill, Alabama. The USDA also initiated a full-scale survey to delimit spread of the IFA throughout the south. It was this survey that first brought to everyone's attention the extent of the IFA problem and the part that the sale of nursery plants played in the spread of IFA (see Fig. 2). In fact, once the direct link between nurseries and IFA spread became obvious, the survey was limited to nurseries because they were easy to locate and inspect. At the conclusion of the survey, the IFA had been found in 102 counties in 10 states (Alabama, Arkansas, Georgia, Florida, Louisiana, Mississippi, North Carolina, South Carolina, Tennessee, and Texas). Infestations had been located west as far as Houston, Texas; east and north to the Carolinas; and south to central Florida (Culpepper 1953; Fig. 2).

IMPORTED FIRE ANT INFESTATION
1949-1953

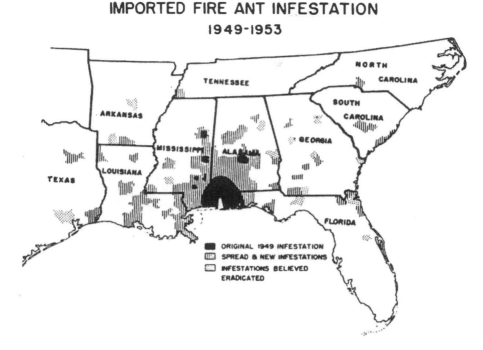

FIGURE 2. Distribution of imported fire ants in 1953 after an intensive 4-year nursery survey in all southern states in the U.S. (Culpepper 1953).

The extent of the IFA spread at this time was truly amazing considering that the red form, which occurred in all areas except northeast Mississippi and northwest Alabama, had only been known to occur in the United States for about 10 years. To date, no additional states are known to be infested and the small infestation in Tennessee was eradicated. In 1981, IFA were found in Puerto Rico (Buren 1982). The dramatic spread of the IFA was, undoubtedly, attributable to their high reproductive capacity (3 to 5 thousand queens per colony per year; Lofgren and Weidhaas 1972) and their propensity for invading soil with nursery plants. Lack of spread farther north is probably attributable to their limited cold tolerance (see Francke and Cokendolpher, Chapter 9). The southernmost infestations of the black form in Argentina are about 35° to 38° latitude which compares to the northernmost spread of either the black or red forms in the U.S. (Buren et al. 1974).

With their "beachheads" established throughout the south, the IFA were free to spread to surrounding areas by both manmade and natural means. Colony densities of 125 to 250 mounds per ha became common in prime pasture land. Intense concern spread among farmers as expressed by demands at the USDA Spring Hill Laboratory for over 5,000 information bulletins in one spring (Fig. 3). Finally, in 1957 the Southern Association of Commissioners of Agriculture passed a resolution recognizing the IFA as an economic pest. The resolution also petitioned the U.S. Congress, which was in session, to provide funds to the USDA to carry out a uniform control program without delay. On August 28, 1957, the U.S. Congress appropriated $2.4 million for IFA control and eradication (Canter 1981).

FIGURE 3. Demonstration to Alabama farmers by George H. Culpepper of imported fire ant control with chlordane drench. Picture from files of USDA in 1952.

In October of 1957, the Plant Pest Control Division, ARS, USDA, and the Southern Plant Board (an association of regulatory and control officials of the southern states) met in Memphis, Tennessee to develop guidelines for the program. The plans for control and eradication were based on the use of aerial or ground applications of granular heptachlor or dieldrin. In addition, a quarantine was proposed on shipment of sand and gravel, grass sod and nursery plants, stumpwood or timbers with soil attached, unless treated with chemicals to kill any associated ants. The quarantine became effective May 6, 1958 (ARS-USDA 1958).

At the same time the control program was initiated, it was recognized that a Methods Development Laboratory was needed to develop improved control techniques. Mr. W. F. Barthel (chemist) and I were selected to organize a USDA Methods Development Laboratory at Gulfport, Mississippi. Two primary goals were set: (1) find means to reduce the amount residual insecticide needed to achieve control and (2) develop a toxic bait.

Almost as soon as the Federal-State program started, problems were encountered. Some of the first applications of heptachlor (2.24 kg/ha) were made during the cold wet winter of 1957-1958. Shortly thereafter, mortality of wildlife, and in some cases cattle, were reported (George 1958). The actual truth of how much damage was due to insecticide and how much to the severe environmental conditions will never be known. The program continued during 1958; but in the spring of 1959, the dosage of heptachlor was reduced to 1.4 kg/ha and in February 1960, to two applications of 0.28 kg/ha spaced 3 to 6 months apart (Lofgren et al. 1961; Lofgren et al. 1965). Both changes were prompted by intense criticism by conservationists (Brown 1961) and were based on studies conducted by the Methods Development Laboratory. In 1960, another critical problem arose when FDA residue tolerances for heptachlor were reduced to zero on harvested crops (Canter 1981). This made the goal of eradication at that time impractical.

In 1961, the research efforts of the Methods Development Laboratory resulted in the formulation of mirex bait (Lofgren et al. 1963; Lofgren et al. 1964; Stringer et al. 1964). The bait consisted of the toxicant mirex dissolved in soybean oil and impregnated on corn cob grits. Various mirex concentrations and bait application rates were tested in 1961 and 1962. In 1963, the application rate was standardized at 2.8 kg/ha (8.4 g AI/ha); and in 1965, the rate was reduced to 1.4 kg/ha (4.2 g AI/ha). The switch to a bait caused a new problem—the lack of a chemical residue allowed the IFA to quickly reinfest the treated areas. This emphasized the need for repeated applications. However, the low application rate of mirex coupled with studies that revealed no apparent harm to wildlife relieved the immediate concerns of environmentalists about the control program (Baker 1963).

From 1964 to 1966, Federal funding decreased and it appeared that the Federal-State control program might be discontinued. However, the IFA continued to spread and to annoy and alarm farmers and urbanites alike. Consequently, the Southern Plant Board proposed to the USDA and Congress a stepped-up eradication effort using two to four applications of mirex bait. Since there was no research to verify the effectiveness of this treatment regimen for eradication, the Agricultural Appropriations Subcommittee of the United States Senate requested a study to evaluate its feasibility. Funds were transferred to the Insects Affecting Man and

Animals Research Laboratory, Gainesville, Florida, to initiate large-scale tests. Three different sites (Savannah, Georgia; Tampa-St. Petersburg, Florida; and Columbus-Starkville, Mississippi) were selected with the sizes varying from 103,600 to 862,750 ha. The results of the studies were published by Banks et al. (1973). They concluded that "we feel that technical problems we did encounter are surmountable and, therefore, total elimination of IFA from large isolated areas may be technically feasible."

While the eradication trials were in progress (1967-1970), applications of bait were being made in most states. During the 9-year period of 1967 to 1975, approximately 45,281,380 ha were treated (USDA, unpublished report). Since the treatment regimen involved three applications, the actual territory receiving mirex bait was about one-third this amount. Amazingly, the cost of some treatments was only $0.74 per ha including bait, aircraft, and the electronic guidance system.

All was not peaceful, however, during this time period. In the late 1960s, it became evident that even with an application rate of 4.2 g per ha, mirex residues were appearing in a variety of nontarget organisms (Markin et al. 1974; Spence and Markin 1974). It was soon obvious that mirex had a very disadvantageous characteristic—it biodegraded very slowly in the environment. This finding aroused the fears of environmental groups once more. In 1970, a court injunction to halt the use of mirex was requested by the Environmental Defense Fund (EDF) and the Committee for Leaving the Environment of America Natural (CLEAN) and filed in the U.S. District Court in the District of Columbia. Also in 1970, the U.S. Department of Interior banned the use of mirex on public lands they managed (Canter 1981). While the injunction was denied, a series of similar requests for injunctions were filed over the next few years; and in 1973, the newly formed Environmental Protection Agency (EPA) called for a public hearing to determine whether uses of mirex should be cancelled or amended. These hearings lingered on for 3 years. Finally, the Allied Chemical Corporation, the manufacturers of mirex bait, decided to discontinue its formulation. Their manufacturing plant at Prairie, Mississippi was sold to the State of Mississippi. A nonprofit agency (Mississippi Authority for Control of Fire Ants) operated the plant until 1977 when agreement was reached between this agency and EPA to cancel all mirex registrations (Johnson 1976). The public hearing on mirex initiated by EPA in 1973 was terminated and no judicial ruling was issued.

While all of the above activities were occurring, some very important research was underway. Dr. William F. Buren, a scientist with the U.S. Public Health Service, but with an avocation for ant taxonomy, began a study of the red and black forms of the IFA. He obtained specimens from their entire range in the U.S. and concluded that they were two separate species based upon (1) the lack of evidence for hybridization and (2) the lack of phenotypic variability (Buren 1972). He supported his conclusion with comparisons with similar specimens of the red form from the state of Mato Grasso, Brazil, and the black form from Uruguay and Argentina. In his revision (Buren 1972), he assigned the currently accepted scientific names: Solenopsis invicta Buren (the red imported fire ant) and Solenopsis richteri Forel (the black imported fire ant).

Buren's revision appeared to clarify the red and black form problem once and for all; but as has happened all too often during the long drama of the IFA versus man, his revision may not be the final word. Recent chemical data on venom, hydrocarbons, and trail pheromones give conclusive evidence for hybridization in the areas of Alabama and Mississippi where the two color forms interface (Vander Meer, Chapter 26; Vander Meer et al. 1985).

The early 1970s also saw a great upsurge in research activity on all phases of IFA biology, physiology, behavior, and control. This was spurred on by the one-time release of over $1 million by the USDA for cooperative research with university scientists. Over 16 separate research groups were involved in studies on everything from the toxicology of IFA venom, to the effects of juvenile hormones, and to the beneficial behavior of IFA as predators of pest insects. Several of the states, especially Texas, Mississippi, and Florida, provided funds to continue long-term research. With the demise of mirex in 1977, the USDA also expanded its research program on toxicants, insect growth regulators (IGRs), pheromones, biocontrol, biology, ecology, and economics. The current status of all of this research is reviewed in other chapters of this book. Bibliographies of research on IFA have been published by Banks et al. (1978), Wojcik and Lofgren (1982), and Wojcik (in press). A symposium on IFA was held January 1982 in conjunction with the Southeastern Branch Meeting of the Entomological Society of America, Mobile, Alabama. The papers presented were published in The Florida Entomologist, volume 66.

While all of the control efforts and research were making the IFA the most studied ant species in the world, the IFA went about "doing what they do best"—reproducing. In a period of about 60 years, they expanded from a small foothold at Mobile, Alabama to about 10,930,000 ha at the time the Federal-State program started in 1957 and over 93,120,000 ha in 1985 (Fig. 4 and 5). When the acreage estimates are plotted (Fig. 6), it is evident that the population increased dramatically from 1955 to 1980. More recently, the rate of expansion has declined since the IFA are reaching their ecological limits in the currently infested areas (see Francke and Cokendolpher, Chapter 9). Any additional major expansion could occur only if the IFA become established along the West Coast of the U.S.

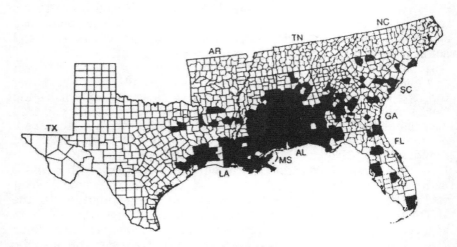

FIGURE 4. Distribution of imported fire ants at time of initiation of Federal-State Cooperative Imported Fire Ant Control Program in 1957.

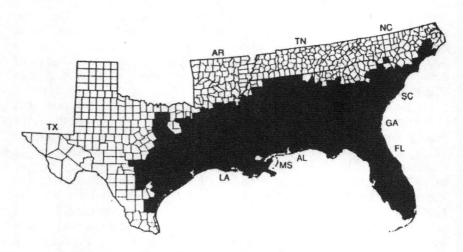

FIGURE 5. Distribution of imported fire ants in 1984.

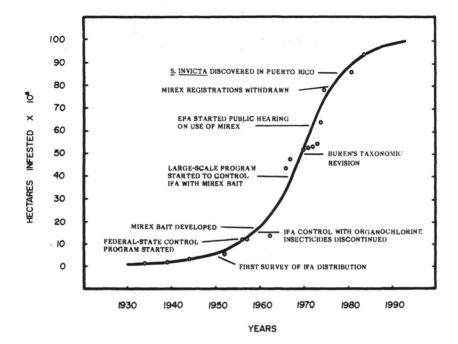

FIGURE 6. Graphic portrayal of the increase in area infested by IFA since 1935 with dates of significant events. Population estimates based on papers by Bruce et al. (1949), Culpepper (1953), Wilson (1951), and unpublished USDA reports.

Our conflicts with the IFA will continue for years to come. Their high reproductive capabilities, efficient foraging behavior, and ecological adaptability make it certain they will be here to perplex and harass us for years to come. It remains for us to develop means to live in accommodation with them. Hopefully, conferences such as this one will provide new insights for the achievement of this accommodation.

REFERENCES CITED

Agricultural Research Service, U.S. Department of Agriculture. 1958. Domestic quarantine notices. Subpart—imported fire ant quarantine and regulations.

Baker, M. F. 1963. New fire ant bait. Highlights Agric. Res. Ala. Agric. Exp. Stn. 10: 16.

Banks, W. A., B. M. Glancey, C. E. Stringer, D. P. Jouvenaz, C. S. Lofgren, and D. E. Weidhaas. 1973. Imported fire ants: Eradication trials with mirex bait. J. Econ. Entomol. 66: 785-789.

46

Banks, W. A., D. P. Wojcik, and C. S. Lofgren. 1978. A bibliography of imported fire ants and the chemicals and methods used for their control. U.S. Dep. Agric., Agric. Res. Serv. ARS-S-180. 25 pp.

Brown, W. L., Jr. 1961. Mass insect control programs: Four case histories. Psyche 68: 75-109.

Bruce, W. G., J. M. Coarsey, Jr., M. R. Smith, and G. H. Culpepper. 1949. Survey of the imported fire ant, Solenopsis saevissima var. richteri Forel. Spec. Rep. S-15 Bur. Entomol. Pl. Quar., U.S. Dep. Agric. 25 pp. (Unpublished).

Buren, W. F. 1972. Revisionary studies on the taxonomy of the imported fire ants. J. Ga. Entomol. Soc. 7: 1-26.

Buren, W. F. 1982. Red imported fire ant now in Puerto Rico. Fla. Entomol. 65: 188-189.

Buren, W. F., G. E. Allen, W. H. Whitcomb, F. E. Lennartz, and R. N. Williams. 1974. Zoogeography of the imported fire ants. J. N. Y. Entomol. Soc. 82: 113-124.

Canter, L. W. 1981. Cooperative imported fire ant programs—final programmatic impact statement. USDA, APHIS-ADM-81-01-F. 240 pp.

Creighton, W. S. 1930. The New World species of the genus Solenopsis (Hymenoptera: Formicidae). Proc. Am. Acad. Arts Sci. 66: 39-151, plates 1-8.

Culpepper, G. H. 1953. Status of the imported fire ant in the southern states in July 1953. U.S. Dep. Agric., Agric. Res. Admin., Bur. Entomol. Pl. Quar. E-867. 8 pp.

Eden, W. G., and F. S. Arant. 1949. Control of the imported fire ant in Alabama. J. Econ. Entomol. 42: 976-979.

George, J. L. 1958. The program to eradicate the imported fire ant. Report to the Conserv. Found. and the N. Y. Zool. Soc. 39 pp.

Johnson, E. L. 1976. Administrator's decision to accept plan of Mississippi Authority and order suspending hearing for the pesticide chemical mirex. Fed. Regist. 41: 56694-56704.

Loding, W. P. 1929. (An observation dated July 15, Alabama) U.S. Dept. Agr., Insect Pest Survey Bull. 9, p. 241.

Lofgren, C. S., V. E. Adler, and W. F. Barthel. 1961. Effects of some variations in formulation or application procedure on control of the imported fire ant with granular heptachlor. J. Econ. Entomol. 54: 45-47.

Lofgren, C. S., F. J. Bartlett, and C. E. Stringer. 1963. Imported fire ant toxic bait studies: Evaluation of carriers for oil baits. J. Econ. Entomol. 56: 62-66.

Lofgren, C. S., F. J. Bartlett, C. E. Stringer, and W. A. Banks. 1964. Imported fire ant toxic bait studies: Further tests with granulated mirex-soybean oil bait. J. Econ. Entomol. 57: 695-698.

Lofgren, C. S., C. E. Stringer, Jr., F. J. Bartlett, W. A. Banks, and W. F. Barthel. 1965. Dual low dosage applications of heptachlor for control of the imported fire ant. Fla. Entomol. 48: 265-270.

Lofgren, C. S., and D. E. Weidhaas. 1972. On the eradication of imported fire ants: A theoretical appraisal. Bull. Entomol. Soc. Am. 18: 17-20.

Lyle, C., and I. Fortune. 1948. Notes on an imported fire ant. J. Econ. Entomol. 41: 833-834.

Markin, G. P., H. L. Collins, and J. Davis. 1974. Residues of the insecticide mirex in terrestrial and aquatic invertebrates following a single aerial application of mirex bait—Louisiana—1971-72. Pestic. Monit. J. 8: 131-134.

Smith, M. R. 1949. Report on imported fire ant investigations. Unpubl. Rep. Bur. Entomol. Pl. Quar., USDA. 42 pp.

Spence, J. H., and G. P. Markin. 1974. Mirex residue in the physical environment following a single bait application—1971-72. Pestic. Monit. J. 8: 135-139.

Stringer, C. E., C. S. Lofgren, and F. J. Bartlett. 1964. Imported fire ant toxic bait studies: Evaluation of toxicants. J. Econ. Entomol. 57: 941-945.

Vander Meer, R. K., C. S. Lofgren, and F. M. Alvarez. 1985. Biochemical evidence for hybridization in fire ants. Fla. Entomol. 68: 501-506.

Wilson, E. O. 1951. Variation and adaptation in the imported fire ant. Evolution 5: 68-79.

Wilson, E. O. 1959. Invader of the South. Nat. Hist. 68: 276-281.

Wilson, E. O., and J. H. Eads. 1949. A report on the imported fire ant, Solenopsis saevissima var. richteri Forel in Alabama. Ala. Dep. Conserv. Spec. Rep., 53 pp., 13 plates (Mimeographed).

Wojcik, D. P. Bibliography of imported fire ants and their control: Second supplement. Fla. Entomol. 69: (In press).

Wojcik, D. P., and C. S. Lofgren. 1982. Bibliography of imported fire ants and their control: First supplement. Bull. Entomol. Soc. Am. 28: 269-276.

5

Agricultural and Medical Impact of the Imported Fire Ants

C. T. Adams

AGRICULTURE

The earliest suggestion of economic damage by imported fire ants (IFA), Solenopsis invicta and S. richteri (=S. saevissima var. richteri), was made in the spring of 1935 when a corn crop was damaged at Fairhope, Alabama; however, it was not reported until more than ten years later by Eden and Arant (1949). They also said damage to several other vegetable crops in 1937 was of sufficient concern that local, state, and Federal agencies attempted control of the ants with cyanogas dust on 2000 acres. Later, Lyle and Fortune (1948) concluded that IFAs were a major crop pest in Mississippi. Wilson and Eads (1949) conducted surveys in Baldwin and Mobile counties in Alabama and found that IFAs fed on corn, peanut and bean seeds, and the roots, stems and occasionally leaves of corn, beans, Irish potatoes, sweet potatoes and cabbage. Crop damage for these two counties for 1948 was estimated to be in excess of $500,000.

The absence of data on the economic impact of IFA for the ensuing 20 years may be correlated with the widespread use of chlorinated hydrocarbon insecticides to control a number of soil and crop insects. Though the use of these chemicals was cancelled by the Environmental Protection Agency during the 1960s, the residue remaining in the soil was probably not depleted until the mid-1970s. Thus, a resurgence of IFA populations in agricultural lands was not apparent until the late 1970s (Lofgren et al. 1975; Adams et al. 1976; Adams et al. 1977). It is commonplace at the present time to find soybean and corn fields supporting as many as 100 to 125 colonies/ha. Potato fields in central Florida have recently been found infested with IFA at damaging levels, and young citrus in the Indian River area of Florida currently support extremely heavy infestations (>200/ha), especially in groves with micro-mist irrigation systems.

The impact of IFA populations on the growth and yield of

48

various crops has been studied in recent years so that cost-benefit ratios could be estimated. Adams et al. (1976, 1977) compared the yield of soybeans from paired fields in Georgia and North Carolina in which one field in each pair was infested with IFA (108 to 184 mounds/ha), and the other field was rendered essentially IFA-free through the use of mirex bait. Mechanical losses due to variations in combine or harvesting techniques ranged from $6.00 to $11.56/ha. However, Lofgren and Adams (1981) found that the lower yield of soybeans in infested fields could not be attributed solely to mound interference with the harvester. They found that yields from heavily infested fields invariably averaged 14.5% less than yields from lightly infested fields. Small plot studies conducted at Gainesville, Florida supported these findings with a statistically significant 15% reduction in yield from the IFA-infested plot.

Smittle et al. (1982) labeled corn, okra and soybean plants with radioisotopes (^{32}P) and found evidence of feeding by the IFA on all three crops. Ants were not observed feeding directly on soybean plants, but ants collected around the roots of these plants were radioactive, implying that they were feeding on plant roots.

Adams et al. (1983) reported that the plant density and yield of soybeans from field plots in Gulfport, Mississippi and Gainesville, Florida infested with IFA were significantly different from comparable non-infested plots. The plot in Gulfport was infested at a rate of ca 115 mounds/ha, while the plot at Gainesville was infested at ca 108 mounds/ha. The reductions in plant density in the infested plots at Gulfport and Gainesville were 20% (3.5 plants/m) and 35% (14.1 plants/m), respectively. Yields were reduced by 20% ($P < 0.05$) in Gulfport and 33% ($P < 0.001$) in Gainesville. Furthermore, the Gainesville study showed a significant decrease (P 0.02) in both plant height (4.9 cm) and in the number of bean pods per meter of row (215.3) in the IFA-infested plots.

Apperson and Powell (1983) studied the factors affecting soybean production in North Carolina in 1979 and 1980 and found a negative correlation between IFA and soybean seed yield. A reduction of 403 to 672 kg/ha of seed was attributed to IFA activity (C. Apperson, personal communication). However, IFA impact in soybeans, like other factors, can vary seasonally and geographically.

Adams (1983) reported the loss of ca 50% of an eggplant crop of 12 ha in Marion County, Florida. The loss was directly attributed to extremely heavy predation by IFA on the newly set seedlings. The growth tips of many surviving plants were damaged, and dirt was carried up and deposited in leaf axils and the crowns of the plants.

Our current research includes an investigation of the impact of IFA on production of Irish potatoes and citrus. The study on potatoes was initiated in 1983 at the Gainesville laboratory. Two plots, each consisting of 10 rows 30 m long, were planted with two varieties of potatoes (Sebago and Russett Centennial) in alternating rows. One plot was infested with IFA at a rate of 108 mature colonies/ha and the other was maintained IFA-free. In 1983, reductions of 27.1 and 27.9% of marketable potatoes were recorded from the Sebago and Russett Centennial varieties, respectively; in 1984 comparable reductions of 20.5 and 20.9% were noted.

Results from the 1984 small plot studies were compared with data collected from field studies of a commercial field in St. Johns County, Florida. Data collected from 25 randomly selected sites (1 meter of row ea.) in a 10 ha field heavily infested with IFA, and 25 similar sites in an ant-free field, showed a 35.0% reduction of marketable potatoes in the IFA-infested field (var. Sebago).

During the mid-1970s, reports were received regarding IFA shredding bark from newly planted, non-bearing citrus groves, four years or younger, in six counties in Florida. Recent surveys indicate that 22 of the citrus-producing counties in Florida are affected to some degree. A survey of young groves in Hendry County indicated that all groves were affected, and one grower claimed 25% tree loss in a newly set one-year-old grove of 240 ha. A second grower was preparing to re-set 3500 trees, representing 5.4% of the trees in a 400 ha grove. A random survey of the affected trees by the author and a Florida Extension Service Citrus Specialist indicated that half of the trees died of a fungus disease (Phythophora sp.) while the remaining trees showed classic symptoms of girdling by IFA. Replacement cost of the trees (1982 prices) was $1.75/tree plus $2.50/tree for planting, fertilization and labor. Grove maintenance was estimated at $10.00/tree/year.

Damage to young trees (one to five years old) apparently begins in late fall at the onset of cold weather. The IFA colony migrates to the base of the young tree and, using the trunk of the tree for support, constructs a mound that may be 30 to 50 cm in height. The ants forage actively in the tree, sometimes tending honeydew-producing scales or aphids. They also appear to be attracted to plant sap exuding from the tree. Other damage occurs when the ants shred the bark within the confines of the mound. In the early spring, citrus flowers are attacked by foraging worker ants, apparently feeding on the nectar. Feeding around the calyx results in a greater than 50% decrease in flower maturation. Newly set fruit from these flowers are readily attacked. With oranges, this occurs at the stylar end, while in grapefruit the stem end is attacked. More than 70% of the trees inspected in three groves were heavily infested with IFA in March coinciding with the appearance of the early flowers. Heavy damage (40.8%) to newly set fruit was apparent in early May, and only 15% of the original fruit set remained on the tree in late May. Following fruit set, IFA attacked new, succulent, flush growth, essentially pruning the young branches from the tree.

Surveys of infested counties showed that IFA populations in young citrus groves were extremely high, ranging from 160 mounds/ha in Hendry County to 556/ha in Indian River County. These high levels of infestation are attributed, in part, to the manner of land preparation prior to planting and to the fact that a majority of the newly planted groves are on reclaimed pasture land already heavily infested with IFA.

Laboratory studies confirmed that IFA readily feed on citrus flowers, sap and various tissues (unpublished data). This was evident from studies in which trees were labeled with ^{32}P and exposed to queen-right colonies, as well as the discovery of fresh plant tissue in infrabuccal pellets. In these tests with potted greenhouse trees, all other arthropods were excluded. About 30% of the ants collected the first day after exposure were radioactive. The percentage of radioactive IFA increased daily through five days reaching a maximum of ca 70%.

MEDICAL AND PUBLIC HEALTH

The word "fire" in the common name of S. invicta and S. richteri is descriptive of the reaction of people to their sting. It is not surprising then, that their medical and public health impact is of primary concern. Presently about 40 million people live in IFA-infested areas of the southeastern U.S. and Puerto Rico. Public health surveys by Clemmer and Serfling (1975) and Adams and Lofgren (1981) suggest that 29 to 35% of the population or 11 to 14 million individuals may be stung one or more times each year. Paull (1984) estimates from published data that 0.61% of people who are stung experience generalized systemic anaphylaxis. Assuming these estimates to be correct, 67 to 85 thousand individuals per year require the care of physicians and/or emergency treatment for anaphylaxis. Severe hyperallergic reactions are a constant concern for these persons. As Paull (1984) states, "Although deaths from fire ant stings are rare, many patients who have sustained allergic reactions to a fire ant sting live in daily fear of subsequent stings." Undoubtedly fears of this type are one of the driving forces behind demands for IFA control.

The serious potential to public health associated with IFA stings is also emphasized by the extreme potency of the protein-aceous allergenic compounds of the venom (Lockey 1974; James 1976). The average volume of venom of the IFA is 0.07 to 0.10 ul, of which less than 1% (0.001 ug) represents the proteinaceous components (Rhoades 1977). By comparison, the protein content of the average bee sting is ca 50 ug.

Additional concern about hypersensitivity to IFA stings is evident from the medical literature. Mueller (1959) suggests that unrecognized fatal cases of insect stings occur with greater frequency than is realized. Schwartz (1984) also emphasized the possibility that insect sting attacks may explain some deaths. His conclusion was based on a study of the blood samples of 95 individuals who died of unknown causes. Antibodies indicative of a reaction to insect stings were found in 23% of his study population while less than 1% of his control group of 216 healthy blood donors contained such antibodies. Further, he suggests that as much as 10% of the population may have allergic reactions to insect attacks and that the number of deaths from insect stings—officially listed as less than 100 per year—is grossly underestimated because of a lack of diagnostic tests by physicians and coroners.

Triplett (1973) reported generalized urticaria and angio-edema as the most common complaints of people to IFA stings. More severe complaints include respiratory, gastro-intestinal and shock symptoms. Rhoades (1977) further defined the symptomatology to include cardiovascular and neurologic symptoms. Adams and Lofgren (1982) reviewed the medical records of 329 patients from Fort Stewart, Georgia and found that edema (81%) and urticaria (51%) were the most common symptoms encountered while respiratory distress (7.1%) was the least common, though the most severe. These less serious reactions, edema and urticaria, though not life-threatening, may be no less debilitating to certain individuals. They further reported that 6.8% of the patients seeking aid for IFA encounters suffered secondary infections that required multiple visits to a doctor or hospital for medical assistance.

MISCELLANEOUS

Forestry

Direct loss of longleaf pine seedlings, Pinus palustris, resulted when 32.8% of germinating seeds were eaten by IFA (Campbell 1974). Wilkinson and Chellman (1979) reported a 40% reduction of mean tree height in a 50-ha pine plantation infested with native pine tortoise scale that was tended and propagated by IFA. The authors cited a high potential for damage should the scale-ant association develop and persist in the future.

Mechanical

Damage to secondary roads, resulting from IFA excavating soil from beneath the roadway, was reported from Carteret County, North Carolina, in 1975 (R. Grothaus, personal communication). This excavation, which reached a depth of 4 ft, weakened the road foundation and, subsequently, vehicular traffic caused a subsidence and eventual breaking of the surface of the road. In a survey of 40 miles of roadway in March 1977, fully mature IFA colonies averaged

20/mile with a corresponding subsidence at each colony location. Cost of repair to the highway, furnished by the North Carolina Department of Transportation, averaged $100/subsidence for material and labor. Similar damage to secondary roads reportedly occurs in all coastal counties in North Carolina.

Damage of a different type has been observed by the author on Interstate 75 in the vicinity of Tampa, Florida. In this case, a new silicone sealant used to seal expansion joints in the concrete highway was damaged when the ants chewed portions of the styrofoam backer rod and penetrated the sealant strip, thus permitting rainwater to seep into the joints. Random examination of 8,267 linear feet of seal revealed 226 sites of penetration by IFA. Each mile of highway contained about 36,500 linear feet of seal. Thus, 996 sites per mile of highway were damaged by IFA. Replacement costs, excluding inspection and labor, was reported to be $3.00/linear foot of sealant.

Wildlife

Fire ant predation on wildlife populations has been verified for 6 avian species. These include quail (Travis 1938, 1943), wood duck and the roseate spoonbill (Ridlehuber 1982), barn swallow (Kroll et al. 1973), Mississippi kite (Parker 1977), and black-bellied whistling duck (Delnicki and Bolen 1977). Mount et al. (1981) reported predation by IFAs on the eggs of the lizard Cnemidophorus sexlineatus using the radioisotope ^{32}P, and Mount (1981) reported that nine species of reptiles that were abundant in the Alabama coastal plain in 1968 were now either "locally common" (8 species) or "infrequent." He suggested that predation by IFA on the eggs and new-born young was primarily responsible for the observed lowered population trends. The author further states that a lapse of 10 to 20 years is to be expected between the time of initial IFA infestation and the time the impact becomes obvious to the field naturalist.

DISCUSSION AND SUMMARY

The preceding review graphically illustrates that IFA have a wide-ranging impact on man, animals, plants, and the general ecology of their habitats. Despite this, their economic impact has been a point of controversy for many years with their status reported as ranging from beneficial or minor nuisance to major economic pest. Published data suggest that their omnivorous diet includes a wide range of plant and animal life including pest and beneficial species (see Reagan, Chapter 6). Much of their economic impact can be attributed to their opportunistic feeding behavior in concert with their life cycle. For example, spring is a time of active planting of commercial crops which coincides with a surge in fire ant production of worker and sex brood. It is not surprising, then, that foraging fire ant workers take advantage of germinating seeds, seedlings, and new plant growth as food. Thus, soybeans, young citrus trees, and potato tubers become susceptible to attack. This plant-feeding aspect of IFA behavior has been neglected since the early report of Wilson and Eads (1949), possibly because soil residual insecticides were controlling the IFA.

Recent laboratory studies demonstrated the need of IFA colonies for carbohydrates (Williams et al. 1980). Plant sap is a logical source of this nutrient since feeding on honey-dew produced by insects is infrequently observed for fire ants. The importance of plant sap as an energy source was shown in detailed studies of the leaf-cutting ant, Atta cephalotes, by Quinlan and Cherrett (1979). They found that carbohydrates in plant sap, not the ants' fungus symbiont, provide the major energy requirements for the worker caste.

All of these data support the need for comprehensive studies of the nutritional and energy requirements of IFA and their potential sources in nature. Studies of this type would greatly enhance our ability to elucidate the potential harmful and beneficial aspects of IFA feeding behavior.

The public health problems of IFA stings have been the subject of many studies, but the actual monetary costs associated with them are, at best, poor estimates. Aside from costs, however, mental anguish coincident with hyperallergic reactions and dangers to very young children while they play outdoors are very real problems for many persons. Therefore, it is only proper that many allergists are alerting their fellow physicians to the fact that IFA stings are an increasing problem, one which they need to recognize and be prepared to treat (Lockey 1980; Paull 1984; deSchazo et al. 1984). For example, we have received three calls in the past six months regarding legal cases in which IFA have invaded nursing homes and attacked bed-ridden elderly patients. Hopefully, additional studies by medical authorities will continue to provide cost and injury data.

Miscellaneous activities of IFA associated with their mound-building and tunnel construction, aggressiveness, and their general obtrusive behavior in defending their territories can lead eventually to a variety of problems. Examples of this are the problems cited with ants tunneling under highways and their predation and harass-ment of wildlife.

For the future it is hoped that others will join in the documen-tation of both the harmful and beneficial aspects of IFA ecology. However, emphasis must be placed on the study of their total ecological impact. The fact that IFA are a "weedy" species (see Tschinkel, Chapter 7) insures that they will be man's constant companion as he lives, plays, and competes for space with them. It will behoove us to know our competitor so that simple, effective methods for maintaining the upper hand will always be available to control them in areas where they cause the most agricultural, medical, and environmental impact.

REFERENCES CITED

Adams, C. T. 1983. Destruction of eggplants in Marion County, Florida, by red imported fire ants (Hymenoptera: Formicidae). Fla. Entomol. 66: 518-520.

Adams, C. T., and C. S. Lofgren. 1981. Red imported fire ants (Hymenoptera: Formicidae): Frequency of sting attacks on residents of Sumter County, Georgia. J. Med. Entomol. 18: 378-382.

Adams, C. T., and C. S. Lofgren. 1982. Incidence of stings or bites of the red imported fire ant (Hymenoptera: Formicidae) and other arthropods among patients at Ft. Stewart, Georgia, USA. J. Med. Entomol. 19: 366-370.

Adams, C. T., W. A. Banks, C. S. Lofgren, B. J. Smittle, and D. P. Harlan. 1983. Impact of the red imported fire ant, Solenopsis invicta (Hymenoptera: Formicidae), on the growth and yield of soybeans. J. Econ. Entomol. 76: 1129-1132.

Adams, C. T., J. K. Plumley, W. A. Banks, and C. S. Lofgren. 1977. Impact of the red imported fire ant, Solenopsis invicta Buren, on soybean harvest in North Carolina. J. Elisha Mitchell Sci. Soc. 93: 150-152.

Adams, C. T., J. K. Plumley, C. S. Lofgren, and W. A. Banks. 1976. Economic importance of the red imported fire ant, Solenopsis invicta Buren. I. Preliminary investigations of impact on soybean harvest. J. Ga. Entomol. Soc. 11: 165-169.

Apperson, C. S., and E. E. Powell. 1983. Correlation of the red imported fire ant (Hymenoptera: Formicidae) with reduced soybean yields in North Carolina. J. Econ. Entomol. 76: 259-263.

Campbell, T. E. 1974. Red imported fire ant a predator of direct-seeded longleaf pine. U.S. For. Serv. Res. Note No. SO-179. 3 pp.

Clemmer, D. I., and R. E. Serfling. 1975. The imported fire ant: Dimensions of the urban problem. South. Med. J. 68: 1133-1138.

Delnicki, D. E., and E. G. Bolen. 1977. Use of black-bellied whistling duck nest sites by other species. Southwest. Nat. 22: 275-277.

deSchazo, R. D., C. Griffing, T. H. Kwan, W. A. Banks, and H. F. Dvorak. 1984. Dermal hypersensitivity reactions to imported fire ants. J. Allergy Clin. Immunol. 74: 841-847.

Eden, W. G., and F. S. Arant. 1949. Control of the imported fire ant in Alabama. J. Econ. Entomol. 42: 976-979.

James, F. K. 1976. Fire ant sensitivity. J. Asthma Res. 13: 179-183.

56

Kroll, J. C., K. A. Arnold, and R. F. Gotic. 1973. An observation of predation by native fire ants on nestling barn swallows. The Wilson Bull. 85: 478-479.

Lockey, R. F. 1974. Systemic reaction to stinging ants. J. Allergy Clin. Immunol. 54: 132-146.

Lockey, R. F. 1980. Allergic and other adverse reactions caused by the imported fire ant, pp. 441-448. In O. Oehling, E. Methov, I. Glazer, and C. Orbesman (eds.), Advances in allergology and clinical immunology. Pergamon Press, New York. 795 pp.

Lofgren, C. S., and C. T. Adams. 1981. Reduced yield of soybeans in fields infested with the red imported fire ant, Solenopsis invicta Buren. Fla. Entomol. 64: 199-202.

Lofgren, C. S., W. A. Banks, and B. M. Glancey. 1975. Biology and control of imported fire ants. Ann. Rev. Entomol. 29: 1-30.

Lyle, C., and I. Fortune. 1948. Notes on the imported fire ant. J. Econ. Entomol. 41: 833-834.

Mueller, H. L. 1959. Further experiences with severe allergic reactions to insect stings. New England J. Med. 261: 374-377.

Mount, R. H. 1981. The red imported fire ant, Solenopsis invicta (Hymenoptera: Formicidae), as a possible serious predator on some southeastern vertebrates: Direct observations and subjective impressions. J. Ala. Acad. Sci. 52: 71-78.

Mount, R. H., S. E. Trauth, and W. H. Mason. 1981. Predation by the red imported fire ant, Solenopsis invicta (Hymenoptera: Formicidae) on eggs of the lizard Cnemidophorus sexlineatus (Squamata: Teiidae). J. Ala. Acad. Sci. 52: 66-70.

Parker, J. W. 1977. Mortality of nestling Mississippi kites by ants. Wilson Bull. 89: 176.

Paull, B. R. 1984. Imported fire ant allergy: Perspectives on diagnosis and treatment. Postgrad. Med. 76: 155-161.

Quinlan, R. J., and J. M. Cherrett. 1979. The role of fungus in the diet of the leaf-cutting ant, Atta cephalotes L. Ecol. Entomol. 4: 151-160.

Rhoades, R. B. 1977. Medical aspects of the imported fire ant. The University Presses of Florida, Gainesville, FL. 75 pp.

Ridlehuber, K. T. 1982. Fire ant predation on wood duck ducklings and pipped eggs. Southwest. Nat. 27: 220.

Schwartz, H. J. 1984. Insect bites may explain some deaths. USAEHA Pest Manag. Bull. 84: 16.

Smittle, B. J., C. T. Adams, and C. S. Lofgren. 1982. Red Imported fire ants: Detection of feeding on corn, okra and soybeans with radioisotopes. J. Ga. Entomol. Soc. 19: 78-82.

Travis, B. V. 1938. The fire ant (Solenopsis sp.) as a pest of quail. J. Econ. Entomol. 31: 649-652.

Travis, B. V. 1943. Notes on the biology of the fire ant, Solenopsis geminata (F.) in Florida and Georgia. Fla. Entomol. 24: 15-22.

Triplett, R. F. 1973. Sensitivity to the imported fire ant: Successful treatment with immunotherapy. South. Med. J. 66: 477-480.

Wilkinson, R. C., and C. W. Chellman. 1979. Toumeyella scale, imported fire ant, reduce slash pine growth. Fla. Entomol. 62: 71-72.

Williams, D. F., C. S. Lofgren, and A. Lemire. 1980. A simple diet for rearing laboratory colonies of the red imported fire ant. J. Econ. Entomol. 73: 176-177.

Wilson, E. O., and J. H. Eads. 1949. A report on the imported fire ant, Solenopsis saevissima var. richteri Forel, in Alabama. Spec. Rpt. Ala. Dept. Conserv. Mimeo. 54 pp.

6

Beneficial Aspects
of the Imported Fire Ant:
A Field Ecology Approach

T. E. Reagan

The imported fire ant, Solenopsis invicta, is a predaceous and omnivorous insect that can be beneficial or detrimental depending upon which organisms they utilize for food. Consequently, research on S. invicta should be oriented towards its total biology and ecology before classifying it as a pest or a beneficial insect in any particular situation. In simplest terms, S. invicta's status depends upon whether it is viewed from the perspective of a sugarcane farmer taking advantage of their predatory nature to help him produce a crop or from that of a golf course manager confronted with their unsightly mounds on the fairways. A prime objective of this chapter is to stress the importance of a sound field ecology approach to the study of the economic aspects of S. invicta because it is influenced by a multiplicity of biotic and abiotic factors.

Our ongoing research program on the importance of S. invicta as a component of sugarcane pest management systems in Louisiana provides the background for this paper. Sugarcane is a perennial crop, and as such the benefits derived from the ground level complex of predators, particularly S. invicta, are greater than in most annual field crops (AliNiazee et al. 1979).

BENEFITS FROM PREDATION

Long et al. (1958) reported an increase in damage caused by the sugarcane borer (SCB), Diatraea saccharalis, in fields treated with heptachlor for S. invicta control in St. Mary, Iberia, and Lafayette parishes of southern Louisiana. The result was a 48% increase in crop damage (60% SCB bored internodes in heptachlor-treated vs. 40.5% in untreated fields). Subsequent research by Hensley et al. (1961), Negm (1966, 1968), and Negm and Hensley (1967, 1969) showed that soil applications of heptachlor or chlordane for S. invicta control suppressed a complex of predators including other ants, spiders, carabids, earwigs, and staphylinids. Charpentier et al. (1967) compared the roles of these predator groups and found that ants were most abundant, and that S. invicta was the most important predator. Carroll (1970) compared the abundance of various ant species in Florida sugarcane, speculated on their collective role as predators of SCB, and reported for the first time that S. invicta had invaded this ecosystem. Negm and Hensley (1969, 1972) correlated the numbers of arthropod predators with the degree of SCB egg and larval mortality. S. invicta was found to be an important predator of SCB eggs and larvae; however, spiders and carabid beetle larvae were the most effective predators, respectively, of SCB eggs and larvae.

Studies were initiated in the early 1970s to define more clearly the role of S. invicta in sugarcane management systems and the effect of chemicals for control of this ant on the total predator complex. Reagan et al. (1972) conducted a test to determine the impact of S. invicta control with mirex on SCB damage in a 22-ha field using a randomized complete-block design with a 2 x 2 factorial arrangement of treatments. SCB infestations and sugarcane damage increased 53% and 69%, respectively, following an application of mirex bait that reduced the S. invicta population by about 94%. Also, azinphosmethyl, a chemical recommended for SCB control, was less effective when S. invicta was suppressed.

Research on the effects of S. invicta in sugarcane during the late 1970s to the present time concentrated primarily on the interrelationships of this species with other arthropods. Reagan et al. (1972) suggested that the aggressive nature of S. invicta might result in competitive displacement of other ant species since S. invicta was the only one collected in their pitfall traps. White (1980) showed that as levels of predation on SCB increased in ratoon crops, significantly stronger negative correlations were noted between S. invicta and other ants. Adams et al. (1981) found a negative correlation between the incidence of S. invicta and other ants in Florida sugarcane fields; they also documented an increase in population resurgence of S. invicta following mirex applications in contrast to lower levels with other ant species.

S. invicta has been reported as a predator of numerous other agricultural pests including the rice stink bug, Oebalus pugnax (Newsom et al. 1960); the striped earwig, Labidura riparia (Gross and Spink 1969); aphids (Wilson and Oliver 1969); the Nantucket pine tip moth, Rhyacionia frustrana (Wilson 1969; Wilson and Oliver 1970); the boll weevil, Anthonomus grandis (Sterling 1978); the soybean looper, Pseudoplusia includens (Whitcomb et al. 1972); the velvetbean caterpillar, Anticarsia gemmatalis (Buschman et al. 1977; Whitcomb et al. 1972); the southern green stink bug, Nezara viridula (Whitcomb et al. 1972; Stam 1978; Krispyn and Todd 1982); the green cloverworm, Plathypena scabra (Snodgrass 1976); Heliothis spp. (McDaniel and Sterling 1979; Eger et al. 1983); the cotton leafworm,

Alabama argillacea (Gravena and Sterling 1983); the hornfly, Haematobia irritans (Howard and Oliver 1979); the pecan weevil, Curculio caryae (Dutcher and Sheppard 1981); immature tabanids (Johnson and Hays 1973); springtails and the greenhouse whitefly, Trialeurodes vaporariorum (Morrill 1977); and Diaprepes abbreviatus larvae (Whitcomb et al. 1982; Richman et al. 1983).

The research on the lone star tick, Amblyomma americanum, by Harris (1971), Harris and Burns (1972), and Burns and Melancon (1977) provides one of the few cases where levels of predation on an arthropod pest were compared before and after their habitat was invaded by S. invicta. They found that as S. invicta infestations advanced through north-central and northwestern Louisiana parishes, there was a permanent reduction in lone star tick populations that correlated with increased levels of S. invicta. Fleetwood et al. (1984) also cited substantial A. americanum reductions associated with S. invicta in east Texas pastures. Oliver et al. (1979), on the basis of data from Burns and Melancon (1977) and the U.S. Public Health Service, suggested an association between the reduction in ticks caused by S. invicta and the decline in incidence of tularemia disease; however, they pointed out that other arthropods also transmit tularemia.

Research in pasture habitats has been conducted to quantify the relationships of S. invicta with other arthropods, soil fertility, and pasture management practices. Herzog et al. (1976) showed that the mound-building activity of S. invicta increased concentrations of elemental phosphorus and potassium and reduced soil acidity. They also found an increase in several nutrients in the surface grass on mounds and pointed out the value of dispersing mound soil for pasture improvement. Howard (1975) and Howard and Oliver (1978, 1979) substantiated that S. invicta was a significant predator of the horn fly and observed that they were aggressive to several other ant species. Summerlin et al. (1984) reported that a predator complex which included S. invicta caused a greater impact on horn fly populations than the ants alone. Their studies also showed that S. invicta reduced populations of native scarabs and an imported scarab, Onthophagus gazella, that colonized fresh cattle manure. However, they found that O. gazella was able to make better utilization of its food source in the presence of large numbers of worker ants. Roth et al. (1983) used specialized screening techniques to exclude certain predators or competitors as a means to separate out the effects of predation and competition on horn fly survival. These workers demonstrated the complex relationships among flies breeding in cattle dung and the complex successional niche changes involved. Their study, and that of Blust et al. (1982), led to the development of cultural management practices that utilize the benefits of S. invicta in pasture-livestock production (Sauer et al. 1982).

Predation by S. invicta on major cotton insect pests in Texas was reported by Sterling (1978), Jones and Sterling (1979), McDaniel and Sterling (1982), and Gravena and Sterling (1983). Other studies defined the interspecific association of S. invicta with aphids and various predaceous arthropods in cotton (Reilly and Sterling 1983; Sterling et al. 1979). This work failed to substantiate a negative effect of fire ants on other predaceous insects but did cite a relatively minor interference with spider predation in cotton fields. Fillman et al. (1983) evaluated sampling techniques for determining population levels of S. invicta needed for effective predation in the cotton ecosystem. Subsequently, Sterling et al. (1984) have advocated selective use of insecticides in cotton fields so that the maximum benefits from S. invicta predation can be realized.

Most field ecology studies of S. invicta have underscored the insect's generalist feeding habits. S. invicta has been credited with being both a specialized predator and a polyphagous scavenger (Sterling et al. 1979). Because of the multiplicity of items in the diet of S. invicta, the usual time-lag relationships associated with efficacy of classical biological control organisms do not apply to predation by this insect on arthropod pests. Additional reviews and citations of the beneficial aspects of S. invicta are presented by Wojcik (in press), Wojcik and Lofgren (1982), Banks et al. (1978), and Apperson and Adams (1983).

ECOLOGICAL TECHNIQUES FOR DETERMINING FOOD HABITS

Because of the complex interactions of S. invicta with other components of their environment, a careful and comprehensive selection of methodologies and procedures must be utilized to determine the true impact of S. invicta in various agricultural systems. Due to its large foraging territories, particularly as related to different habitats and food needs, the selection of appropriately sized plots may be critical for data collection. Such data is exemplified from pre- and post-treatment pitfall trap collections (previously unpublished) following the application of mirex bait to 10 of 20 plots (0.6 to 1.4 ha; 23 to 37 rows spaced 1.8 m apart) in a 22.3 ha sugarcane field (Reagan et al. 1972). The number of ants collected per trap (two per plot placed on the center row of each plot) two weeks later dropped from 247 to 2.4 in the treated plots and 273 to 22.0 in the check plots. This reduction (ca 10-fold) of ants collected in untreated plots was attributed either to foraging by the ants into the baited plots or collection of bait which had been retrieved and discarded by other ants. Thus, emphasis is placed on the importance of using large plots in habitat comparison studies, especially where insecticidal baits are being used. Lofgren (1968) also emphasized the problem of obtaining accurate evaluations of the effectiveness of bait formulations due to foraging habits of

S. invicta.

Another important consideration when insecticides are used to study S. invicta predation is that these pesticides also exert effects on nontarget organisms. In the study mentioned previously (Reagan et al. 1972), mirex caused significant reductions in crickets and carabid and staphylinid beetles, the latter two being SCB predators.

Methods of collecting data for feeding preference studies also deserve very careful scientific scrutiny, particularly due to the complexity of ant/plant associations. S. invicta may feed on plant exudates, very small arthropods such as aphids and mites (Hays and Hays 1959), and certain plant foliage or fruit material (Smittle et al. 1983). Because of the variety of interactions, data from studies that utilize radio-labeling or other tracer techniques should be verified with additional methodologies to eliminate possible erroneous inter-pretations. It must be recognized that when S. invicta is stressed for food, its feeding habits can be modified and directed toward other available food sources (Reagan 1982). Thus, it is conceivable that test techniques could force S. invicta to feed or not feed on a particular animal or plant with obvious misleading results.

A useful technique for identifying and quantifying S. invicta's diet in the field has been described recently by Ali et al. (1984). This method, which has been used previously by other researchers, involves collecting foraging workers on their trails with an aspirator or forceps as they are returning to the nest with food particles. This material is identified later in the laboratory with the aid of a micro-scope. When replication is adequate and the data is stratified both in time and space, reliable results can be obtained as shown in the histograms in Fig. 1. The results should be validated with other sampling methodologies such as sweep nets and/or D-vacs for addi-tional quantification of the arthropods. Identification of liquids ingested by workers (nectar and other plant fluids) requires other collection techniques that incorporate analyses by various chromato-graphic or other biochemical procedures. Such techniques have been reported previously by Collins (1983) in studies of the interaction of S. invicta with weeds in soybean fields.

From the pest management perspective, research also must address the possibility of S. invicta predation on key beneficial organisms. This was illustrated by Reagan (1982) when he compared samples of foraging materials collected in August and September simultaneously from weedy and weed-free field plots of sugarcane (see Table 1). It is apparent that in weed-free habitat the ants directed their feeding more toward important beneficial groups (earthworms, spiders, other ants). It was additionally noted that the overall ant populations in the weed-free fields declined, and the percentage of abandoned nests increased.

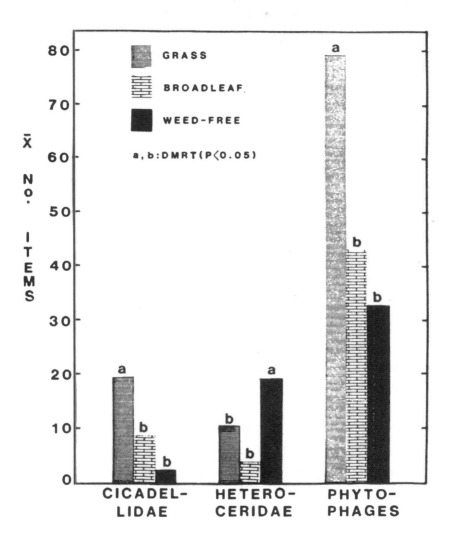

FIGURE 1. Relative abundance of selected groups of $\underline{S}.$ $\underline{invicta}$ prey in grass, broadleaf, and weed-free sugarcane habitats (means of items collected over a sampling time of 9 h in each habitat). Adapted from Ali et al. (1984).

TABLE 1. Relative proportions of invertebrate organisms collected by S. invicta workers in weedy and weed-free sugarcane field plots, Bunkie, Louisiana.[a]

Weed habitat		Weed-free habitats	
Invertebrate order	% of total	Invertebrate order	% of total
Homoptera	26	Annelida	19
Gryllidae	17	Araneae	15
Lepidoptera	14	Heteroceridae	15
Scarabaeidae	12	Homoptera	11
Unidentified fragments	12	Formicid pupae	11

[a]Adapted from Reagan (1982).

PEST MANAGEMENT SYSTEMS IN SUGARCANE

The complexity of S. invicta predation relative to various aspects of a pest management system have been emphasized in our studies of sugarcane production in Louisiana. This research has included observations on the interrelationships between weeds, arthropod predators, other invertebrates, and sugarcane (Reagan 1982; Ali et al. 1984; Ali and Reagan 1985). The more important conclusions are as follows:

1. As the habitat and arthropod fauna become more diverse with succeeding crops of sugarcane, predation on SCB progressively increases and almost doubles by the second ratoon crop.

2. S. invicta are polyphagous feeders and respond positively to prey abundance.

3. Annual broadleaf and shallow-rooted grass weeds at levels sub-competitive with sugarcane help maintain high S. invicta populations by enhancing their predation on insects associated with these weeds.

4. S. invicta predation on entomophagous insects was greatest in weed-free habitats where food limitation enhanced "scramble" competition.

5. Maximum predation of SCB occurs under conditions of moderate weed control which suggests the necessity for determining economic thresholds for weed populations in sugarcane.

More recent studies (Ali and Reagan 1985) have shown that applications of azinphosmethyl, commonly used for SCB control, may destabilize the predator system through food web disruption by significantly decreasing coleopterans (51%), phytophagous insects (35%), and spiders (67%). Also, these studies have demonstrated that a subcompetitive stand of shallow-rooted annual broadleaf weeds can enhance predator populations (including S. invicta) early in the season. Later in the season, these weeds die back as the sugarcane canopy closes, thus eliminating competitive losses to the crop. Economic returns from such broadleaf habitats were 19% higher than in the weed-free fields. Thus, stabilization of the predator system through floral, and consequently faunal, diversification has proven to be a judicious and economically superior approach to integrated pest management in Louisiana sugarcane (Ali and Reagan 1985).

While this predator system may function well in Louisiana, such may not be the case in Florida sugarcane. In Florida, two introduced entomophagous parasites, Apanteles flavipes and Agathis stigmatera have exerted the predominant control of SCB (Hall 1982). Additionally, a recently discovered exotic plant hopper pest, Perkinsiela saccharicida, has become established (Sosa 1985). This homopteran pest may have a mutualistic association with S. invicta at certain times of the year similar to that described for other Homoptera by Scarborough (1984). This relationship may be strongest during periods of extensive brood rearing by S. invicta. Thus, the possibility of S. invicta interacting negatively with the SCB parasites and positively with the homopteran pest could pose a significant problem. Such interactions when more clearly defined may present a contrast to the beneficial role played by the ants in Louisiana sugarcane.

SUMMARY AND CONCLUSIONS

The research discussed in this review has involved the use of numerous and varied methods for quantifying the abundance and beneficial impact of S. invicta, as well as its interaction with other arthropods and flora. These methods include the use of continuous sampling devices such as pitfall traps, various baits or modifications thereof (i.e., index cards as used by Dutcher and Sheppard 1981), D-vac machines, whole plant destructive sampling, direct observation of foliage, mound numbers and size stratification, and ant collections for diet determination. All of these yield useful data for field ecology studies. Because they do not provide counts of total organisms in a particular area, most of these methods mentioned can be used only for replicated comparisons of differences between habitats or other treatments. It is important for researchers to realize, however, that all of these sampling methods have at least some degree of bias (Southwood 1978). Therefore, effective interpretation of results from field ecology studies requires multiple sampling approaches used simultaneously to minimize bias.

In conclusion, the potential benefits of predation of S. invicta on pest arthropods has been reviewed, and an emphasis has been placed on the importance of comprehensive and thorough field research to the study of interactions of S. invicta with the flora and fauna in its various habitats. Whenever pesticides are used as a research tool, we need to be continually aware that they may affect the data through various degrees of niche modification. Likewise, it is also likely that efforts to eradicate S. invicta with broad spectrum residual pesticides have resulted in increases in subsequent infestation levels (see Tschinkel, Chapter 7; Lofgren and Williams 1985). Such programs may have maintained S. invicta in the status of a recent colonizer with the potential for high population growth. One result of eradication programs can be ecosystem simplification; however, S. invicta may also play a role in simplification, especially through its competition with other ants (Newsom 1978). In areas where S. invicta has been established for a long time, population levels seem to have subsided substantially (L. D. Newsom, personal communication).

Additionally, the point is made that "pest" is a man-made term and that an insect such as S. invicta should not be described as such until the research to define its relationships has been conducted. Finally, it is possible that because of the complexity of this ant's food and niche relationships, the greatest potential for its control in areas where it is a pest could be made through cultural modifications and habitat management (Sauer et al. 1982).

ACKNOWLEDGMENT

Gratitude is expressed to Drs. W. L. Sterling (Texas A & M University), W. R. Tschinkel (Florida State University), and L. D. Newsom (Louisiana State University) for their constructive reviews and helpful suggestions on this manuscript.

REFERENCES CITED

Adams, C. T., T. E. Summers, C. S. Lofgren, D. A. Focks, and J. C. Prewitt. 1981. Interrelationship of ants and the sugarcane borer in Florida sugarcane. Environ. Entomol. 10: 415-418.

Ali, A. D., and T. E. Reagan. 1985. Vegetation manipulation impact on predator and prey populations in Louisiana sugarcane ecosystems. J. Econ. Entomol. 78: 1409-1414.

Ali, A. D., T. E. Reagan, and J. L. Flynn. 1984. Influence of selected weedy and weed-free sugarcane habitats on diet composition and foraging activity of the imported fire ant (Hymenoptera: Formicidae). Environ. Entomol. 13: 1037-1041.

AliNiazee, M. T., L. A. Andres, J. W. Beardsley, G. W. Bishop, D. W. Davis, K. S. Hagen, S. C. Hoyt, L. Moore, J. A. McMurtry, E. R. Oatman, R. G. Simpson, and T. F. Watson. 1979. Biological control and insect pest management. Univ. Calif. Div. Agric. Sci. Publ. 4096.

Apperson, C. S., and C. T. Adams. 1983. Medical and agricultural importance of red imported fire ant. Fla. Entomol. 66: 121-126.

Banks, W. A., C. S. Lofgren, and D. P. Wojcik. 1978. A bibliography of the imported fire ants and the chemicals and methods used for their control. U.S. Dept. Agric., Agric. Res. Serv., Bull. ARS-S-180. 35 pp.

Blust, W. E., G. H. Wilson, K. L. Koonce, B. D. Nelson, and J. E. Sedberry, Jr. 1982. The red imported fire ant, Solenopsis invicta Buren: Cultural control and effects on hay meadows. Louisiana State Univ. Agric. Exp. Sta. Bull. No. 738. 27 pp.

Burns, E. C., and D. G. Melancon. 1977. Effect of imported fire ant (Hymenoptera: Formicidae) invasion on lone star tick (Acarina: Ixodidae) populations. J. Med. Entomol. 14: 247-249.

Buschman, L. L., W. H. Whitcomb, R. C. Hemenway, D. L. Mays, N. Ru, N. C. Leppla, and B. J. Smittle. 1977. Predators of velvetbean caterpillar eggs in Florida soybean. Environ. Entomol. 6: 403-407.

Carroll, J. F. 1970. Role of ants as predators of the sugarcane borer, Diatraea saccharalis. M.S. Thesis, University of Florida, Gainesville, FL. 70 pp.

Charpentier, L. J., W. J. McCormick, and R. Mathes. 1967. Beneficial arthropods inhabiting sugarcane fields and their effects on borer infestations. Sugar Bull. 45: 276-277.

Collins, F. L. 1983. Interactions of soybean arthropods and crop-weeds: Diversity, moth nutrition and stability. Ph.D. Dissertation, Louisiana State University, Baton Rouge, LA. 98 pp.

Dutcher, J. D., and D. C. Sheppard. 1981. Predation of pecan weevil larvae by red imported fire ants. J. Ga. Entomol. Soc. 16: 210-213.

Eger, J. E. Jr., W. L. Sterling, and A. W. Hartstack, Jr. 1983. Winter survival of Heliothis virescens and Heliothis zea (Lepidoptera: Noctuidae) in College Station, Texas. Environ. Entomol. 12: 970-975.

Fillman, D. A., W. L. Sterling, and D. A. Dean. 1983. Precision of several sampling techniques for foraging red imported fire ant (Hymenoptera: Formicidae) workers in cotton fields. J. Econ. Entomol. 76: 748-751.

Fleetwood, S. C., P. D. Teel, and G. Thompson. 1984. Impact of imported fire ant (Hymenoptera: Formicidae) on lone star tick (Acari: Ixodidae) mortality in open and canopied pasture habitats of East Central Texas. Southwest. Entomol. 9: 158-163.

Gravena, S., and W. L. Sterling. 1983. Natural predation on the cotton leafworm (Lepidoptera: Noctuidae). J. Econ. Entomol. 76: 779-784.

Gross, H. B., Jr., and W. T. Spink. 1969. Response of the striped earwig following application of heptachlor and mirex and predator-prey relationships between imported fire ants and earwigs. J. Econ. Entomol. 63: 686-689.

Hall, D. G. 1982. A biological-chemical IPM program for the sugarcane borer. Proc. II Inter-American Sugarcane Seminar—Insect and Rodent Pests - 1981, Florida International Univ. 2: 89-95.

Harris, W. G. 1971. The relationship of the imported fire ant, Solenopsis saevissima (F. Smith), to populations of the lone star tick, Amblyomma americanum (Linnaeus) and the effects of mirex on populations of arthropods. Ph.D. Dissertation. Louisiana State Univ., Baton Rouge, LA. 112 pp.

Harris, W. G., and E. C. Burns. 1972. Predation on the lone star tick by the imported fire ant. Environ. Entomol. 1: 362-365.

Hays, S. B., and K. L. Hays. 1959. Food habits of Solenopsis saevissima richteri Forel. J. Econ. Entomol. 52: 455-457.

Hensley, S. D., W. H. Long, L. R. Roddy, W. J. McCormick, and E. J. Concienne. 1961. Effects of insecticides on the predaceous arthropod fauna of a Louisiana sugarcane field. J. Econ. Entomol. 54: 146-149.

Herzog, D. C., T. E. Reagan, D. C. Sheppard, K. M. Hyde, S. S. Nilake, M. Y. B. Hussein, M. L. McMahan, R. C. Thomas, and L. D. Newsom. 1976. Solenopsis invicta Buren: Influence on Louisiana pasture soil chemistry. Environ. Entomol. 5: 160-162.

Howard, F. W. 1975. Arthropod population dynamics in pastures treated with mirex-bait to suppress red imported fire ant populations. Ph.D. Dissertation, Louisiana State Univ., Baton Rouge, LA. 162 pp.

Howard, F. W., and A. D. Oliver. 1978. Arthropod populations in permanent pastures treated and untreated with mirex for red imported fire ant control. Environ. Entomol. 7: 901-903.

Howard, F. W., and A. D. Oliver. 1979. Field observations of ants (Hymenoptera: Formicidae) associated with red imported fire ants, Solenopsis invicta Buren, in Louisiana pastures. J. Ga. Entomol. Soc. 14: 259-262.

Johnson, A. W., and K. L. Hays. 1973. Some predators of immature Tabanidae (Diptera) in Alabama. Environ. Entomol. 2: 1116-1117.

Jones, D., and W. L. Sterling. 1979. Manipulation of red imported fire ants in a trap crop for boll weevil suppression. Environ. Entomol. 8: 1073-1077.

Krispyn, J. W., and J. W. Todd. 1982. The red imported fire ant as a predator of the southern green stinkbug in Georgia. J. Ga. Entomol. Soc. 17: 19-26.

Lofgren, C. S. 1968. Toxic bait studies with the imported fire ant, Solenopsis saevissima richteri: Forel. Ph.D. Dissertation, Univ. of Florida, Gainesville, FL. 67 pp.

Lofgren, C. S., and D. F. Williams. 1985. Imported fire ants: Population dynamics following treatment with insecticidal baits. J. Econ. Entomol. 78: 863-867.

Long, W. H., E. A. Concienne, E. J. Concienne, R. N. Dobson, and L. D. Newsom. 1958. Fire ant eradication increases damage by the sugarcane borer. Sugar Bull. 37: 62-63.

McDaniel, S. G., and W. L. Sterling. 1979. Predator determination and efficiency on Heliothis virescens eggs in cotton using ^{32}P. Environ. Entomol. 8: 1083-1087.

McDaniel, S. G., and W. L. Sterling. 1982. Predation of Heliothis virescens (F.) eggs on cotton in East Texas. Environ. Entomol. 11: 60-66.

Morrill, W. L. 1977. Red imported fire ant foraging in a greenhouse. Environ. Entomol. 6: 416-418.

Negm, A. A. 1966. Studies on natural control of the sugarcane borer, Diatraea saccharalis (Fabricius), and the effect of certain insecticides on predaceous arthropods. M.S. Thesis. Louisiana State Univ., Baton Rouge, LA. 71 pp.

Negm, A. A. 1968. Ecological studies on certain mortality factors of the sugarcane borer, Diatraea saccharalis (F.), in Louisiana. Ph.D. Dissertation, Louisiana State Univ., Baton Rouge, LA. 84 pp.

Negm, A. A., and S. D. Hensley. 1967. The relationship of arthropod predators to crop damage inflicted by the sugarcane borer. J. Econ. Entomol. 69: 1503-1506.

Negm, A. A., and S. D. Hensley. 1969. Evaluation of certain biological control agents of the sugarcane borer in Louisiana. J. Econ. Entomol. 62: 1008-1013.

Negm, A. A., and S. D. Hensley. 1972. Role of predaceous arthropods of the sugarcane borer, Diatraea saccharalis (F.), in Louisiana. Proc. XIV Cong. Int. Soc. Sugarcane Tech. pp. 445-453.

Newsom, L. D. 1978. Eradication of plant pests - Con. Bull. Entomol. Soc. Amer. 24: 35-40.

Newsom, L. D., W. T. Spink, and S. Warter. 1960. Effect of the fire ant eradication program on the fauna of rice fields. Insect Cond. La. 2: 8-14.

Oliver, A. D., T. E. Reagan, and E. C. Burns. 1979. The fire ant - an important predator of some agricultural pests. Louisiana Agric. 22: 6, 7, 9.

Reagan, T. E. 1982. Sugarcane borer pest management in Louisiana: Leading to a more permanent system. Proc. II Inter-American Sugarcane Seminar—Insect and Rodent Pests - 1981, Florida International Univ. 2: 100-110.

Reagan, T. E., G. Coburn, and S. D. Hensley. 1972. Effects of mirex on the arthropod fauna of a Louisiana sugarcane field. Environ. Entomol. 1: 588-591.

Reilly, J. J., and W. L. Sterling. 1983. Dispersion patterns of the red imported fire ant (Hymenoptera: Formicidae), aphids and some predaceous insects in a cotton agroecosystem. Environ. Entomol. 12: 541-545.

Richman, D. B., W. H. Whitcomb, and W. F. Buren. 1983. Predation on neonate larvae of Diaprepes abbreviatus (Coleoptera: Curculionidae) in Florida and Puerto Rico citrus groves. Fla. Entomol. 66: 215-222.

Roth, J. P., G. T. Fincher, and J. W. Summerlin. 1983. Competition and predation as mortality factors of the horn fly, Haematobia irritans (L.) (Diptera: Muscidae), in a central Texas pasture habitat. Environ. Entomol. 12: 106-109.

Sauer, R. J., H. L. Collins, G. Allen, D. Campt, T. D. Canerday, G. Larocca, C. Lofgren, T. E. Reagan, D. L. Shankland, M. Trostle, W. R. Tschinkel, and S. B. Vinson. 1982. Imported fire ant management strategies—Panel VI, pp. 91-110. In S. L. Battenfield (ed.), Proceedings of a Symposium on the Red Imported Fire Ant, EPA/USDA(APHIS) 0-389-890/70, Washington, D.C. 255 pp.

Scarborough, T. A. 1984. Mutualism of the red imported fire ant, Solenopsis invicta Buren, with honeydew producing Homoptera. Ph.D. Dissertation. Texas A & M Univ., College Station, TX. 102 pp.

Smittle, B. J., C. T. Adams, and C. S. Lofgren. 1983. Red imported fire ants: Detection of feeding on corn, okra and soybeans with radioisotopes. J. Ga. Entomol. Soc. 18: 78-82.

Snodgrass, G. L. 1976. An evaluation of the black imported fire ant, Solenopsis richteri as a predator in soybeans in northeast Mississippi. M.S. Thesis, Miss. State Univ., Mississippi State, MS. 86 pp.

Sosa, O. Jr. 1985. The sugarcane delphacid, Perkinsiella saccharicida (Homoptera: Delphacidae), a sugarcane pest new to North America detected in Florida. Fla. Entomol. 68: 357-360.

Southwood, T. R. E. 1978. Ecological methods, with particular reference to the study of insect populations. Chapman and Hall (N.Y. and London). 524 pp.

Stam, P. A. 1978. Relation of predators to population dynamics of Nezara viridula (L.) in a soybean ecosystem. Ph.D. Dissertation. Louisiana State Univ., Baton Rouge, LA. 220 pp.

Sterling, W. L. 1978. Fortuitous biological suppression of the boll weevil by the red imported fire ant. Environ. Entomol. 7: 564-568.

Sterling, W. L., D. A. Dean, D. A. Fillman, and D. Jones. 1984. Naturally-occurring biological control of the boll weevil. Entomophaga 29: 1-9.

Sterling, W. L., D. Jones, and D. A. Dean. 1979. Failure of the red imported fire ant to reduce entomophagous insect and spider abundance in a cotton agroecosystem. Environ. Entomol. 8: 976-981.

Summerlin, J. W., H. D. Petersen, and R. L. Harris. 1984. Red imported fire ant (Hymenoptera: Formicidae): Effects on the hornfly (Diptera: Muscidae) and coprophagous scarabs. Environ. Entomol. 13: 1405-1410.

Whitcomb, W. H., H. A. Denmark, A. P. Bhatkar, and G. L. Greene. 1972. Preliminary studies on the ants of Florida soybean fields. Fla. Entomol. 55: 129-142.

Whitcomb, W. H., T. D. Gowan, and W. F. Buren. 1982. Predators of Diaprepes abbreviatus (Coleoptera: Curculionidae) larvae. Fla. Entomol. 65: 150-158.

White, E. A. 1980. The effects of stubbling and weed control in sugarcane on the predation of the sugarcane borer, Diatraea saccharalis (F.). M.S. Thesis. Louisiana State Univ., Baton Rouge, LA. 216 pp.

Wilson, N. L. 1969. Foraging habits of the fire ant, Solenopsis saevissima richteri Forel, on some arthropod populations in southeastern Louisiana. Ph.D. Dissertation. Louisiana State Univ., Baton Rouge, LA. 80 pp.

Wilson, N. L., and A. D. Oliver. 1969. Food habits of the imported fire ant in pastures and pine forest areas in Southeastern Louisiana. J. Econ. Entomol. 63: 1268-1271.

Wilson, N. L., and A. D. Oliver. 1970. Relationship of the imported fire ant to Nantucket pine tip moth infestations. J. Econ. Entomol. 63: 1250-1252.

Wojcik, D. P. Bibliography of imported fire ants and their control: Second supplement. Fla. Entomol. (In press).

Wojcik, D. P., and C. S. Lofgren. 1982. Bibliography of imported fire ants and their control: First supplement. Bull. Entomol. Soc. Am. 28: 269-276.

7

The Ecological Nature
of the Fire Ant: Some Aspects
of Colony Function and Some
Unanswered Questions

W. R. Tschinkel

Presented here is a conceptual interpretation of the available facts about fire ants that will be used to paint a general picture of their biological nature and function. Considering the gaps in our knowledge, the future may very well see changes of interpretation. The facts upon which I base my review are the results of the work of many people over more than three decades. Information on history and spread of Solenopsis invicta is discussed by Lofgren in Chapter 4.

ECOLOGICAL NATURE OF S. INVICTA

Ecologically, S. invicta is a weed species and, as such, shows many of the biological properties of weeds. Weeds are animal or plant species that are adapted for the opportunistic exploitation of ecologically disturbed habitats. Naturally, these are created by flood, fire, and storm and consist of new sandbanks, slumps, and landslides, burns, and windfalls. Man, however, creates vast areas of disturbed habitat by clearing forest for agricultural, domestic, and other uses. The plant and animal communities that occupy such disturbed habitats are called early secondary or early succession communities because, if left alone, they will gradually revert to dominant climax communities. In the southeast, this is mostly deciduous forests. Because such early succession communities are ephemeral and underexploited, the weed species utilizing these habitats are adapted for very rapid, scramble-type exploitation with an emphasis on high reproductive rates and efficient dispersal rather than competition with other members of the community (Ito 1980).

The weed-like properties of the fire ant are as follows: First, the fire ant is clearly and dramatically associated with ecologically disturbed habitats created mostly by man both in the United States and South America (Banks et al. 1985). Thus, S. invicta is abundant in old fields, pastures, lawns, roadsides, and any other open, sunny habitats. It shares these habitats with many other weedy plant and animal species, from man's crops to lawn and pasture grasses, goldenrods, and dog-fennel. Man is the fire ant's greatest friend, even though the sentiment may not be returned.

On the other hand, the fire ant is absent or rare in late succession or climax communities such as mature deciduous or pine forest (personal observation). When it is found in these communities, it is usually associated with local disturbances such as seasonal flooding and roads. In North Florida, on transects through longleaf pine-wiregrass-turkey oak forest, I found S. invicta strictly associated with temporary ponds or pond margins, graded dirt roads, and the margins of paved roads. All other areas were occupied by S. geminata, if any fire ants were present at all. This was true even in recently clearcut and replanted areas. Increased insolation associated with disturbance is thus not sufficient explanation for S. invicta's distribution. Unfortunately, there are almost no data on the biotic and abiotic causes of its distribution. Nevertheless, it is clear that S. invicta, like other weeds, is associated with open, disturbed habitats. This also appears to be true in its native homeland in Southern Brazil (Buren, personal communication; Banks et al. 1985).

A high reproductive rate is a second weedy property that is associated with the sporadic, unpredictable and ephemeral availability of suitable habitat (in the absence of man-made disturbance). Success in such a situation goes to the animals and plants that "gits thar the fustest with the mostest" (attributed to Civil War General Lee DeForrest in response to being asked how to win a battle) with little attention paid to competition within the community. Fire ants, like other weeds, achieve a high reproductive rate in part by very high investment of resources in reproductives. From the meager data available (Markin et al. 1973; Morrill 1974), I estimate that S. invicta allocates 30 to 40% of its annual biomass production to sexuals. This is similar to the energy allocation to seeds found in weedy species of goldenrod and much higher than non-weedy goldenrods adapted to competition in late-succession communities (Ito 1980) (Fig. 1). Thus, the average fire ant colony in North Florida produces about 4500 sexuals per year (Morrill 1974). Although very little information is available for comparison, this seems high for ants in general and is almost certainly an adaptation to its weedy habits.

A third weedy property is effectiveness of dispersal and colonization. In the absence of man-made disturbance, secondary habitat is scattered and unpredictable. Its exploitation depends upon the ability to scatter propagules (sexuals or seeds) over wide areas on the chance that a few will find their way to an appropriate site and colonize it. The fire ant achieves this by producing a large number of sexuals who take part in high-altitude, dispersive mating

74

flights throughout a large portion of the year (Fig. 2). The queens often fly or are wind-carried one-fourth to one-half mile or more before settling to the ground although most settle at shorter distances (Markin et al. 1971).

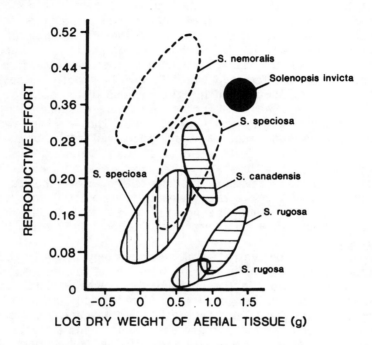

FIGURE 1. Reproductive effort (proportion of production invested in seeds or sexuals) in relation to size of organism or colony for goldenrods (Solidago spp.) and fire ants (black circle). Each enclosed area is represented by the individuals of a single population on a dry site (enclosed in dotted curve), a wet site (with horizontal shading), or a hardwood site (vertical shading) (modified from Ito 1980).

Fire ant queens do not have to depend on general habitat disturbances to enhance their ability to establish a colony; even a specific disturbance to the ant community can provide the necessary conditions for success. This phenomenon was first observed by Summerlin et al. (1977) in mirex-treated plots in which S. invicta was a minor component of the ant community. The mirex killed almost all of the ground-nesting ants; but after recolonization, S. invicta and another weedy species, Conomyrma insana, had greatly increased their dominance over all other species (Fig. 3). Many of the native species did not reappear in the course of this two-year study.

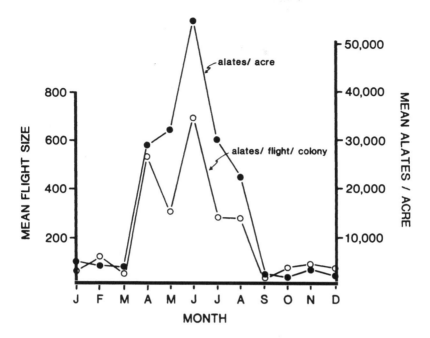

FIGURE 2. Occurrence and size of mating flights throughout
the year in North Florida (from Morrill 1974).

The implication of these studies is clear: Large-scale,
unspecific control programs utilizing chemical baits in areas of low
fire ant populations may aid rather than hinder the establishment of
the fire ant and accentuate its dominance over native ants. The
reversal of this dominance appears to be very slow, if it occurs at
all. Clearly, the relationship of the fire ant to other native ants in
its community is an area of utmost importance about which we know
very little. No program to manage fire ants can hope to succeed
without such knowledge.

The fourth and fifth weedy properties of S. invicta are rapid
growth and early reproduction. Fire ants achieve rapid growth by
cooperation rather than competition during the founding and incipient colony period. Thus, a number of newly mated queens may share
excavation of the founding nest and rearing of the first brood of
minims. This cooperation, called pleometrosis, increases the
chances of surviving the founding period and results in incipient
colonies with about three times as many minim workers as colonies
founded by a single queen (haplometrosis) (Tschinkel and Howard
1983). Because of exponential colony growth, this initial boost is
maintained throughout the early growth period, so that after four

months, pleometrotically-founded colonies are still three times as large as haplometrotic ones (Fig. 4), a clear advantage in competition and winter or drought survival.

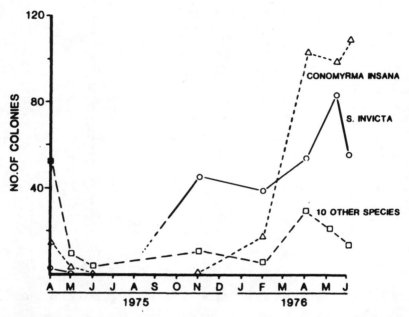

FIGURE 3. Changes in an ant community after treatment with mirex. The two weed species, S. invicta and Conomyrma insana, are shown separately (from Summerlin et al. 1977). Ten other species are lumped together.

Once past the incipient period, colony growth is rapid, requiring about three to four years to reach upper colony size limit of 150,000 to 200,000 (Fig. 5). At the end of one year, 50% of the colonies produced some sexuals while 100% produced sexuals at the end of two years. Reduction in the generation time has a great effect on increasing the rate of population increase.

We have little understanding of the population dynamics within a colony during founding, growth, maturity, and senescence. Descriptive data are very crude and knowledge of mechanisms is almost entirely absent. How intracolonial population dynamics might be influenced by habitat differences has received little attention, yet it is of considerable importance to any management program. This subject is one of the more obvious voids in our knowledge and desperately needs attention.

An interesting aspect of fire ant colony development is that several newly-mated queens may cooperate in colony founding;

however, as the colony matures a single queen survives, and the colony becomes competitive with conspecifics (Tschinkel and Howard 1983). The most obvious aspect of this change is that colonies become territorial (see Jaffe, Chapter 18) as they grow (unpublished data). That is, they defend a plot of ground against all other fire ant colonies (Wilson et al. 1971) and possibly to some extent against some other ant species. The size of these colony territories is probably proportional to the number of workers in the colony and the food resources within the territory. A typical figure for pasture land is about 20 to 25 mature colonies per acre (Morrill 1974). Establishment of new fire ant colonies within territories occupied by mature colonies is probably impossible, because the resident workers kill newly-mated queens that land in their territory (Tschinkel and Howard 1983).

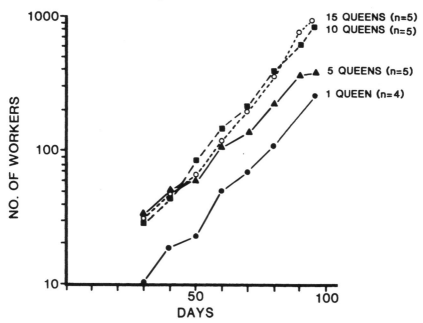

FIGURE 4. Early growth in worker number in colonies founded by various numbers of queens. Pleometrotic colonies begin growth with about three times as many workers as haplometrotic ones and maintain this advantage throughout (Tschinkel and Howard 1983).

A final brief point concerning the ecological nature of the fire ant is its seasonality. S. invicta is a tropical ant lacking true hibernation, yet it shows strong seasonality in many aspects of colony life. This seasonality is probably imposed by seasonal fluctuations in temperature and, to some degree, rainfall (Markin and Dillier 1971).

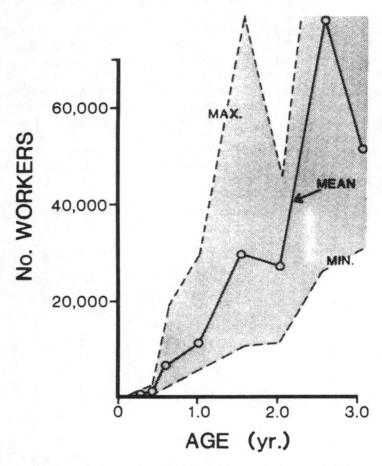

FIGURE 5. Growth in colonies of S. invicta. Shaded area includes the extremes (data from Markin et al. 1973).

The influence of temperature on ecology and life cycle may be another underappreciated factor in fire ant biology (see Francke and Cokendolpher, Chapter 9). We normally associate temperature with the rates of various processes, but there is evidence that temperature may also act as a threshold or trigger for fire ants. Examples of processes whose occurrence as well as rate is controlled by temperature include worker brood production, alate brood production, foraging (Markin and Dillier 1971), mating flights (Morrill 1974), and successful colony founding (Markin et al. 1972). Such knowledge is of obvious importance to our understanding of the potential and realized ecological range of the fire ant.

In summary, S. invicta is ecologically an opportunistic weed species depending upon habitat disturbance for its continued existence and success.

COLONY FUNCTION

Colony function refers to the general internal workings and social relationships within the colony, including behavioral, physiological and morphological aspects. A useful organizing principle is to consider colony life from the point of view of production. Organisms (and colonies) carry out the production of new biomass in such a way as to maximize reproductive success, this being merely a restatement of the principles of natural selection. Production of any kind requires some degree of control; so let us look at colony function from the perspective of production and its control.

Production is the procurement of raw material in the form of food, water, and air and their conversion into new biomass. There are two primary localities for the production of new biomass in the colony: (1) The queen produces new biomass in the form of eggs. A queen in a large colony is capable of producing her own weight in eggs every 24 hours (unpublished data). (2) The larvae produce new biomass as they grow to become new individuals, either workers or sexuals. Under some conditions, colonies are capable of tripling their biomass (mainly workers) in a month (unpublished data).

Under the heading of control are a variety of behavioral and physiological phenomena, many of which have been investigated. As related to production, control deals with problems of the routing of materials (where does it go, and when?), rates of production (how fast?), and allocation of resources, including labor (how much to each subcomponent?).

A partial list of control mechanisms would include: (1) pheromones (chemical signals), (2) various behavioral interactions, (3) trophic relations (food flow and use), and (4) thermoregulation. This list is not complete but will suffice for this general discussion. Pheromones are discussed by Fletcher (Chapter 15), Vander Meer (Chapter 17), and Glancey (Chapter 19); therefore, I will proceed directly to behavioral interactions.

Behaviorally and morphologically, colony function is dominated by an intricate division of labor based upon caste (queen or worker) and worker size and age. A basic division of labor for social insects is that of reproduction—generally, the queen lays all the eggs and the workers do all the work. Within the worker caste, however, labor is further subdivided on the basis of worker size and age (Fig. 6). In a mature colony, workers range in size such that the largest weighs about 15 times as much as the smallest. Colonies begin life with only small workers; but as they grow, the proportion and size of larger workers gradually increases. This phenomenon adds a developmental-time dimension to any discussion of division of labor by size that has heretofore not been recognized. Several general differences in behavior are apparent between large and small workers—large workers carry and handle larger particles and brood, they are more likely to cut up insect prey, less likely to feed on liquid food, and less likely to function in brood and queen care (Mirenda and Vinson 1981).

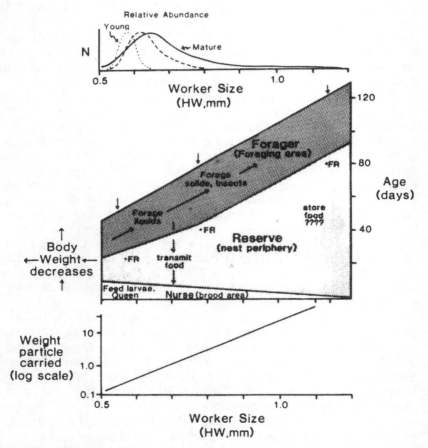

FIGURE 6. Division of labor by worker size and age in S. invicta. Top: Size-frequency distribution of workers in colonies during growth and maturity (after Wood and Tschinkel 1981). Middle: Age, size, task, and location in nest. Age on vertical axis; size (head width) on horizontal axis (data from Mirenda and Vinson 1981). Bottom: Size of particle carried by worker in relation to worker size. Note log scale (data from Wilson 1978).

Large workers live much longer than do small workers (up to 3-fold), but all workers pass through a series of changes in the labor they carry out as they grow older (Mirenda and Vinson 1981). This polyethism is superimposed over and modified by the division of labor by size already discussed. Early in their adult lives, workers act as nurses taking care of the queen and brood and are, thus, found mostly in the brood area. Small workers spend a larger proportion of their lives as nurses than do large workers. The largest essentially do not function as nurses. As the workers age, they leave the brood area and move to the peripheral nest areas to take up their roles as "reserves." As such, they receive food from foragers returning to the nest, transfer it to the nurses, and take part in the many other nest functions including construction, sanitation, and defense. During this period, large workers are more likely to store liquid food in their crops for longer periods than are small workers. Only during the last 25 to 40% of their lives do workers ever leave the nest to forage. During this final period, small workers are more likely to forage on liquids and larger workers on solids and insects (Mirenda and Vinson 1981).

It should be noted that none of these behavioral differences with age and size are sharp and that there appears to be a good deal of built-in flexibility (Mirenda and Vinson 1981). Most of these patterns have been worked out for small colonies, and there may be both quantitative and qualitative changes as the colony grows and ages.

Turning to trophic relationships, most of the fire ant's natural diet is insects and other small invertebrates although carrion, honeydew, and some plant material are also taken (Hays and Hays 1959). Only about 10 to 20% of the workers act as foragers at any given time, so that the food they collect must be shared with the other 80 or 90% of the colony. Workers themselves utilize only a small proportion of the colony's food. The bulk of utilization is by the queen and the larvae, both of which get the lion's share of the protein (Howard and Tschinkel 1981). Once foragers have brought the food back to the nest, it enters a complex web of exchanges, conversions, and processes (Fig. 7) which we are finally beginning to unravel (Howard and Tschinkel 1981; Sorensen and Vinson 1981). Foragers pass the food to the younger reserves who, in turn, transport it from the nest periphery to the brood and queen area where they share it with the nurse workers. These nurses are the youngest class of workers and carry out the functions of brood and queen care. They pass liquid food and probably glandular secretions to the second and third stage larvae and to the queen, all of whom subsist entirely on a liquid diet. Solid food is passed to the 4th-stage larvae (Petralia and Vinson 1978); they process it, utilize part, and probably pass some of it back to the workers in partly digested form. They may also pass back glandular secretions of an unknown nature in addition to excreted material (unpublished data). The queen may also pass back excreted or secreted material, but her major role is to produce eggs. We have recently found that late 4th-instar larvae are needed for continued egg-laying by the queen. Removal of such larvae causes egg-laying to cease within 24 hours while their re-addition causes egg-laying to commence once more in relation to the log of their number (Fig. 8). We have shown that a bulk substance moves from the larvae to the workers to the queen and into the eggs (unpublished). The image that is emerging with respect to the flow and use of food is that the various members of the colony are all pieces of a metabolic system in which each plays a distinct metabolic and behavioral role and all are interdependent. As yet, we have only a dim idea of how this metabolic labor is subdivided and specialized among colony members, but its beauty and sophistication are becoming apparent.

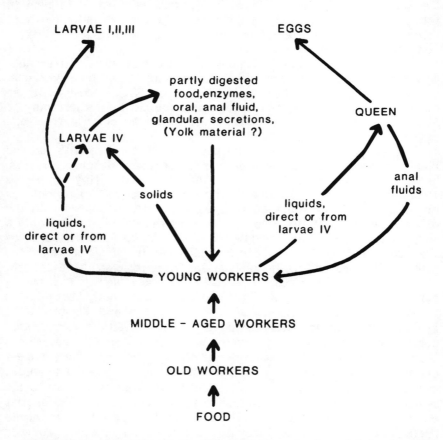

FIGURE 7. Trophic relationships within colonies of S. invicta. Arrows indicate the direction of flow of materials.

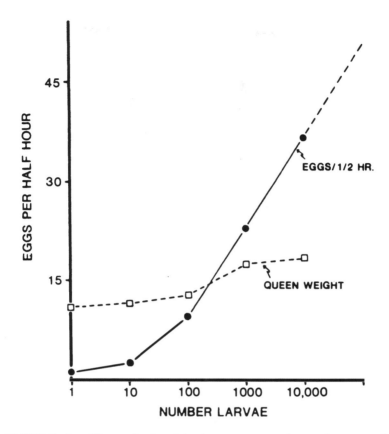

FIGURE 8. The relationship between number of larvae in colonies and the queen's egg-laying rate. Note the log scale. The queen's weight increases in parallel to the ovarian development required for increased oviposition (Tschinkel, original data).

SOME IMPORTANT UNANSWERED QUESTIONS

There are many important unanswered questions concerning the biology of S. invicta. Some of the most significant are as follows:

1. Certainly one of the largest holes in our knowledge of fire ants is the area of colony population dynamics. In Fig. 9, I present a hypothetical survivorship curve based upon my impressions and very sketchy data. The curve incorporates the following characteristics: (a) very high mortality during the mating flight and colony founding period with perhaps as little as 0.1% survival; (b) reduced, but still substantial and steady, colony mortality during the growth period; and (c) relatively low mortality during the colony maturity stage.

We have absolutely no idea of what happens to old colonies, so I have entered this phase as a dotted line. Is there increased mortality during this period? We have no quantitative data on the age-specific mortality during any of the life cycle phases. This kind of "life-table" information is absolutely necessary for an ecological understanding of the fire ant and the eventual development management programs. Ultimately, we will need this kind of information for each of the different kinds of environment the fire ant occupies if we are to manage it intelligently.

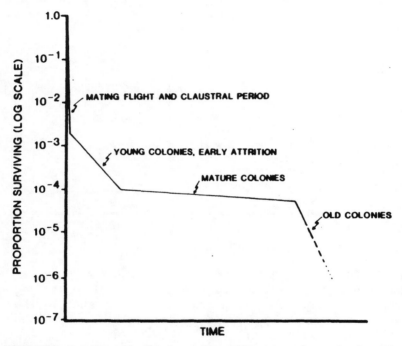

FIGURE 9. Hypothetical survivorship curve for colonies of S. invicta. Note that the proportion surviving is on a log scale.

2. The relationship to native ants has already been mentioned. Although an exotic in the USA, the fire ant is a member of a biological community relating in some manner to other members of that community. I have already provided evidence that its relationship to other ants is a key component in our understanding of fire ant ecology; however, only scattered, generally unquantitative information is available.

3. While we know generally what fire ants eat, we have only poor quantitative knowledge of the impact their feeding has on the community in which they live. Furthermore, individual colonies

show idiosyncratic food preferences, the meaning and origin of which we do not understand (Glunn et al. 1981). Surely, this is an important aspect of any bait-based fire ant control method.

4. The process of colonization of newly available habitat is poorly understood. How much control do queens exercise over the sites in which they settle after a mating flight? How are these sites selected? Which site and other physical characteristics influence success of colony foundation and how much? The founding period is obviously one of the most vulnerable stages of colony life so that, from the management point of view, knowledge of this period seems imperative.

5. The number of mated queens during various parts of the life cycle is an important question. I have mentioned that colonies are often founded by groups of mated queens in cooperation (Tschinkel and Howard 1983). We have some idea of what the advantages of such behavior might be, though none are as yet quantitatively characterized. We know that during the post-founding period workers in laboratory nests kill all but one queen so that the colony enters the growth phase with a single queen. It is, therefore, interesting that mature colonies often have more than one inseminated queen capable, under the right conditions, of laying eggs (Tschinkel and Howard 1978). In most of the fire ant's range, these extra queens are inhibited from laying eggs but can take the place of the colony queen-mother if she is lost or dies. In some populations scattered throughout the range, all of the mated queens in a colony may lay eggs (polygyny), and a colony may contain hundreds or even thousands of laying queens (Glancey et al. 1973). What brought about this profound change and why? What is the effect of queen number during various parts of the colony cycle on the ant's ecology and possibly on its potential range? Are they adopted after mating flights? If so, why are they not killed? Does mating take place in the nest? If so, what is the genetic importance, and what effect does it have on the westward spread? If mating flights are unnecessary in polygynous colonies, can the ant survive in areas where there is no summer rain, such as California?

6. Finally, the fact that, in comparison with those of Brazil, the fire ants in the USA are all much more closely related to one another because they are descended from a single pair or small number of individuals has received practically no attention (Tschinkel and Nierenberg 1983). This situation could well bring about major differences between the biology of the fire ant in the USA and its biology in Brazil. For example, it might affect the degree of cooperation and thus success during the founding period, it might affect the degree of colony distinctiveness and competition, and it might even affect the mode of reproduction.

SUMMARY AND FINAL REMARKS

In summary, the fire ant is a weed species whose continued existence is favored by human ecological meddling. Its success as a weed is based upon high reproductive rate, excellent dispersal and colonizing ability, rapid colony growth and early reproduction. What we don't know about its ecology can indeed hamper efforts to control them with respect to large-scale, insecticide-based control programs. Colony life is dominated by a complex and subtle division of labor and an intricate set of trophic relationships. These and various behavioral control mechanisms regulate colony function to maximize reproductive success.

Seen from the biologist's point of view, the fire ant represents a wonderful research opportunity almost unique in the history of myrmecology. The ant's abundance, ease of maintenance, and general habits make it an outstanding subject for research while our society's relatively high need for knowledge of this ant gives us the opportunity to carry out this research. This potential knowledge may some day more than repay us for the real and imagined damage and aggravation the fire ant caused along the way.

Whatever we, as a society, decide to do about the fire ant, whether it be to intervene or leave it alone, we are obligated to do it intelligently with a sound foundation of biological knowledge. If the decision is to "do something" about the fire ant, even if only in certain situations, we must know a great deal more of the fire ant's secrets before we can ever hope for success.

REFERENCES CITED

Banks, W. A., D. P. Jouvenaz, D. P. Wojcik, and C. S. Lofgren. Observations on fire ants, Solenopsis spp. in Mato Grosso, Brazil. Sociobiology 11: 143-152.

Glancey, B. M., C. H. Craig, C. E. Stringer, P. M. Bishop, and B. B. Martin. 1973. Multiple fertile queens in colonies of the imported fire ant, Solenopsis invicta. J. Ga. Entomol. Soc. 8: 327-328.

Glunn, F. J., D. F. Howard, and W. R. Tschinkel. 1981. Food preference in colonies of the fire ant, Solenopsis invicta. Insect. Soc. 28: 217-222.

Hays, S. B., and K. L. Hays. 1959. Food habits of Solenopsis saevissima richteri Forel. J. Econ. Entomol. 52: 455-457.

Howard, D. F., and W. R. Tschinkel. 1981. The flow of food in colonies of the fire ant, Solenopsis invicta: A multi-factorial study. Physiol. Entomol. 6: 297-306.

Ito, Y. 1980. Comparative ecology. Cambridge Univ. Press, Cambridge, England.

Markin, G. P., and J. H. Dillier. 1971. The seasonal life cycle of the imported fire ant, Solenopsis saevissima richteri on the Gulf Coast of Mississippi. Ann. Entomol. Soc. Am. 64: 562-565.

Markin, G. P., H. L. Collins, and J. H. Dillier. 1972. Colony founding by queens of the red imported fire ant, Solenopsis invicta. Ann. Entomol. Soc. Am. 65: 1053-1058.

Markin, G. P., J. H. Dillier, and H. L. Collins. 1973. Growth and development of colonies of the red imported fire ant, Solenopsis invicta. Ann. Entomol. Soc. Am. 66: 803-808.

Markin, G. P., J. H. Dillier, S. O. Hill, M. S. Blum, and H. R. Hermann. 1971. Nuptial flight and flight ranges of the imported fire ant, Solenopsis saevissima richteri. J. Ga. Entomol. Soc. 6: 145-156.

Mirenda, J. T., and S. B. Vinson. 1981. Division of labor and specification of castes in the red imported fire ant, Solenopsis invicta. Anim. Behav. 29: 410-420.

Morrill, W. L. 1974. Production and flight of alate red imported fire ants. Environ. Entomol. 3: 265-271.

Petralia, R. S., and S. B. Vinson. 1978. Feeding in larvae of the imported fire ant, Solenopsis invicta: Behavior and morphological adaptations. Ann. Entomol. Soc. Am. 71: 643-648.

Sorensen, A. A., and S. B. Vinson. 1981. Quantitative food distribution studies within colonies of the imported fire ant, Solenopsis invicta. Insect. Soc. 28: 129-160.

Summerlin, J. W., A. C. F. Hung, and S. B. Vinson. 1977. Residues in non-target ants, species simplification and recovery of populations following aerial applications of mirex. Environ. Entomol. 6: 193-197.

Tschinkel, W. R., and D. F. Howard. 1978. Queen replacement in orphaned colonies of the fire ant, Solenopsis invicta. Behav. Ecol. Sociobiol. 3: 297-310.

Tschinkel, W. R., and D. F. Howard. 1983. Colony founding by pleometrosis in the fire ant, Solenopsis invicta. Behav. Ecol. and Sociobiol. 12: 103-113.

Tschinkel, W. R., and N. C. E. Nierenberg. 1983. Possible importance of relatedness in the fire ant, Solenopsis invicta, in the United States. Ann. Entomol. Soc. Am. 76: 989-991.

Wilson, E. O. 1978. Division of labor in fire ants based on physical castes. J. Kans. Entomol. Soc. 51: 615-636.

Wilson, N. L., J. H. Dillier, and G. P. Markin. 1971. Foraging territories of imported ants. Entomol. Soc. Am. 64: 660-665.

8

Observations on the Biology and Ecology of Fire Ants in Brazil

D. P. Wojcik

The biology and ecology of the red imported fire ant (Solenopsis invicta) and the black imported fire ant (S. richteri) have been studied extensively in the United States (see reviews by Lofgren et al. 1975; Tschinkel 1982 and Chapter 7; Wojcik 1983), but comparable studies have not been conducted in South America. Silveira–Guido et al. (1968, 1973) reported on S. richteri and other S. saevissima group species present in Uruguay and Argentina; however, these studies must be reevaluated because of taxonomic changes in this group (Buren 1972). In addition many new species need to be described, the status of older names clarified, and the distribution limits of each species demarcated (J. Trager, Univ. of Florida, Gainesville, Florida, is currently revising the genus). There is little published information on S. invicta in South America; this paper presents observations accumulated during a series of trips made by me and other scientists of the Insects Affecting Man and Animals Research Laboratory, ARS, USDA, Gainesville, Florida to the State of Mato Grosso, Brazil over the past 11 years. Where appropriate, comparisons are made with populations of S. invicta in the United States.

DISTRIBUTION

The distribution of S. invicta given by Buren et al. (1974) is essentially unchanged despite recent data from several collecting trips made between 1973 and 1983 by W. F. Buren and associates and USDA personnel to other areas of Brazil, Uruguay, Argentina, Paraguay, and Bolivia.

Also, no new information has been discovered relating to the introduction of S. invicta into the United States since the extensive zoogeographical study by Lennartz (1973). It has been assumed S. invicta entered on products shipped through the Rio de la Plata region of Argentina and Uruguay; however, S. invicta has been found as far north as Porto Velho, Rondonia, Brazil (Buren et al. 1974). It is not known if its distribution extends further down the Madeira River from Porto Velho towards the Amazon basin or into north-eastern Brazil. S. invicta has never been collected from either the Amazon basin or the Rio de la Plata region, although there are collection records of other Solenopsis spp. The presence of S. invicta in the former area could, of course, provide an alternate explanation for importation of S. invicta into the United States.

MOUND SIZE

One of the most obvious manifestations of S. invicta colonies in the United States is their large conspicuous mounds. The absence of these large mounds, however, does not mean that ant colonies are not present, since the presence and/or persistence of the mound is usually dependent on soil type and moisture conditions (Wojcik 1983). It should be stressed that colony size (number of ants) and mound size (height and diameter of the tumulus) are not necessarily related (Banks et al. 1985). Observations in Brazil indicate similar phenomena. During the extensive dry season (May through September) (Coutinho 1982) or other dry periods, S. invicta often do not or cannot construct a mound. However, large mounds can readily be found if appropriate conditions prevail, particularly during the rainy season (Fig. 1a) (Wojcik 1983; Banks et al. 1985). Most of the S. invicta colonies collected in Mato Grosso during the wet season (January to February 1985) had well-developed mounds of average size (20-30 cm high and 30-40 cm wide) that were typical of their mounds in the United States (Anonymous 1958).
During the 1984 dry season in Mato Grosso, the road shoulders of BR-070 from Cuiaba to Caceres (ca 215 km) were cleared by burning. Fire ant mounds (mostly S. invicta) were found along almost the entire length of the highway. These mounds (Fig. 1a, b) were often as large as the largest mounds found in the United States (Anonymous 1958). They were characterized by a coarse surface, often accompanied by vegetative growth. They could readily be distinguished from termite mounds, which had a smoother surface and rarely supported any vegetation. Also, fire ant mounds were much softer than the hard, rock-like termite mounds.
Fire ant colonies collected in January to February 1985 (wet season) were large and mature as indicated by the presence of alates and large numbers of workers and brood of all sizes (Wojcik and Jouvenaz, unpublished). In contrast, few small colonies have been found, a condition that suggests populations in Brazil are in a steady-state condition with a low turnover rate (Wojcik, unpublished; Banks et al. 1985).
In 1984 and 1985, numerous abandoned mounds were found in densities approximating those found at times in the United States

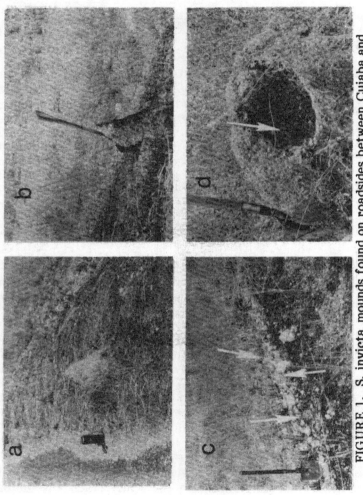

FIGURE 1. S. invicta mounds found on roadsides between Cuiaba and Caceres, Mato Grosso, Brazil during the 1984 dry season. a) 41 cm high mound on roadside. b) 51 cm high mound on roadside in moister area. c) Abandoned mounds (arrows) near active S. invicta mound. d) Active S. invicta colony (arrow) in termite mound.

(Wojcik, unpublished; Hays et al. 1982). This condition was not observed during previous trips, perhaps because mounds abandoned on roadsides were hidden by vegetation and mounds on lawns were rapidly destroyed by mowing machinery. The abandoned mounds observed along roadsides in the 1984 dry season (Fig. 1c) were numerous and of various sizes; in the 1985 wet season, there were about one-third as many abandoned mounds as active mounds. Indeed, when an abandoned mound was located, an active colony was usually found nearby. These active colonies may have migrated from the abandoned mounds.

These observations and those of Banks et al. (1985) negate, for the most part, published statements about the lack of large mounds in South America (Allen et al. 1974; Buren et al. 1974; Williams and Whitcomb 1974) and suggest that mound building behavior is related to temperature, moisture, and soil type rather than behavioral or population differences (see Francke and Cokendolpher, Chapter 9).

NUMBERS OF ANTS PER COLONY

Although estimates of the number of ants per colony in mounds have not been made in Brazil, as in the United States (Markin et al. 1973), some general observations can be made. Based on years of experience in excavating S. invicta colonies in the United States and in South America, it appeared that the number of workers, alates, and immatures in large colonies in South America were about equal to the numbers in large colonies in the United States (Wojcik and Banks, unpublished). Recently, Banks et al. (1985) confirmed this observation when they estimated S. invicta colony size in Brazil with a population index method (Harlan et al. 1981) and found that 65% of the colonies examined had over 50,000 workers and 16% had between 10,000 and 50,000 workers.

NUMBERS OF FERTILE QUEENS PER COLONY

S. invicta had been considered a monogynous species until the reports by Glancey et al. (1975) of a colony that contained hundreds of fertile queens. Subsequently, several other cases of polygyny have been recorded in the U.S. (Lofgren and Williams 1984). We have found no evidence to date for polygyny in any of our surveys in South America.

MATING FLIGHTS

Mating flights are the primary means of natural spread and reinfestation for S. invicta and fire ants in general. When conditions are suitable, unmated males and females fly from the same or different nests to a height of 100 to 300 m and mate (Markin et al.

1971; Bass and Hays 1979). Flights can occur in all months (Morrill 1974) or weeks (Banks et al., unpublished) of the year when conditions are suitable. In the United States, the major peak of flights occurs in the spring with a smaller peak in the fall. In Mato Grosso, alates have been observed in colonies every month of the year (Banks et al. 1985; Wojcik, unpublished). Although no counts were made, Wojcik and Banks (unpublished) found that the numbers and sex of alates in the nests varied considerably just as they do in the United States (Anonymous 1958). Mating flights of S. invicta have been observed in Mato Grosso in January, February, April, May, and November (Banks et al. 1985), March (Williams 1980), and in August (Allen et al. 1974). J. Trager (personal communication) reported that small flights of S. invicta started with the commencement of rains in Cuiaba in October 1984 and continued through November 1984. After continuous heavy rains in early December 1984, very large mating flights of S. invicta occurred throughout the remainder of December 1984.

One apparent difference has been noted in mating flight behavior between North American and Brazilian populations of S. invicta. North American populations are generally reported to fly from 11 AM to 4 PM (Markin et al. 1971; Roe 1973; W. A. Banks, personal communication), although Rhoades and Davis (1967) observed one flight at 9 AM and Hays (1959) stated that flights took place in the morning and were usually completed by noon. In Brazil, S. invicta is reported to fly from 9 AM to 12 PM (W. A. Banks and D. F. Williams, personal communication). While the actual mechanisms that initiate mating flights are not well understood (Lofgren et al. 1975), the importance of the time of day has not been established. The success of mated queens in colony founding has not been studied in South America.

FIRE ANTS IN TERMITE MOUNDS

Fire ant colonies are supposed to be common in termite mounds in Mato Grosso (Whitcomb 1974). I have found this to be true in only one location in southwestern Mato Grosso, one-half way between Varzea Grande and Ilha do Piraim (near Joselandia). In this large grassland that is periodically flooded, 14 of 16 termite mounds in a 2-acre area contained fire ant nests, mostly S. invicta. In contrast, in other areas of Mato Grosso, fire ant colonies were found only occasionally in termite mounds (Fig. 1d). The fire ants rework the tunnels so that the section of the mound they occupy is easily distinguished by differences in tunnel architecture.

It has been suggested (C. S. Lofgren, personal communication) that the presence of S. invicta in termite mounds is a response to periodical flooding (example given above). However, the mound in Fig. 1d was located on an embankment along a highway where there was no possibility of flooding. Another study by Redford (1984) of termite mounds in Brazil (in cerrado in Goias) has shown other Solenopsis spp. to be common inhabitants of termite mounds. Redford's study was conducted in an area where flooding was not a consideration.

NATURAL FOOD SOURCES

The main food sources of S. invicta in Brazil appear to be other arthropods (Wojcik, unpublished). I have seen S. invicta feeding on carrion, fresh-caught fish, and garbage. The ants readily feed on beef and honey baits (Wojcik 1983). In Brazil, S. invicta is evidently an opportunistic predator and scavenger, as it is in the United States (Lofgren et al. 1975). In the 1984 dry season, I found fire ant mounds along highway BR-174 west of Caceres that contained large amounts of plant material, mostly fragments of rice grains which had fallen or blown off passing trucks. There were no rice fields in the immediate vicinity of the mounds sampled. Previously W. A. Banks and D. P. Jouvenaz (personal communication) had observed large amounts of mascerated plant material (leaf and stem fragments, seeds, and seed husks) in S. invicta mounds near Cuiaba. Thus, in some circumstances (mechanisms unknown), S. invicta will feed on plant material. This behavior has been noted in the United States (Lofgren et al. 1975).

In the United States, S. invicta has been reported tending Homoptera on a variety of plants (Anonymous 1958; Nielsson et al. 1971) including citrus (Adams, Chapter 5). Solenopsis species other than S. invicta have been observed tending Homoptera on citrus in South America (Williams et al. 1975; Bartoszeck 1976). Recently S. invicta were observed tending aphids on citrus (C. S. Lofgren and W. A. Banks, personal communication; Wojcik, unpublished) and feeding on extrafloral nectaries of a plant (family Leguminosae) (Wojcik, unpublished) on the EMPA Research Station at Caceres. These associations have not been previously reported in the literature, probably because of the allopatric distributions of S. invicta and entomologists in South America.

COLONY DENSITIES

Recent data on fire ant populations in Mato Grosso (Wojcik, unpublished; Wojcik 1983; Banks et al. 1985) indicate that published statements (Buren et al. 1978; Buren 1980; Whitcomb 1980) implying that S. invicta populations in Mato Grosso are lower than populations in the United States are not true. These latter authors did not make population counts and based their statements only on observations. General comparisons between North and South America are difficult because of differences in ecology and patterns of land use. The only

94

valid comparisons are those using comparable habitats. It must be stressed, however, that few if any of the habitats in South America are strictly comparable to any of the habitats in North America (Wojcik 1983). Fire ants are ants of disturbed areas (Buren et al. 1978) and over half of the area of Mato Grosso is classified as cerrado (Santos et al. 1977) which is mostly undisturbed except for periodic grazing and burning (Coutinho 1982). In Mato Grosso, cerrado (floristically described by Eiten 1982) is characterized by fire climax vegetation and soils which have nutrient deficiencies, acid ph, aluminum and/or manganese toxicity, and lateritic hardpans (Coutinho 1982). It is generally agreed that such a moderately disturbed area does not constitute a primary habitat for S. invicta in Brazil (Allen et al. 1974; Wojcik 1983; Banks et al. 1985). Allen et al. (1974) did not find any S. invicta in cerrado; however, Banks et al. (1985) surveyed three small plots (0.031 ha each) in cerrado and found two S. invicta nests. In severely disturbed cerrado such as roadsides, S. invicta nests can be very common (Table 1). For example, roadside populations of S. invicta colonies in Mato Grosso at times exceed average counts from the United States (Banks et al. 1973; Williams and Lofgren 1982). Conversely, many areas in Brazil are without fire ants or have low populations (Whitcomb 1980), a condition that is also found in the United States (Wojcik and Banks, unpublished).

Another way to assess ant population densities is with baits, which have the advantage of being a non-biased collection method (non-biased in reference to minimum human collecting error; see Wojcik et al. 1975 for methods). Bait surveys along transects in the Cuiaba area (Coxipo and Federal University of Mato Grosso) in 1981 showed high populations of S. invicta attracted to meat and honey baits even though medium populations of other ants were also found (see Wojcik 1983, Fig. 2, 3). Bait surveys in the United States from a highly infested area (Baldwin, Florida) and a lightly infested area (Gainesville, Florida) showed similar results.

PATHOGENS AND INQUILINES

Some of the arthropods associated with fire ants in Uruguay and Argentina were described by Silveira-Guido et al. (1973) and in Brazil (mainly in the Sao Paulo area) by Williams (1980). Many of these same or related arthropods have been found in fire ant nests in Mato Grosso. Jouvenaz (1983; Chapter 27) summarized the known data on fire ant pathogens. These references do not provide data on seasonal distribution; thus, in both the 1984 dry season (July to September) and the 1985 wet season (January to February), I collected inquilines and pathogens from the same three plots. Standard samples of 2 1/2 liters of tumulus were collected in 5 liter buckets and the ants and inquilines separated from the soil by water floatation (Jouvenaz et al., in press).

TABLE 1. Colony densities of fire ant nests in Mato Grosso, Brazil and the southern United States.[a]

| Habitat | Ant nests per acre | | | |
	Brazil	Florida	Alabama	Texas
Highway (age)[b]				
1-2 yr	66[ce]	2[g]	16[g]	10[g]
3-4 yr	25[cf]			
	9[df]			
4-5 yr	13[e]			
	8[df]			
6-7 yr	23[e]			
10+ yr		10[g]	15[g]	13[g]
		109[j]		
Forest[h]	13[e]	15[i]		
		3[g]	3[g]	2[g]
Lawn	14[e]	22[i]		
	9[f]			

[a]Modified from Wojcik 1983.
[b]The Brazilian highway counts are all from roadsides in disturbed cerrado.
[c]The same highway
[d]The same plot.
[e]Modified from Banks et al. 1985.
[f]D. P. Wojcik and D. P. Jouvenaz, unpublished data.
[g]USDA, 1967, unpublished data, 2% random samples of 64,000 acre blocks.
[h]Brazil data from cerrado; United States data from pine forests.
[i]R. E. Brown, 1979, unpublished data, systematic random sampling over entire state of Florida.
[j]Modified from Williams and Lofgren 1982.

Thelohania solenopsae was not detected in either the 1984 or 1985 collections, although it was commonly found in Mato Grosso on previous trips (Jouvenaz et al. 1980). Parasitic ants, Solenopsis (=Labauchena) spp., were not found in Mato Grosso. Five species have been reported from fire ant colonies (Wilson 1971; Williams 1980; Banks et al. 1985), but none from S. invicta colonies. A parasitic ant would have potential in a biological control program for S. invicta. No phorid flies (Diptera: Phoridae) were collected in 1984 and 1985 even though Williams (1980) collected them from S. invicta colonies and I have collected them at other times in Mato Grosso.

Chrysomelid beetle larvae (Coleoptera: Chrysomelidae) (Kistner 1982) were not collected from S. invicta nests; these case-bearing larvae had been collected once on an earlier trip (Wojcik, unpublished). Other inquilines associated with Solenopsis spp. colonies in South America and not collected in Mato Grosso are: Acarina (Reichensperger 1927); Lepidoptera (Bruch 1926); Hemiptera (San Martin 1966a, 1966b); Strepsiptera (Teson and de Remes Lenicov 1979); Hymenoptera, Diapriidae (Borgmeier 1939), Bethylidae (Bruch 1917); and Coleoptera, Staphylinidae (Frank 1977; Wojcik 1980; Kistner 1982), Pselaphidae (San Martin 1968a, 1968b), Tenebrionidae (Steiner 1982). In addition, many other groups of arthropods are known to be inquilines in the nests of other ants (Wilson 1971; Kistner 1982).

The diseases and inquilines found in S. invicta colonies in the three plots checked in the 1984 dry season and the 1985 wet season are listed in Table 2. The percentage of total collections as shown do not add up to 100%, because it was common to have more than one disease and/or inquiline in each colony; e.g., one colony contained over 100 larvae, pupae, and adults of an Orasema sp. (Eucharitidae), and 12 1/2% of the adult worker ants were infected with an undescribed nematode (Jouvenaz et al., in press). The data on the influence and effects of pathogens on S. invicta is limited (Jouvenaz, Chapter 27). The incidences of the diseases were higher but similar to those given by Jouvenaz et al. (1980). The nematode mentioned above is the first record of these parasitic organisms in Solenopsis in South America. There are several species of myrmecophilous Scarabaeidae (Coleoptera) present in Mato Grosso (Chalumeau 1983), one of which is known to be predaceous on S. invicta (Wojcik 1975). Nothing is known about the hister beetle (Coleoptera: Histeridae), the Thysanura are sometimes predaceous (Wojcik, unpublished), and the Diplopoda are scavengers (Wojcik, unpublished). The Orasema spp. (Hymenoptera: Eucharitidae) are ectoparasites of ant larvae and pupae (Williams 1980), which cause malformation and eventual death of the pupae (Wojcik, unpublished). Any promise that these parasites may have for a biological control program is reduced by the fact that, in all cases where oviposition behavior is known, the eggs are laid in plant tissue, including that of many economically important plants (Clausen 1940; Tocchetto 1942; Nicolini 1950). In at least one case, damage to bananas by an Orasema sp. required bait treatment of the host ants to eliminate the problem (Ostmark and Evers 1976).

Although we have little data on the effects of diseases and/or inquilines on colony longevity, it is reasonable to assume that these organisims reduce colony fitness. It is known that some diseases reduce colony vitality and cause greater mortality under stress conditions (Allen and Knell 1980; Jouvenaz, Chapter 27). Inquilines that eat fire ant immatures (e.g., some scarab beetles and Thysanura) or cause the malformation and death of fire ant immatures (e.g., Orasema spp.) produce additional stress on fire ant colonies. This combination of additive mortalities and draining of resources has to have some effect on fire ant colony vigor and/or survival. Since the effects of a biological control agent in its homeland are not necessarily indicative of its utility when introduced into a new environment (Doutt and DeBach 1964), the effectiveness of these organisms must be evaluated before they can be discarded from an IPM program for fire ant control.

TABLE 2. Percentage of fire ant nests examined that contained diseases and/or inquilines in Mato Grosso, Brazil.[a]

	1984[b]		1985[c]	
	With other species	By itself	With other species	By itself
Diseases				
Protozoa				
Thelohania	0.0	0.0	0.0	0.0
Vairimorpha	6.7	1.9	5.7	1.6
Mattesia	7.8	3.8	1.9	1.6
Nematode	0.0	0.0	5.7	3.1
Inquiline				
Scarabaeidae	6.3	2.8	31.3	9.4
Histeridae	0.0	0.0	18.7	6.3
Thysanura	0.0	0.0	3.0	0.0
Diplopoda	0.0	0.0	30.3	18.8
Eucharitidae	28.1	16.2	45.5	25.0
Without disease or inquiline	53.1		12.3	

[a]Results from colonies collected on the same 3 plots.
[b]64 colonies collected; 34 contained disease and/or inquilines; 23 colonies contained only a single species of disease or inquiline.
[c]114 colonies collected; 86 contained diseases and/or inquilines; 40 colonies contained only a single species of disease or inquiline.

COLONY LIFE SPAN

Buren et al. (1978) have hypothesized a population model for fire ants in South America based on the assumption that enclaves of fire ant colonies are ephemeral and die out as a result of diseases, parasites, or predators. While these biotic factors certainly have an effect, other biotic factors, such as habitat, and abiotic factors, particularly the prolonged dry season, cannot be ignored. There are no data on the effects of diseases and/or inquilines on the length of colony life. None of the previously published studies on these organisms followed specific colonies or groups of colonies through several seasons. The statement by Allen and Knell (1980) that mounds over 2 yr of age are seldom found in South America is simply not supported by the facts. From studies by Markin et al. (1973), we know it takes approximately 2 to 3 yr for a colony to amass over 100,000 workers in the United States. Large colonies of this size are seen regularly in Brazil (Banks et al. 1985). The recent data on S. invicta populations in Brazil indicate it is unlikely that the hypothesis of Buren et al. (1978) reflects the real situation in Brazil.

CONCLUSIONS

While many aspects of S. invicta biology in Brazil remain unstudied and unknown, the following points can be made:

1. No new data are available on the origin or method of introduction of S. invicta into the United States.

2. S. invicta builds mounds in Brazil equal in size to the large mounds built in the United States, provided similar soil types are compared.

3. The numbers of ants found in colonies in Brazil are at least equal to the numbers found in colonies in the United States.

4. Polygynous colonies have not been found to date in Brazil.

5. The numbers of alates in mounds in Brazil are similar to numbers in the United States. The importance of the differences between Brazilian and North American populations concerning the time of day of mating flights has not been established.

6. S. invicta is not common in termite mounds in Mato Grosso.

7. S. invicta feed on arthropods, carrion, garbage, meat and honey baits, and extrafloral nectaries, and tend aphids as they do in the United States.

8. Colony densities in Brazil can be similar to those in the United States where similar habitats are considered.

9. The presence of pathogens and/or inquilines is one of many factors influencing S. invicta populations in Brazil.

10. S. invicta colonies in Brazil probably have a life span equal to the life span of colonies in the United States.

The many similarities between North and South American populations of S. invicta should not discourage the seeking of biological control agents in South America for use in North America. It is reasonable to assume that any biological control agents present are exerting pressure on S. invicta populations. These agents must be discovered and studied before we can determine how they can fit into a viable IPM program for fire ant control. The recent establishment of a cooperative USDA-EMBRAPA fire ant research program at the EMPA laboratory in Caceres, Mato Grosso, Brazil includes studies of the ecology and biology of S. invicta and their natural control agents, so that many unanswered questions about these pests can be resolved.

REFERENCES CITED

Allen, G. E., W. F. Buren, R. N. Williams, M. de Menezes, and W. H. Whitcomb. 1974. The red imported fire ant, Solenopsis invicta: Distribution and habitat in Mato Grosso, Brazil. Ann. Entomol. Soc. Am. 67: 43–46.

Allen, G. E., and J. D. Knell. 1980. Pathogens associated with the Solenopsis saevissima complex in South America. Proc. Tall Timbers Conf. Ecol. Anim. Control Habitat Manage. 7: 87–94.

Anonymous. 1958. Observations on the biology of the imported fire ant. USDA-ARS. ARS-33-49. 21 pp.

Banks, W. A., B. M. Glancey, C. E. Stringer, D. P. Jouvenaz, C. S. Lofgren, and D. E. Weidhaas. 1973. Imported fire ants: Eradication trials with mirex bait. J. Econ. Entomol. 66: 785–789.

Banks, W. A., D. P.Jouvenaz, D. P. Wojcik, and C. S. Lofgren. 1985. Observations on fire ants, Solenopsis spp., in Mato Grosso, Brazil. Sociobiology 11: 143–152.

Bartoszeck, A. B. 1976. Afideos de laranjeira (Citrus sinensis Osb.) e minoseira (Citrus reticulata B.), seus predadores e parasitas. Acta Biol. Parana, Curitiba 5: 15–48.

Bass, J. A., and S. B. Hays. 1979. Nuptial flights of the imported fire ant in South Carolina. J. Ga. Entomol. Soc. 14: 158–161.

Borgmeier, T. 1939. Sobre alguns Diapriideos myrmecophilos, principalmente do Brasil (Hym. Diapriidae). Rev. Entomol. 10: 530–545.

Bruch, C. 1917. Nuevas capturas de insectes mirmecofilos. Physis (Buenos Aires) 3: 458–465.

Bruch, C. 1926. Orugas mirmecofilas de Hameris epulus signatus Stich. Rev. Soc. Entomol. Argentina 1: 1–9.

Buren, W. F. 1972. Revisionary studies on the taxonomy of the imported fire ants. J. Ga. Entomol. Soc. 7: 1–26.

Buren, W. F. 1980. The importance of fire ant taxonomy. Proc. Tall Timbers Conf. Ecol. Anim. Control Habitat Manage. 7: 61–66.

Buren, W. F., G. E. Allen, W. H. Whitcomb, F. E. Lennartz, and R. N. Williams. 1974. Zoogeography of the imported fire ants. J. N. Y. Entomol. Soc. 82: 113–124.

Buren, W. F., G. E. Allen, and R. N. Williams. 1978. Approaches toward possible pest management of the imported fire ants. Bull. Entomol. Soc. Am. 24: 418–421.

Chalumeau, F. 1983. Batesiana et Martinezia, nouveaux generes d'Eupariini (Coleoptera: Scarabaeidae: Aphodiinae) du nouveau monde. Bull. Mensuel Soc. Linn. Lyon 52: 142-153.

Clausen, C. P. 1940. The oviposition habits of the Eucharidae (Hymenoptera). J. Wash. Acad. Sci. 30: 504-516.

Coutinho, L. M. 1982. Ecological effects of fire in Brazilian cerrado, pp. 273-291. In B. J. Huntley and B. H. Walker (eds.), Ecology of tropical savannas. Springer-Verlag, Berlin. 669 pp.

Doutt, R. L., and P. DeBach. 1964. Some biological control concepts and questions, pp. 118-142. In P. DeBach (ed.), Biological control of insect pests and weeds. Reinhold Publ. Co., NY. 844 pp.

Eiten, G. 1982. Brazilian "Savannas," pp. 25-47. In B. J. Huntley and B. H. Walker (eds.), Ecology of tropical savannas. Springer-Verlag, Berlin, 669 pp.

Frank, J. H. 1977. Myrmecosaurus ferrugineus, an argentinian beetle from fire ant nests in the United States. Fla. Entomol. 60: 31-36.

Glancey, B. M., C. E. Stringer, C. A. Craig, and P. M. Bishop. 1975. An extraordinary case of polygyny in the red imported fire ant. Ann. Entomol. Soc. Am. 68: 922.

Harlan, D. P., W. A. Banks, H. L. Collins, and C. E. Stringer. 1981. Large area tests of AC-217,300 bait for control of imported fire ants in Alabama, Louisiana, and Texas. Southwest. Entomol. 6: 150-157.

Hays, K. L. 1959. Ecological observations on the imported fire ant, Solenopsis saevissima richteri Forel, in Alabama. J. Ala. Acad. Sci. 30: 14-18.

Hays, S. B., P. M. Horton, J. A. Bass, and D. Stanley. 1982. Colony movement of imported fire ants. J. Ga. Entomol. Soc. 17: 266-274.

Jouvenaz, D. P. 1983. Natural enemies of fire ants. Fla. Entomol. 66: 111-121.

Jouvenaz, D. P., W. A. Banks, and J. D. Atwood. 1980. Incidence of pathogens in fire ants, Solenopsis spp., in Brazil. Fla. Entomol. 63: 345-346.

Jouvenaz, D. P., D. P. Wojcik, M. A. Naves, and C. S. Lofgren. Observations on a parasitic nematode of fire ants, Solenopsis spp., from Mato Grosso. Pesquisa Agropecuaria Brasileira. (In press).

Kistner, D. H. 1982. The social insects' bestiary, pp. 1-244. In H. R. Hermann (ed.), Social insects, Vol 3. Academic Press, NY. 459 pp.

Lennartz, F. E. 1973. Modes of dispersal of Solenopsis invicta from Brazil into the continental United States—a study of spatial diffusion. M.S. Thesis, Univ. Florida, Gainesville, Fla. 242 pp.

Lofgren, C. S., W. A. Banks, and B. M. Glancey. 1975. Biology and control of imported fire ants. Annu. Rev. Entomol. 20: 1-30.

Lofgren, C. S., and D. F. Williams. 1984. Polygynous colonies of the red imported fire ant, Solenopsis invicta (Hymenoptera: Formicidae) in Florida. Fla. Entomol. 67: 484-486.

Markin, G. P., J. H. Dillier, and H. L. Collins. 1973. Growth and development of colonies of the red imported fire ant, Solenopsis invicta. Ann. Entomol. Soc. Am. 66: 803-808.

Markin, G. P., J. H. Dillier, S. O. Hill, M. S. Blum, and H. R. Hermann. 1971. Nuptial flight and flight ranges of the red imported fire ant, Solenopsis saevissima richteri (Hymenoptera: Formicidae). J. Ga. Entomol. Soc. 6: 145-156.

Morrill, W. A. 1974. Production and flight of alate red imported fire ants. Environ. Entomol. 3: 265-271.

Nicolini, J. A. 1950. La avispita costurera y la tuberculosis del olivo. Rev. Agron. Porto Alegro (35-36): 20.

Nielsson, R. J., A. P. Bhatkar, and H. A. Denmark. 1971. A preliminary list of ants associated with aphids in Florida. Fla. Entomol. 54: 245-248.

Ostmark, H. E., and C. Evers. 1976. Manipulation of ant populations to control Orasema costaricensis in bananas. Paper presented at the XV Intern. Congr. Entomol., Washington, D.C. 23 August 1976.

Redford, K. H. 1984. The termitaria of Cornitermes cumulans (Isoptera: Termitidae) and their role in determining a potential keystone species. Biotropica 16: 112-119.

Reichensperger, A. 1927. Eigenartiger nestbefund und neue gastarten neotropischer Solenopsis-arten. Folia Myrmecologica et Termitologica 1: 48-51.

Rhoades, W. C., and D. R. Davis. 1967. Effects of meterological factors on the biology and control of the imported fire ant. J. Econ. Entomol. 60: 554-558.

Roe, R. A., II. 1973. A biological study of Solenopsis invicta Buren, the red imported fire ant, in Arkansas, with notes on related species. M.S. Thesis, Univ. of Arkansas, Fayetteville, Ark. 135 pp.

San Martin, P. R. 1966a. Nota sobre Anommatocoris coleopteratus Kormilev, 1955 (Vianaidina, Tingidae, Hemiptera). Rev. Bras. Biol. 26: 327-328.

San Martin, P. R. 1966b. Notas sobre Neoblissus parasitaster Bergroth, 1903 (Blissinae, Lygaeidae, Hemiptera). Rev. Bras. Biol. 26: 247-251.

San Martin, P. R. 1968a. Notas sobre Fustiger elegans Raffray (Coleoptera, Pselaphidae) en el Uruguay y la Argentina. Physis 28: 59-64.

San Martin, P. R. 1968b. Nuevo hallazgo de Metopioxys gallardoi Bruch, 1917 (Coleoptera, Pselaphidae, Metopiinae). Rev. Bras. Biol. 28: 27-28.

Santos, L. B. D., N. R. Innocencio, and M. R. de Silva Guimaraes. 1977. Vegetacao, pp. 59-84; Geografia do Brasil, Regiao Centro-oeste, Vol. 4. Fundacao Inst. Brasil. Geogr. Estat., Rio de Janeiro, Brasil. 364 pp.

Silveira-Guido, A, J. Carbonell, and C. Crisci. 1973. Animals associated with the Solenopsis (fire ants) complex, with special reference to Labauchena daguerrei. Proc. Tall Timbers Conf. Ecol. Anim. Control Habitat Manage. 4: 41-52.

Silveira-Guido, A., J. Carbonell-Bruhn, C. Crisci, and P. San Martin. 1968. Labauchena daguerrei Santschi como parasito social de la hormiga Solenopsis saevissima richteri Forel. Agron. Trop. (Maracay, Venez.) 18: 207-209.

Steiner, W. E., Jr. 1982. Poecilocrypticus formicophilus Gebien, a South American beetle established in the United States (Coleoptera: Tenebrionidae). Proc. Entomol. Soc. Wash. 82: 232-239.

Teson, A, and A. M. A. de Remes Lenicov. 1979. Estrepsipteros parasitoides de Hymenopteros (Insecta - Strepsiptera). Rev. Soc. Entomol. Argentina 38: 115-122.

Tocchetto, A. 1942. Bicho costureiro. Rev. Agron. Porto Alegre 6: 587-588.

Tschinkel, W. R. 1982. History and biology of fire ants, pp. 16-35. In S. L. Battenfield (ed.), Proc. Symp. Imported Fire Ant. USDA-APHIS, USEPA, Washington, DC. 255 pp.

Whitcomb, W. H. 1974. Natural populations of entomophagous arthropods and their effects on the agroecosystem, pp. 150-169. In F. G. Maxwell and F. A. Harris (eds.), Proc. Summer Inst. Biol. Control Plant Insects Diseases. Univ. Press of Mississippi, Jackson, Miss. 647 pp.

Whitcomb, W. H. 1980. Expedition into the Pantanal. Proc. Tall Timbers Conf. Ecol. Anim. Control Habitat Manage. 7: 113-122.

Williams, D. F., and C. S. Lofgren. 1982. Aerial application of AC-217,300 baits for control of red imported fire ants, 1979. Insect. Acar. Tests. 7: 269.

Williams, R. N. 1980. Insect natural enemies of fire ants in South America with several new records. Proc. Tall Timbers Conf. Ecol. Anim. Control Habitat Manage. 7: 123-134.

Williams, R. N., M. de Menezes, G. E. Allen, W. F. Buren, and W. H. Whitcomb. 1975. Observacoes ecologicas sobre a formiga lava-pe, Solenopsis invicta Buren, 1972 (Hymenoptera: Formicidae). Rev. Agric. (Piracicaba) 50: 9-22.

103

Williams, R. N., and W. H. Whitcomb. 1974. Parasites of fire ants in South America. Proc. Tall Timbers Conf. Ecol. Anim. Control Habitat Manage. 5: 49-59.
Wilson, E. O. 1971. The insect societies. Belknap Press, Cambridge, Mass. 548 pp.
Wojcik, D. P. 1975. Biology of Myrmecaphodius excavaticollis (Blanchard) and Euparia castanea Serville (Coleoptera: Scarabaeidae) and their relationships to Solenopsis spp. (Hymenoptera: Formicidae). Ph.D. Dissertation, Univ. of Florida, Gainesville, Fla. 74 pp. Dissert. Abstr. Intern. B36: 5962.
Wojcik, D. P. 1980. Fire ant myrmecophiles: Behavior of Myrmecosaurus ferrugineus Bruch (Coleoptera: Staphylinidae) with comments on its abundance. Sociobiology 5: 63-68.
Wojcik, D. P. 1983. Comparison of the ecology of red imported fire ants in North and South America. Fla. Entomol. 66: 101-111.
Wojcik, D. P., W. A. Banks, and W. F. Buren. 1975. First report of Pheidole moerens in Florida (Hymenoptera: Formicidae). Coop. Econ. Insect Rep. 25: 906.

9
Temperature Tolerances
of the Red Imported Fire Ant

O. F. Francke
and J. C. Cokendolpher

      The geographic distribution of organisms is determined by historical and ecological factors. Historical factors which have influenced the distribution of the fire ant, Solenopsis invicta Buren, include those which led to isolation from its nearest relatives in South America and its introduction into North America during this century. The ecological factors which influence the distribution of S. invicta are all those which confine its current range in South America (Buren et al. 1974) and which eventually will determine its final distribution in North America. The ecological factors can be biotic; i.e., the influence of other organisms, such as predators, parasites, pathogens, and competitors including man, and abiotic factors such as climate, photoperiod, and soil properties.

      The role that climate plays in influencing the distribution of organisms has long been recognized by biogeographers. Climatic factors have been implicated in influencing the final distribution of S. invicta in North America (Buren et al. 1974; Hung and Vinson 1978; Pimm and Bartell 1980; Moody et al. 1981). How does S. invicta adapt to temperature and what are its limits of tolerance to heat and cold? How will these limits relate to its final distribution in North America? These questions are addressed in this contribution.

TOLERANCE TO HEAT

      The resistance exhibited by insects to high temperatures depends on the interaction of at least three factors: length of exposure, relative humidity (RH), and behavior. Sudden exposure to high temperatures causes death due to denaturization of enzymes and other proteins, whereas gradual exposure to hot temperatures usually does not cause such severe biochemical effects. However, longer exposures to sub-acute temperatures can cause dehydration and ultimately lead to death. Thus, insects can presumably tolerate longer exposures to warm temperatures when the RH of the air is high (reducing evapotranspiration). The significance of RH is size dependent: small insects dehydrate more rapidly than larger insects because of their surface area/mass relationships.

O'Neal and Markin (1975) determined rates of brood development of S. invicta in the laboratory, starting with newly mated queens. Queens held at 25, 30, 32, and 35°C laid eggs from which minim workers developed successfully. At 38°C all the larvae had died by the fourth instar, presumably due to dehydration. Finally, queens held at 40°C died within 24 h of isolation, before laying any eggs.

Braulick (1982) determined the role played by dehydration in temperature-induced mortality in S. invicta. Major and minor workers were exposed to 26.7, 32.3, and 37.8°C at 0% RH, and the time of death was determined. The results (Fig. 1) confirm that minor workers dehydrate faster than major workers, and the warmer the temperature the faster the ants dehydrate. Ninety-nine percent of the minor workers died in 9 h at 27.6°C, 6 h at 32.3°C, and 4 h at 37.8°C.

Francke et al. (1985) determined the maximum temperatures that S. invicta minor workers could tolerate for 1 h. Ants were acclimated to 12, 22, and 32°C for at least 2 wk prior to the experiments; the tests were performed at 0% and 100% RH and mortality was determined after 24 h. Analysis of variance indicated that experimental temperature [$F_{(1,258)}=264.2$, $p < 0.001$], experimental RH [$F_{(1,258)}=72.2$, $p < 0.001$], and prior acclimation history [$F_{(2,258)}=3.2$, $0.05 > p > 0.01$] significantly affected mortality rates in minor workers. Comparison of the temperatures required to kill 50% and 95% of the minor workers (Table 1) at both 0% and 100% RH indicated that the ants did not die of dehydration and that survival was probably slightly enhanced at 0% RH (presumably by evaporative cooling). All ants acclimated at either 12 or 22°C died after 1 h exposure to 46°C, whereas a 1 h exposure to 48°C was necessary to kill all ants acclimated at 32°C (Fig. 2).

TABLE 1. Temperature tolerance (LD_{50} and LD_{95}) values for S. invicta after 1 h exposure to upper lethal temperatures (°C). Based on data from Francke et al. (1985).

	Acclimation regime					
	12°C		22°C		32°C	
	0% RH	100% RH	0% RH	100% RH	0% RH	100% RH
LD_{50}	43.2	40.8	42.4	42.3	44.0	—
LD_{95}	44.7	43.1	45.2	43.8	47.3	—

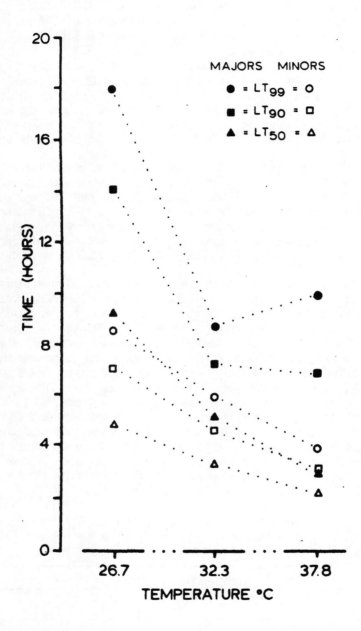

FIGURE 1. Lethal times (LT) for minor and major workers of S. invicta exposed to 26.7, 32.3, and 37.8°C at 0% RH (based on data from Braulick 1982).

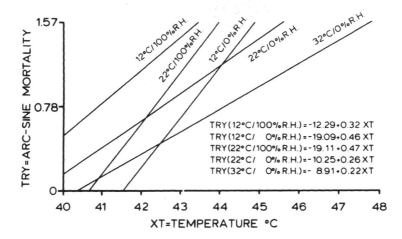

FIGURE 2. Regression models (lines and equations) predicting mortality values (Y-axis depicts arc-sine values for mortality: LD_{50}=0.78 rad., LD_{95}=1.49 rad.) for S. invicta acclimated at three different temperatures and tested at two different RH (redrawn after Francke et al. 1985).

The temperatures necessary to kill fire ants, directly or by dehydration, might appear to be rarely attained under natural conditions in North America. However, during the summer, temperatures in sunny areas at ground level can easily surpass the ants' limits of tolerance. For example, Hadley (1970) reported air temperatures (1 m high) and soil surface temperatures of 46 and 80.5°C, respectively, in the hot deserts of Arizona in July.

Two important behaviors influence fire ant heat tolerance. The first is the avoidance of direct exposure to the sun by foraging at night or by seeking shade during the warmest months of the year. The second behavioral modification is to move deeper into the soil as air and soil temperatures increase during the day and to move upwards into the mound at night.

Pinson (1980) monitored soil and S. invicta mound temperatures in central Texas during two consecutive summers. Results (Fig. 3) indicate that temperatures became more stable with increasing soil depth, and temperatures in a fire ant mound have a wider range of variation than do temperatures in the undisturbed soil column adjacent to a mound. Pinson (1980) also monitored the movement of eggs, larvae, pupae, and adult ants within those mounds and found that there was a daily progression from shallow-to-deep and back-to-shallow in the nests (presumably to escape high temperatures). In 1978, the temperature even at a depth of 15 cm in the mound was higher than 32°C for at least 6 h. At low RH such conditions would be lethal to 99% of the minor workers in a fire ant colony. During hot and dry conditions brood production ceases in fire ant colonies, and adult ants retreat to greater depths in their mounds.

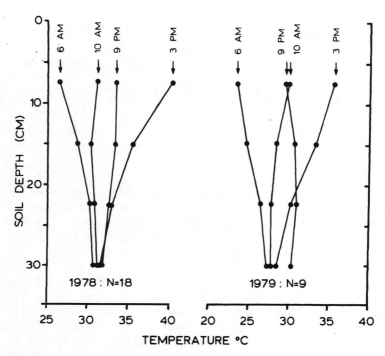

FIGURE 3. Mean mound temperatures of S. invicta at various depths and times during the summers of 1978 and 1979. Based on data from Pinson (1980): 1978 mounds at Boerne, Texas; 1979 mounds at Flatonia, Texas.

TOLERANCE TO COLD

The responses of insects to cold depend on spatial, temporal, and physiological parameters, and interactions among them (Danks 1978). Migration and dormancy are possible adaptive alternatives; however, neither is known to occur in S. invicta. Temporal parameters include phenology and other aspects such as the timing and control of entry into a dormant stage and completion of dormancy; however, S. invicta overwinter as adults which do not enter dormancy.

Physiological adaptations include thermoregulation and cold-hardiness. Thermoregulation assumes that the insect maintains its body temperature between a minimum and maximum for at least a portion of its active period, even though thermal conditions in the environment vary (Heinrich 1981). Thermoregulation in social insects can be achieved by endothermy (the heat that determines body temperature is produced by metabolic activity) and by ectothermy (the heat that determines body temperature is acquired from the environment).

Endothermy requires that rates of heat production be greater than the rates of heat loss. Numerous authors have suggested that endothermy is important in some ants (Coenen-Stass et al. 1980, and citations therein), but no data are available for S. invicta. However, Green (1959) reported that they do not form clusters in the nest during winter.

The placement of the nest (sunny versus shaded areas) and mound structure are important adaptations for ectothermy, and the shape and placement of the mound determines insolation. The mounds of S. invicta have been hypothesized to have a thermoregulatory function (Hubbard and Cunningham 1977) based on their shape and orientation, but this hypothesis awaits testing by measuring winter temperatures in the mounds and adjacent soil columns at different times on sunny and cloudy days. Morrill (1977) measured temperatures in fire ant mounds during January 1976 but did not compare them with adjacent soil temperatures. At 0830 h on January 9 the temperatures at depths of 1, 6, 11, and 25 cm were 0°C, and the air temperature was -7°C. At 1430 h on January 12, the temperatures at the same depths were 21, 20, 16, and 9°C, respectively, and air temperature was 14°C. Thus, at least on some winter days, there is a thermal gradient in the mound which can be exploited by the ants. However, a similar gradient might develop in undisturbed soils, in which case the thermoregulatory function of the mound's shape and position would have to be questioned.

Thermoregulation is also dependent on the composition of the mound. The thermal diffusivity of mound material has also been proposed as important in temperature control (Brandt 1980). Heat production within nest materials by microbial metabolism is also known for some mound building ants (Coenen-Strass et al. 1980). Data on thermal diffusivity of S. invicta mounds are not available, and heat production by microbial activity is minimal as little organic material is available for microbial consumption.

Cold-hardiness in insects encompasses three distinct phenomena: cold acclimation (preparation to avoid injury at temperatures below those at which continued growth is possible), supercooling (ability to avoid freezing at temperatures below the freezing point), and freeze tolerance (ability to withstand bodily freezing) (Danks 1978).

Cold acclimation usually involves seasonal changes in metabolites correlated with depressed supercooling points or freezing tolerance. Some insects become partially dehydrated in winter, and the increased concentration of solutes in the body depresses their freezing points. Other insects increase the concentration of cryoprotectant substances in their bodies (Danks 1978). To our knowledge, seasonal changes in metabolites have not been studied in S. invicta; however, the role of acclimation was investigated by determining the supercooling points of worker ants maintained at 12, 22, and 32°C. There were no significant differences among treatments [$F(2,81)=1.19$, $p < 0.05$] (unpublished data), leading us to conclude that cold acclimation is not an important physiological adaptation by S. invicta to avoid freezing. Preliminary studies (unpublished data) reveal they are not freeze tolerant.

Mean supercooling points vary among castes and developmental stages of S. invicta (Francke et al., in press). Adults freeze at slightly higher temperatures ($\bar{x} \pm$ SD extremes = -9.4 ± 4.2°C for minors to -11.2 ± 3.6°C for males) than larvae (-12.0 ± 3.3°C for worker larvae), and larvae freeze at higher temperatures than pupae (-21.4 ± 2.8°C for worker pupae). Because fire ants overwinter as adults, supercooling is not considered an important physiological adaptation.

Winter survival in fire ants is affected by length of exposure and by colony size (larger colonies have deeper nests and can thus avoid sudden cold spells). Markin et al. (1973) monitored winter survival of new mounds in southern Mississippi and found that older colonies with larger mounds survived better than younger colonies with smaller mounds (Table 2).

TABLE 2. Winter survival of new mounds of S. invicta in southern Mississippi. Based on data from Markin et al. (1973).

Age of colony	Average diameter of mounds	Depth	No. of mounds in October	No. mounds surviving by			
				Dec.	Feb.	Apr.	June
90 days	5-8 cm	18-25 cm	12	5	5	1	0
4 months	8-12 cm	—	15	10	2	1	0
5 months	12-18 cm	40-60 cm	12	9	9	6	2
7 months	18-25 cm	69-90 cm	5	5	4	2	2

Morrill (1977) and Morrill et al. (1978) compared survival of mature S. invicta colonies in central Georgia during normal and severe winters. The winter of 1975-1976 was considered normal, and colonies examined had 100% survival (Morrill 1977). The winter of 1976-1977 was extremely cold in central Georgia, breaking many climatic records, and only 5 of 50 colonies surveyed survived that winter.

DISCUSSION

The red imported fire ant, S. invicta, does not have any known unusual physiological adaptations to cope with temperature extremes and differs little from the native fire ants of North America in heat tolerance and supercooling points (Francke et al. 1985; unpublished data). Their gallery-riddled mounds have been hypothesized to be a thermoregulatory adaptation, but there is scant evidence to support this hypothesis. Pinson (1980) provided data which show that during the summer, mound temperatures vary more widely than, and may even exceed soil temperatures. S. invicta move up and down in the temperature gradient which develops in the nests, but so do many other ant species (including some native fire ants) which do not build mounds. Measurements of mound and surrounding soil temperatures on sunny and cloudy days are needed in order to verify the thermoregulatory nature of the mound.

The large size of Formica spp. nests (1 m high, 2 m diameter), the organic matter incorporated into them, and the thatching presumably act as insulation and retard heat loss (Heinrich 1981). S. invicta generally have smaller mounds and do not cover their nests with thatch. This suggests that heat loss at night or on cloudy days is very rapid in fire ant nests, perhaps even faster than in the soil because of their gallery-riddled nature. Morrill (1977) reported temperatures of 0°C down to a depth of 25 cm in a nest at 0830 h on January 9 in central Georgia. These observations, plus the fact that it never freezes in the homeland of S. invicta strongly suggest that mounds of S. invicta are not adaptations to cold tolerance. However, measurements of mound and soil temperatures during winter and on cold nights are still necessary to test these ideas.

Another important point is the relationship between the mound and soil moisture. The shape of the mounds, their porous outer layer (rather than having one or a few entrances that can be plugged), and their gallery-riddled nature all serve to increase the surface area through which evaporation can occur during hot-dry summer days. The soil column under a fire ant mound probably dessicates faster than undisturbed soil or soil under rocks and other objects under which many desert ants nest. Actually, such a mound would appear to be highly adaptive in areas with considerable rainfall and where the soils are often near their saturation point or undergo periodic flooding. Not surprisingly, much of S. invicta native habitat consists of a swampy area that does undergo yearly flooding.

The potential spread of S. invicta across North America is a subject of much conjecture. Its limited temperature and humidity tolerances, and its habit of building mounds in the open, lead us to postulate that its spread will slow down considerably as it encounters the hotter and drier regions of the Southwest. However, this species can probably become established with the help of man in some areas of the Southwest (particularly in California) in moist refugia such as stream beds, watering holes, irrigated agricultural land, and urban areas. In these areas it could attain pest status. It is doubtful that it can succeed in the hot deserts; and even if it manages to establish itself, colony size and mound density undoubtedly will be considerably less than in the southeastern U.S. and thus it will not attain pest status.

The northern limits of distribution of S. invicta in North America seem to be determined by its tolerance to cold temperatures. Its northward spread virtually stopped several years ago (Pimm and Bartell 1980; Francke et al., in press), and considerable winter mortality of mature colonies has already been reported near its northern limits (Morrill et al. 1978). Furthermore, apparently no mating flights occur when morning soil temperatures (surface to 10 cm deep) are below 18°C (Rhoades and Davies 1967); colony founding is successful only if the soil temperatures at a depth of 5 to 7 cm are equal to or greater than 24°C (Markin et al. 1974); and young colonies, less than four months old, survive winter poorly even in southern Mississippi (Markin et al. 1973). Thus, colder temperatures and the duration of winter, both of which progressively increase further north, have a significant negative impact upon imported fire ant populations.

ACKNOWLEDGMENTS

We thank William P. MacKay, Sherman A. Phillips, Jr., and M. Kent Rylander for critically reviewing the manuscript. This work was supported by the Texas Department of Agriculture.

REFERENCES CITED

Brandt, D. C. 1980. Is the mound of Formica polyctena Foerst. in origin a simulation of a rock. Oecologia 44: 281-282.

Braulick, L. S. 1982. The effect of acute exposure to relative humidity and temperature on the worker caste of four species of fire ants. Texas Tech Univ. 59 p. Thesis.

Buren, W. F., G. E. Allen, W. H. Whitcomb, F. E. Lennartz, and R. N. Williams. 1974. Zoogeography of the imported fire ants. J. N. Y. Entomol. Soc. 82: 113-124.

Coenen-Stass, D., B. Schaarschmidt, and I. Lamprecht. 1980. Temperature distribution and calorimetric determination of heat production in the nest of the wood ant, Formica polyctena (Hymenoptera, Formicidae). Ecology 61: 238-244.

Danks, H. V. 1978. Modes of seasonal adaptation in the insects. 1. Winter survival. Can. Entomol. 110: 1167-1205.

Francke, O. F., L. R. Potts, and J. C. Cokendolpher. 1985. Heat tolerances of four species of fire ants (Hymenoptera: Formicidae: Solenopsis). Southwest Nat. 30: 59-68.

Francke, O. F., J. C. Cokendolpher, and L. R. Potts. Supercooling studies in North American fire ants (Hymenoptera, Formicidae). Southwest. Nat. (In press).

Green, H. B. 1959. Imported fire ant mortality due to cold. J. Econ. Entomol. 52: 347.

Hadley, N. F. 1970. Micrometeorology and energy exchange in two desert arthropods. Ecology 51: 434-444.

Heinrich, B. (ed.) 1981. Insect thermoregulation. John Wiley & Sons, Inc., New York. 328 p.

Hubbard, M. D., and W. G. Cunningham. 1977. Orientation of mounds in the ant Solenopsis invicta (Hymenoptera, Formicidae, Myrmicinae). Insect. Soc. 24: 3-7.

Hung, A. C. F., and S. B. Vinson. 1978. Factors affecting the distribution of fire ants in Texas (Myrmicinae: Formicidae). Southwest Nat. 23: 205-214.

Markin, G. P., J. H. Dillier, and H. L. Collins. 1973. Growth and development of colonies of the red imported fire ant, Solenopsis invicta. Ann. Entomol. Soc. Am. 66: 803-808.

Markin, G. P., J. O'Neal, J. H. Dillier, and H. L. Collins. 1974. Regional variation in the seasonal activity of the imported fire ant, Solenopsis saevissima richteri. Environ. Entomol. 3: 446-452.

Moody, J. V., O. F. Francke, and F. W. Merickel. 1981. The distribution of fire ants, Solenopsis (Solenopsis) in western Texas (Hymenoptera: Formicidae). J. Kans. Entomol. Soc. 54: 469-480.

Morrill, W. L. 1977. Overwinter survival of the red imported fire ant in central Georgia. Environ. Entomol. 6: 50-52.

Morrill, W. L., P. B. Martin, and D. C. Sheppard. 1978. Overwinter survival of the red imported fire ant: Effects of various habitats and food supply. Environ. Entomol. 7: 262-264.

O'Neal, J., and G. P. Markin. 1975. Brood development of the various castes of the imported fire ant, Solenopsis invicta Buren (Hymenoptera: Formicidae). J. Kans. Entomol. Soc. 48: 152-159.

Pimm, S. L., and D. P. Bartell. 1980. Statistical model for predicting range expansion of the red imported fire ant, Solenopsis invicta, in Texas. Environ. Entomol. 9: 653-658.

Pinson, C. K. 1980. The temperature regime in the Solenopsis invicta mound and its effect on behavior. Texas Tech Univ. 74 p. Thesis.

Rhoades, W. C., and D. R. Davies. 1967. Effects of meteoreological factors on the biology and control of the imported fire ant. J. Econ. Entomol. 60: 554-558.

10
Temporal Foraging Patterns of *Solenopsis invicta* and Native Ants of Central Texas

*S. A. Phillips, S. R. Jones,
and D. M. Claborn*

In most biotic communities, ants are conspicuous by their large numbers, but ant population diversity may be limited by resource availability. The niche breadth of any one population should increase if a resource is in short supply (Bernstein 1979); however, interspecific competition between predominant ant species will act to oppose niche broadening. Niche breadth equilibria achieved by populations within the community depends on the efficiency of resource recovery, resource density, dependability, and dispersion. The number and phenotypic makeup of species in the community may also have a profound effect on niche broadening (Bernstein 1979; Davidson 1978). Levins et al. (1973) state that "the adjustment of the niches of each species to local physical and biotic conditions takes place largely through interspecific behavior—the way in which the species interact when they meet in the course of foraging." They also state that for ants niche breadth is approximately proportional to carrying capacity, but the proportion of the carrying capacity that the ants actually utilize depends on the aggressiveness of the species involved.

The likelihood of interspecific aggression is determined by cost–benefit relations between the risk of losing workers, energetic drains of aggressive behavior, and the reward of secured resources. Since loss of foragers may be more costly to small colonies than to large colonies, colony size may relate directly to agonistic fervor (Carroll and Janzen 1973). Solenopsis invicta (red imported fire ant or RIFA), for example, rapidly invades occupied niches through agonistic behavior, possibly correlated with the large size of their colonies (Adkins 1970). However, a small contingent of one ant species encountering a sizable number of another species may become "timid" (Levins et al. 1973). These factors are important in niche delineation and regulation of territorial boundaries for those species which are territorial.

Most ants are omnivorous (Carroll and Janzen 1973); and as a consequence, competitive displacement is a common occurrence (MacKay and MacKay 1982). Competitive avoidance is also common and may be exhibited via territoriality, character displacement, foraging strategy, or diel partitioning of foraging activity (Davidson 1977).

Forager success is most accurately measured by colony fitness (Carroll and Janzen 1973). Niche breadth may indicate the heterogeneity of resources used by a species, but not how efficiently these resources are gathered (Levins et al. 1973). An array of different foraging methods are employed by various ant taxa, the most efficient of which include group or recruitment foraging (Davidson 1977).

The following two-part study was designed to investigate some parameters at the western edge of S. invicta infestation that may be important for the coexistence of S. invicta and the predominant native ant species. The particular parameters of interest in this study involve phenotypic and behavioral aspects which may contribute to species foraging success. Because of (1) difficulty in determining accurately the exact number of colonies of an ant species in the field and (2) the wide assortment of variables that make it hard to separate foraging efficiency from other aspects of niche delineation, we studied interactions between the following ant species in the laboratory: Pheidole dentata Mayr, S. invicta, Forelius foetidus (Buckley), and Monomorium minimum (Buckley). These four species were found to occur sympatrically at high frequencies during a year-long survey conducted in Kerr and Bandera Counties, Texas (Phillips, unpublished data). The objective of the study was to compare the foraging abilities of these four species when the number of ants per colony were equal. In addition, a study was conducted in Bandera County (summer 1984) to determine if territories foraged by S. invicta were recognized and avoided by native ants.

MATERIALS AND METHODS

Laboratory Study

Three colonies of S. invicta and three colonies each of P. dentata, F. foetidus, and M. minimum were collected from Kerr and Lubbock Counties, Texas, respectively. One thousand individuals chosen at random from each of the twelve colonies were placed into separate foraging tray/nest containers, one species per container. These rectangular containers were constructed of 3.2 mm plastic, measuring 27.9 x 55.9 cm with sides 7.6 cm in height. The sides were coated with Fluon® to prevent escape. All colonies contained about the same amount of brood, and each was maintained with only one fertile queen. All colonies were held at constant temperatures

(30°C) and humidity (70%) with a 12:12 light/dark cycle. Numbers of individuals recruited to a variety of food sources were determined. Only foods eliciting the highest recruitment were used (Jones 1985). All baits consisted of a mixture of beef and chicken dog food, grape jelly, oatmeal, and fish food. Although each species was presented a different proportion of these ingredients, the proportions were such that the species recruitment response remained approximately equal to the response for the single most attractive ingredient for that species. All baits were ground in a mortar and placed on 1 x 1 cm cards for testing. Numbers of ants on bait cards were recorded every 15 seconds for 40 minutes from the time of bait discovery. These results were averaged to produce a mean number of ants per minute and again averaged for all colonies within a species. Four colonies (one of each species) were tested daily at randomized times within each day. Each colony was starved for two days between successive trials. Ten replications were conducted per colony resulting in 30 replications per species. A nested ANOVA followed by a Student-Newman-Keul's Test (8 d.f.) (Steel and Torrie 1960) was performed on the following parameters: initiation of recruitment, time to reach peak numbers, numbers recruited, time for bait retrieval, number foraging both before bait placement and after bait retrieval, and distance traveled per minute followed by number of turns executed (36 reps).

Field Study

Two 20 x 20 m sites containing at least two S. invicta mounds were established in grazed pasture land in Bandera County, Texas. Each plot was fenced to prevent disturbance by cattle. A pitfall trap was positioned flush with the ground within each square meter of each site (400 traps/site). Traps consisted of 437-ml plastic drinking cups lined with unscented talc. Uncoated sleeves were placed in the ground permanently, whereas talc coated traps were positioned within the sleeves only during the 3-h sampling intervals.

Two 3-h periods were sampled in each 24-h period. Traps were placed in the field from 1300 h to 1600 h and again from 2000 h to 0100 h. These periods had been identified previously as the minimum and maximum foraging times for S. invicta, respectively (Claborn 1985). Initially, data were to be collected twice each month during the summer at each site; however, due to inclement weather, collections were made five times on site A and three times on site B (Claborn 1985). Ants collected were identified according to sampling period, and the percentage of the area traversed by each species was analyzed by Student's paired t-test (Steel and Torrie 1960). Associations between species were determined by means of Chi-square analysis (Pielou 1974).

RESULTS

Laboratory Study

An evaluation of the data in Fig. 1 reveals that the recruitment patterns of P. dentata and S. invicta are similar as are the patterns of F. foetidus and M. minimum. However, no significant difference in time for initiation of recruitment ($p > 0.10$) and of time to reach peak numbers ($p > 0.10$) was detected among the above species pairs.

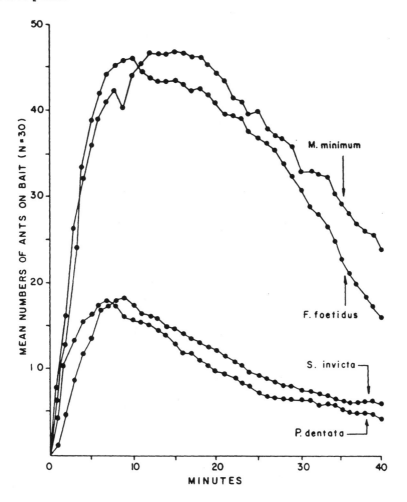

FIGURE 1. Comparative recruitment patterns of four sympatric ant species. Data points represent mean numbers of ants recorded at baits.

Significant differences were found among species (Fig. 2A) for actual peak numbers recruited (p < 0.001) and for bait retrieval times (p < 0.001) (Fig. 2B). F. foetidus and M. minimum recruited significantly higher numbers to the baits, but were significantly slower in bait retrieval than either P. dentata or S. invicta.

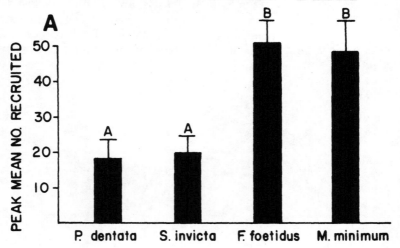

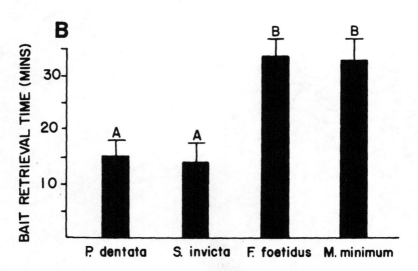

FIGURE 2. Peak numbers of workers recruited (A) and bait retrieval time (B) for four species of ants. Data analyzed by Student-Newman-Keul's Test. Means with the same letter are not significantly different.

No significant difference was detected between species ($p > 0.10$) in number of individuals foraging prior to bait placement (Fig. 3A). However, significant differences were detected between species for numbers of individuals foraging after bait retrieval ($0.025 > p > 0.01$). Significant differences were found between species ($p < 0.001$) for distance covered per minute (Fig. 3B) and number of turns made per minute ($0.05 > p > 0.025$). M. minimum executes approximately twice as many turns per unit distance as do the other three species.

Field Study

Diel fluctuations in percent of area utilized by each of the predominant ant species are significant (Table 1). Although S. invicta and Conomyrma insana each occupies a large percent of the foraging range nocturnally, their foraging area is drastically reduced diurnally. The activity schedule of Iridomyrmex pruinosus is opposite that of S. invicta and C. insana (Table 2). Both of these species, in turn, exhibit a strong negative association with I. pruinosus. These associations were detected only on one site. The scarcity of native ants may have contributed to a lack of significance at the second site.

TABLE 1. Mean percent of area occupied by predominant ant species at each sampling period in Bandera County, Texas (summer 1984). Data analyzed by Student's paired t-test.

Species	Site	Percent		Probability level
		Nocturnal	Diurnal	
S. invicta	A	64.0	8.0	$p < 0.001$
	B	70.0	8.8	$p < 0.05$
I. pruinosus	A	1.2	46.5	$p < 0.001$
	B	10.5	5.6	$p > 0.05$
C. insana	A	31.0	1.7	$p < 0.001$
	B	*	*	*

*Species not found at this site

120

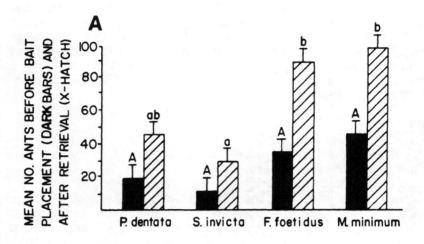

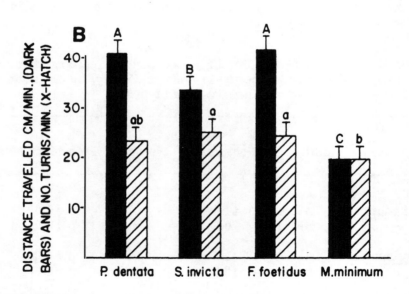

FIGURE 3. Comparison between species of four foraging parameters. Data analyzed by Student-Newman-Keul's Test. Means with the same letter are not significantly different within parameter. (A) mean number of ants before bait placement and after bait retrieval; (B) distance traveled and number of turns made.

TABLE 2. Chi-square analysis of temporo-spatial associations of predominant ant species from two sites in Bandera County, Texas (summer 1984).

Species pairs	Site	Month	x^{2*}	Relationship
C. insana/S. invicta	A	June	128.15	positive
		July	184.67	positive
		August	95.69	positive
I. pruinosus/S. invicta	A	June	271.47	negative
		July	133.21	negative
		August	80.97	negative
C. insana/I. pruinosus	A	June	115.44	negative
		July	55.83	negative
		August	26.73	negative
I. pruinosus/S. invicta	B	June	1.22 NS	none
		July	1.35 NS	none
		August	0.01 NS	none

NS = not significant
*d.f. = 1; critical value = 3.84

DISCUSSION

The ability of S. invicta to displace other ant species may well be a function of its aggressive nature and large colony size (Adkins 1970; Howard and Oliver 1979). This study indicates that when numbers within colonies of different species are equal, S. invicta does not exhibit dramatic superiority in the categories tested. In fact, P. dentata and S. invicta are nearly identical in their foraging abilities under simplified experimental conditions. As S. invicta continues its northward and westward expansion, environmental conditions required for survival become less favorable with respect to temperature (north) and humidity (west). Consequently, a decrease in population density should be expected. Under these conditions, the ability of S. invicta to compete with resident ant species and invade already filled niches should also decrease. In addition, a temporo-spatial separation of activity may be a means of partitioning resources among different ant species (Hansen 1978; Carroll and Janzen 1973). An indication of this decrease in competitiveness by S. invicta in central Texas may be the diel foraging fluctuations that result in resource partitioning. Indeed, this study suggests a need for further investigation of the competitive abilities of these ants in natural communities, and emphasizes the importance of resource preemption in determining the success of a species in a competitive environment.

REFERENCES CITED

Adkins, H. G. 1970. The imported fire ant in the southern United States. Assoc. Am. Geog. 60: 578-592.

Bernstein, R. A. 1979. Evolution of niche breadth in populations of ants. Am. Natur. 114: 533-544.

Carroll, C. R., and D. H. Janzen. 1973. Ecology of foraging by ants. Annu. Rev. Ecol. Syst. 4: 231-257.

Claborn, D. M. 1985. Effect of Solenopsis invicta Buren territoriality on native ants of central Texas. M.S. Thesis, Texas Tech Univ., Lubbock, TX. 63 pp.

Davidson, D. W. 1977. Foraging ecology and community organization in desert seed-eating ants. Ecology 58: 725-737.

Davidson, D. W. 1978. Size variability in the worker caste of a social insect (Veromessor pergandei Mayr) as a function of the competitive environment. Am. Natur. 112: 523-532.

Hansen, S. R. 1978. Resource utilization and coexistence of three species of Pogonomyrmex ants in an upper Sonoran grassland community. Oecologia 35: 109-117.

Howard, F. W., and A. D. Oliver. 1979. Field observations of ants (Hymenoptera: Formicidae) associated with red imported fire ants, Solenopsis invicta Buren, in Louisiana pastures. J. Ga. Entomol. Soc. 14: 259-263.

Jones, S. R. 1985. Foraging and recruitment abilities of Solenopsis invicta Buren, compared with other ant species indigenous to Texas. M.S. Thesis, Texas Tech Univ., Lubbock, TX. 99 pp.

Levins, R., M. L. Pressick, and H. Heatwole. 1973. Coexistence patterns in insular ants. Am. Sci. 61: 463-472.

MacKay, W., and E. MacKay. 1982. Coexistence and competitive displacement involving two native ant species (Hymenoptera: Formicidae). Southwest. Nat. 27: 135-142.

Pielou, E. C. 1974. Population and community ecology: Principles and methods. Gordon and Breach Science Publishers, N.Y., NY. 423 pp.

Steel, R. G. D., and J. H. Torrie. 1960. Principles and procedures of statistics. McGraw-Hill Book Co., Inc., N.Y., NY. 481 pp.

11
Population Dynamics of Leaf-Cutting Ants: A Brief Review

H. G. Fowler, V. Pereira-da-Silva,
L. C. Forti, and N. B. Saes

In spite of the tremendous economic importance of leaf-cutting ants, relatively little is known about their basic biology and ecology, and the published information is widely scattered through a diverse literature—largely in Portuguese and Spanish. These facts attest to the problems of developing more rational control and management strategies for this group of highly important ants. We have, therefore, summarized and synthesized the existent literature on the population dynamics of these ants with the hope that our review will serve as a stimulus for additional research on their basic ecology.

SEXUAL PRODUCTION AND NUPTIAL FLIGHTS

The number of sexuals produced in colonies of <u>Acromyrmex</u> and <u>Atta</u> is highly variable (Table 1), and little is known about the factors that favor the production of males and gynes. Nevertheless, given normal colony densities, it seems reasonable to expect annual production of 5 to 20 x 10^3 gynes/ha. The nuptial flight (=revoada) may be a single annual event or may be composed of several swarming events. During the revoada, the queen must be mated and obtain sufficient sperm to last her life span, generally more than 10 years. Leks of males are probably formed (Amante 1972a) and females probably mate with three or more males. Mated females may contain more than 300 million sperm in their spermatheca after the revoada (Kerr 1961; Moser 1967; Corso and Serzedello 1981). During the revoada, predation, especially by birds, is very intense (Mariconi 1970; Autuori 1950; Warter et al. 1962). Dix and Dix (unpublished) estimated that birds removed 3900 of 7500 swarming <u>Atta</u> <u>cephalotes</u> queens in Guatemala.

123

TABLE 1. The number of males and gynes and their respective sex-ratios in leaf-cutting ant colonies.

Species	Number of		Sex ratio	Source[a]
	males	gynes		
Atta				
sexdens rubropilosa	2406	486	4.95	1
	27622	1696	16.3	2
	7624	4684	1.6	2
	4098	2765	1.4	2
	7873	5177	1.5	2
	35381	1378	25.6	2
	7896	3202	2.4	2
	9138	1411	6.4	2
	160[b]	20[b]	8.0	3
cephalotes	50[b]	6[b]	8.3	3
vollenweideri	7014	657	10.7	4
laevigata	2402	1325	1.8	2
	3986	2308	1.7	2
	2893	712	4.0	2
	1550	1030	1.5	2
	891	85	10.4	2
	5130	411	12.4	2
texana	1080	875	1.2	5
bisphaerica	6393	3688	1.7	2
	4879	2487	1.9	2
	9055	841	10.7	2
	1813	160	11.3	2
	4470	1263	3.5	2
Acromyrmex				
octospinosus	33	8	4.1	6
	273	283	0.9	6
coronatus	3577	1606	2.2	7
	2240	1680	1.3	7
landolti fracticornis	328	184	1.8	8
striatus	428	196	2.2	8

[a]Reference for each species according to number as follows: (1) Pereira–da–Silva 1979, (2) Autuori 1950, (3) Geijskes 1953, (4) Jonkman 1977, (5) Moser 1967, (6) Lewis 1975, (7) Pereira–da–Silva et al. 1981, (8) Fowler, unpublished.
[b]Data from colony fragments only and are presented only for sex ratios.

COLONY-FOUNDING OR POST-NUPTIAL FLIGHT

Colony-founding is an extremely important aspect in understanding the ecology of these ants, yet we know very little about it. Atta texana queens are known to fly 10.5 km (Moser 1967), A. cephalotes 9.6 km (Cherrett 1969), and A. sexdens rubropilosa 11 km (Jutsum and Quinlan 1978). The distribution, duration, and length of the dispersal flights, however, remain unknown. Nevertheless, A. capiguara and A. bisphaerica make more colonization attempts in areas with high densities of mature colonies (Fowler et al. 1984; Fowler and Saes, in press, a). In both of these cases, queens may have mistaken large nest mounds for disturbed areas. In A. sexdens rubropilosa and Acromyrmex muticonodis, such a pattern has not been documented (Fowler and Saes, in press, a, b). Cherrett (1968), using pitfall traps, found more A. cephalotes queens in forested areas and more Ac. octospinosus queens in clearings which corresponded to the observed distribution of mature colonies.

Even though colonies of leaf-cutting ants generally are founded by haplometrosis, it seems reasonable, based on the literature (Weber 1937; Jonkman 1977; Huber 1907; Moser and Lewis 1981; Mariconi and Zamith 1963; Moser 1963; Walter et al. 1938) that some colonies can be founded by pleometrosis or at least they can adopt queens following the death of their own queens. Mintzer and Vinson (1985) report that pleometrosis occurs in A. texana.

The number of queens that escape predation during the revoada and alight on the ground to search for a site to begin nest construction (Table 2) varies tremendously. Even these queens still face predation by birds; vertebrates, especially armadillos and rodents; and ants (Mariconi 1970; Mariconi and Zamith 1963; Pereira-da-Silva 1973; Weber 1972). Dix and Dix (unpublished) found that during this phase of colony establishment, 4% of A. cephalotes queens were lost to vertebrates and 7% to other ants. In Brazil, studies are being conducted on the scarab, Canthon dives, as a control agent for founding queens of Atta. A large scale mass-rearing and field-release program for the same beetle in Paraguay in the 1950s, however, failed to have any effect. It is likely that the effect of this beetle is only to improve the survivorship of the remaining founding queens.

Execution of founding queens by established colonies of leaf-cutting ants is another important source of mortality with the percentage of queens killed varying from 20 to 50 (Cherrett 1968; Rockwood 1973; Fowler 1982; Fowler et al. 1984; Fowler and Saes, in press, a,b; Dix and Dix, unpublished). The high mortality of founding queens attributed to the actions of conspecific workers suggests that competition is intense between colonies, and that in some areas, colony density may be at carrying capacity.

TABLE 2. Densities of post-revoada founding queens on the ground reported for various species of leaf-cutting ants.

Species (source)[a]	Number queens/ha	Locality[b]
Acromyrmex		
landolti fracticornis(1)	1500-7500	Paraguay
muticonodis(2)	50-350	Sao Paulo
Atta		
capiguara(3)	200-400	Sao Paulo
capiguara(4)	100-800	Paraguay
capiguara(5)	3000-13300	Sao Paulo
cephalotes(6)	2000-10000	Guatemala
sexdens rubropilosa(7)	300-800	Sao Paulo
sexdens and laevigata(8)	20000-50000	Sao Paulo
vollenweideri(1)	635-2870	Paraguay
bisphaerica(2)	500-1350	Sao Paulo

[a]Reference for species according to number as follows: (1) Fowler, unpublished, (2) Fowler and Saes, in press, a, (3) Amante 1972b, (4) Fowler et al. 1984, (5) Pereira-da-Silva and Forti, unpublished, (6) Dix and Dix, unpublished, (7) Fowler and Saes, in press, b, (8) Pereira-da-Silva 1973.
[b]Localities in Brazil are listed by city.

EARLY COLONY GROWTH AND SURVIVAL

The growth of incipient colonies of various species of Atta has been investigated by Autuori (1941), Mariconi (1974), and Pereira-da-Silva (1973, 1975, 1979). Egg production by the queen during the first three or four months is apparently correlated with a cycle of activity in the corpora allata (Fig. 1). The patterns of egg, larval, pupal, and worker production during this phase are generally quite consistent (Fig. 2). Once the first worker brood matures, the queen no longer has to depend upon muscle catalysis and the production of trophic eggs for her survival as workers begin to forage for vegetation and the fungus garden begins to grow (Weber 1972).

Survival of the queen and her incipient colony during this phase is still precarious. Pereira-da-Silva and Forti (unpublished) found that of 13,300 A. capiguara queens establishing incipient nests, only 12 were still alive after three months. In predator exclusion experiments, Autuori (1941) found that over a five-year period, the survivorship of incipient colonies during the first three months varied between 18.9 and 6.4%. These colony failures must be attributed to the lack of establishment of a fungal garden, to pathogens, or to the fact that the queen was not fertilized. Autuori (1950) found that of 3558 incipient nests of A. sexdens rubropilosa studied, only 90 (2.5%) were alive at the end of three months. Dix and Dix (unpublished) found a 90% reduction of colonies of A. cephalotes over their first three months of life with 74% of these failures being due to pathogens and 26% to the fact that queens were unmated. Furthermore, they found that after 24 months, the density of young nests decreased from 7.5 to 1.7 nests/ha. They decreased to 0.8 nests/ha at 36 months, and, finally, stabilized at 0.5 nests/ha after five to six years. Jacoby (1944) found only a 6.6% survivorship over the first three months in his studies of A. sexdens rubropilosa, and Mariconi (1974) obtained no survivorship after three months in another study of A. capiguara. During this initial phase, major predators included armadillos, rodents, and, perhaps, occasional attacks by ants, including aggression by established colonies.

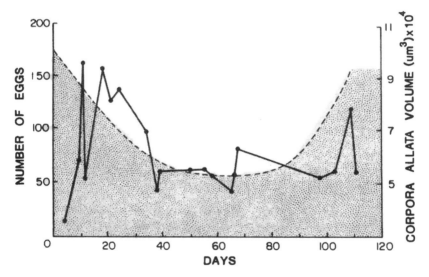

FIGURE 1. Solid line: Egg production in newly founded colonies of A. capiguara (Pereira-da-Silva 1979); broken line: corpora allata volume of colony founding queens of A. sexdens rubropilosa over the same time period (Moraes et al. 1980).

COLONY AND WORKER SURVIVAL

After the colony has passed the critical period of colony establishment; i.e., once the first nest opening is made and workers begin to forage, colony survival tends to be greater. However,

128

colonies are still lost, especially while small, to attacks of conspecific colonies and armadillos. Existent data suggest that queens, and
thus colonies of A. sexdens rubropilosa, live at least 15.5 years
(Autuori 1950). Colony survival for other species is estimated at 11
years for Ac. octospinosus (Weber 1976a), 10 years for A. cephalotes
(Weber 1976b), and at least 7 years for Ac. niger (Forti and Pereira-
da-Silva, unpublished). Mariconi (1970) affirms that it is quite easy
to maintain colonies in the laboratory for 5 to 10 years.

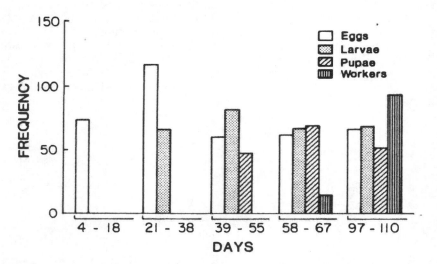

FIGURE 2. Initial colony growth in newly founded colonies
of A. capiguara (Pereira-da-Silva 1979).

Survival under field conditions is not well documented.
Jonkman (1979) has estimated that colonies of A. vollenweideri live
10 to 20 years. Once a colony reaches the age of three years, he
estimates that it has an additional life expectancy of seven more
years, with maturity reached at the age of five to six years. Dix and
Dix (unpublished) also estimated the average age to maturity was
five to six years for A. cephalotes and that colonies probably
survived until the age of 14 to 15 years. Fowler (1984) studied the
population dynamics of A. capiguara and found a Type I (heavy
initial mortality) survivorship. In this species, colonies are estimated to live 10 to 12 years once they pass through the initial phase.
Schade (1973) referred to colony life spans for A. sexdens
rubropilosa on the order of 30 to 140 years, but apparently these
estimates are not based on data. Data for estimated colony populations are given in Table 3 for several species. Graphically presented
in Fig. 3 are the relationships between age, nest dimensions, and
colony populations for several species.

TABLE 3. Estimated colony populations of various species of leaf-cutting ants.

Species (source)[a]	Colony age	Estimated Population (x 10^6)	Locality[b]
Acromyrmex			
landolti balzani(1)	Mature	0.03	Sao Paulo
octospinosus(2)	Mature	0.10	Trinidad
octospinosus(3)	Mature	0.04–0.11	Trinidad
coronatus(4)	Mature	0.05–0.15	Sao Paulo
crassispinus(1)	Mature	0.27	Paraguay
disciger(1)	Mature	0.21	Sao Paulo
Atta			
colombica			
tonsipes(5)	Mature	1.00–2.50	Panama
laevigata(6)	Mature	3.50	Sao Paulo
sexdens			
rubropilosa(6)	Young	0.25	Sao Paulo
sexdens			
rubropilosa(7)	Mature	5.00–80.00	Paraguay
cephalotes(8)	Young	0.60	Trinidad
vollenweideri(9)	Mature	4.00–7.00	Paraguay

[a]Reference for species according to number as follows: (1) Fowler, unpublished, (2) Weber 1972, (3) Lewis 1975, (4) Pereira-da-Silva et al. 1981, (5) Martin et al. 1967, (6) Pereira-da-Silva 1975, (7) Schade 1973, (8) Lewis et al. 1974, (9) Jonkman 1977.
[b]Localities in Brazil are listed by city.

Colony nest dimensions have been used extensively to estimate colony age (Autuori 1941; Bitancourt 1941; Jonkman 1980; Lewis 1975; Fowler 1984). The assumption that these nest dimensions are a function of colony age (Fig. 3) and worker populations has been accepted for a long time. However, when compared with real data (Fig. 3D), there is no correlation between nest dimensions and worker populations. Nevertheless, we feel that nest dimensions approximate maximal yearly worker populations for the following reasons:

1. Nest dimensions are highly correlated with the size of the foraging territories. For example, the area of the nest surface is strongly correlated with nearest neighbor distance in A. sexdens rubropilosa (Forti and Pereira-da-Silva 1979).

2. Nest dimensions are significantly correlated with distances foraged by various species of leaf-cutters (Fowler and Robinson 1979a, b; Everts, unpublished; Forti 1985; Lewis et al. 1974).

130

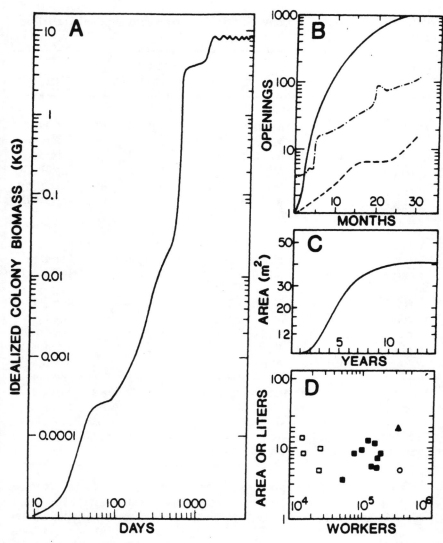

FIGURE 3. A: Idealized <u>Atta</u> colony growth based on Pereira–da–Silva (1975, 1979), Autuori (1941), and Fowler (1984). B: Relation between colony age and the number of nest openings. Top curve, <u>A. sexdens rubropilosa</u> (Autuori 1941); middle curve, <u>A. vollenweideri</u> (Jonkman 1980); lower curve, <u>Ac. octospinosus</u> (Lewis 1975). C: Relationship between colony age and nest surface area for <u>A. vollenweideri</u> (Jonkman 1980). D: Worker populations versus colony surface area. Open squares, <u>A. sexdens rubropilosa</u> (Pereira–da–Silva 1975); open circle, <u>A. laevigata</u> (Pereira–da–Silva 1975); triangle, <u>A. vollenweideri</u> (Jonkman 1977). Closed squares, <u>Ac. coronatus</u>, given as liters of mound volume (Pereira–da–Silva et al. 1981).

3. Nest measurements eventually stabilize and vary little once the colony reaches maturity (Jonkman 1980; Fowler 1984).

All of these reasons suggest that nest dimensions do indeed adequately index colony populations. We suggest that the lack of correlation shown in published excavations are either due to inexact estimates of worker populations or, more probably, to seasonal effects. Colony dimensions probably reach their maximum during the pre-revoada yearly phase after which colony populations decline.

Given these data, we diagrammed an idealized colony growth in Fig. 3A based on pieces of information from various sources. From what we have been able to extract from the literature, it seems that colony growth passes through various exponential phases with the most pronounced between the age of three and five years; i.e., once a colony approaches maturity.

With respect to worker survivorship, even fewer sources of information are available. The only data we have, albeit incomplete, are given in Fig. 4. If these patterns hold, workers probably exhibit a Type IV individual survivorship and live for at least four months as Mariconi (1970) also estimates. Although not much is known about age polyethism, older workers probably become foragers, and their life span may be short.

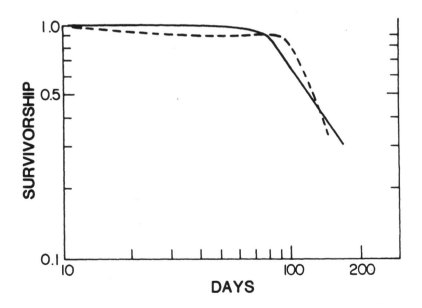

FIGURE 4. Worker survivorship obtained by marking workers as they eclosed. Solid line, A. sexdens rubropilosa (Forti and Pereira-da-Silva, unpublished); broken line, Ac. landolti balzani (Fowler, unpublished).

Forager populations per active trail have been estimated at from 800 to 1500 for Ac. landolti fracticornis (Fowler and Robinson 1979a, b), from 1000 to 1700 for young colonies of A. sexdens rubropilosa (Forti et al. 1985), and 19,800 workers for A. cephalotes (Lewis et al. 1974). Lewis et al. (1974) estimated the foraging force of an A. cephalotes colony with a population of 600,000 workers to be 40,000 or about 6% of the total worker population. A reasonable assumption would be that approximately 10% of a colony's worker population are foragers.

Mean lifespans of foragers have been estimated to be two to three months for A. capiguara (Forti, unpublished), 1.5 months for Ac. landolti balzani (Fowler, unpublished), and two to three weeks for A. colombica (Porter and Bowers 1982).

POLYMORPHISM AND COLONY ONTOGENY

Species of Atta and Acromyrmex have highly polymorphic workers. In Acromyrmex, worker allometry is monophasic, while in Atta, it is diphasic. During early colony growth, the worker population shows little size variation being comprised of a narrow range of worker sizes, all very small (Fig. 5). Even at the mature stages of colony development, major workers (Fig. 6) are apparently produced at the expense of media workers (Fig. 5) as the frequency of minor workers varies little through the colony life span (Pereira-da-Silva 1973, 1975, 1979). In A. sexdens rubropilosa, media workers do not appear until about four months after the first nest opening is made, and major workers only appear some 18 months later (Autuori 1941). Most species of Atta probably have similar worker size distribution (Fig. 5 and 6), although it is still unknown why A. capiguara has relatively fewer major workers than other species of the genus. As the colony grows, the gradual acquisition of larger workers expands the efficiency of colony operations (Wilson 1980). Wilson (1980) suggested that one of the principal adaptive values of polymorphic workers in species of Atta was that it allowed the harvest of many more physical types of plants.

Colonies of Acromyrmex, besides being less polymorphic than Atta, may also have more variable caste ratios (Fig. 5). If we accept the argument of Wilson (1980), then worker size may be subject to more variability, determined primarily by the availability of vegetative types for harvest. This implies that unlike Atta which, due to its more developed polymorphism, can harvest many types of plants that differ physically, Acromyrmex spp. may be limited to harvesting a much smaller range of plants. The type of plant would be determined primarily by the physical characteristics of the leaves and, therefore, the caste ratio of the entire colony would have to be adjusted through differential mortality and production of workers to track this narrower range of resources.

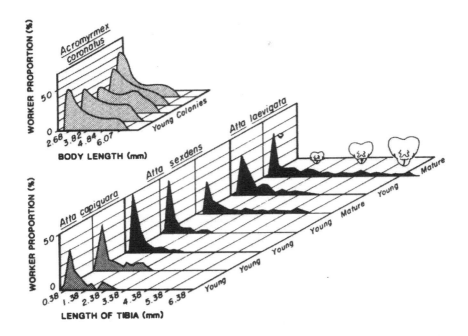

FIGURE 5. Worker polymorphism and biomass in incipient and large colonies of some leaf-cutting ants. Data from Pereira-da-Silva (1975, 1979) and Pereira-da-Silva et al. (1981).

Finally, an additional aspect of colony polymorphism needs to be addressed. Even though minor workers, in numbers, comprise the bulk of the colony population (Fig. 5 and 6), they do not make up the bulk of the biomass which is concentrated in the larger workers (Fig. 7) due to the cubic function between the size of a worker and its weight.

COLONY DENSITIES

Colony densities of leaf-cutting ants, as ascertained from our data and the published literature, vary widely probably in response to local habitats (Table 4). Evidence exists that, at least in some communities, colony density varies little from one year to the next (Dix and Dix, unpublished; Forti, unpublished; Weber 1977), suggesting that populations are near the carrying capacity of the habitat. Colony densities of grass-cutting ants also tend to be larger than colony densities of leaf-cutting ants (Table 4).

134

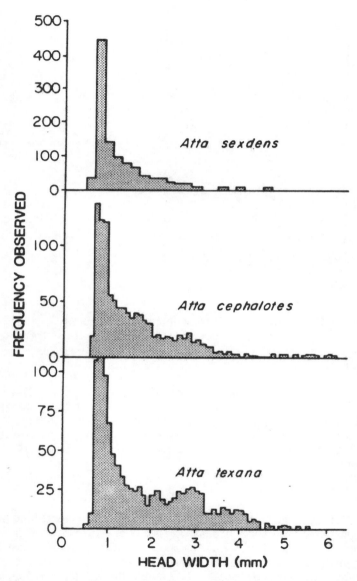

FIGURE 6.   Caste structure in large colonies of <u>Atta</u>, as indexed by head measurements.   Data taken from <u>Stradling</u> (1978), Wilson (1980), and Fowler (1983c).

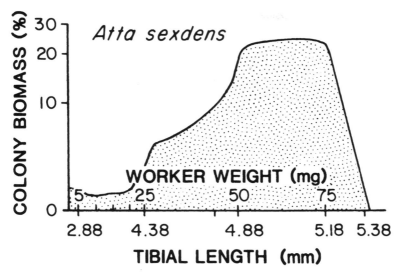

FIGURE 7. Percent colony biomass relative to worker size.

TABLE 4. Reported colony densities of species of <u>Acromyrmex</u> and <u>Atta</u>. Dispersions are given after estimated colony densities: (?) not reported; (R) regular; (N) random; (S) overdispersed.

Species	Locality[a]	Colonies/ha	Source
<u>Atta</u>			
capiguara	Sao Paulo	10–18(?)	Amante 1967a, 1967b
	Sao Paulo	1–2(N)	Forti 1985
	Sao Paulo	2.34(?)	Amante 1972b
	Paraguay	4.2–8.3(R)	Fowler, unpubl.
	Sao Paulo	1–3(R)	Fowler, in press
	Paraguay	8–14(R)	Fowler, unpubl.
cephalotes	Trinidad	0.4–0.6(S)	Cherrett 1968
	Costa Rica	1.25–5.0(S)	Rockwood 1973
	Guatemala	0.5(R)	Dix & Dix, unpubl.
	Bahia	3.0(?)	Leston 1978
colombica	Costa Rica	5–10(S)	Rockwood 1973
<u>laevigata</u>	Guyana	0.6–1.9(S)	Cherrett et al. 1974
	Paraguay	1–3(R)	Fowler, unpubl.
opaciceps	Ceara	7.91(?)	Sales et al. 1983
	Bahia	5.0(?)	Goncalves 1951

136

Table 4 continued.

sexdens			
rubropilosa	Bahia	4-5(?)	Goncalves 1951
	Sao Paulo	2.8(N)	Forti & Pereira-da-Silva 1979
	Sao Paulo	1-5(R)	Fowler, unpubl.
	Paraguay	0.5-3.9(R)	Fowler, unpubl.
	Sao Paulo	4-5(?)	Goncalves 1967
vollenweideri	Argentina	2.0-4.5(?)	Bucher & Zuccardi 1967
	Paraguay	0.4(S)	Jonkman 1977
	Paraguay	0.9(R)	Fowler, unpubl.
	Paraguay	1-4(?)	Robinson & Fowler 1982
spp. (?)	Para	1-30(?)	Ribeiro & Woosner 1979
Acromyrmex			
landolti			
landolti	Venezuela	6000(?)	Labrador et al. 1972
	Guyana	800-900(?)	Weber 1972
	Venezuela	1000(?)	Espina & Timaure 1977
	Guyana	0.96-94.88(S)	Cherrett et al. 1974
landolti balzani	Peru	6000(?)	Everts, unpubl.
	Ceara	120(?)	Goncalves 1967
	Sao Paulo	152-330(S)	Fowler, in press
	Bahia	800-900	Lewis 1975
landolti fracticornis	Paraguay	210-3420(S)	Fowler 1977b
	Paraguay	687-5350(S)	Fowler 1977a
	Argentina	420-890(?)	Fowler, unpubl.
lobicornis	Argentina	1.0-14.7(R,N)	Claver, unpubl.
hystrix	Amazona	4-5(?)	Rodrigues 1966
coronatus	Amazona	0.15-0.30(?)	Rodrigues 1966
muticonodis	Sao Paulo	1-4(S)	Fowler, in press
lundi	Paraguay	3.0(R)	Fowler, in press
octospinosus	Trinidad	2.62-11.68(?)	Cherrett 1968
	Trinidad	3-36(S)	Lewis 1975
crassispinus	Paraguay	0.3-2.4(R)	Fowler, unpubl.
heyeri	Paraguay	3.0-12.5(R)	Fowler, unpubl.
subterraneus	Paraguay	0.4-3.0(R)	Fowler, unpubl.
rugosus	Paraguay	12.8-243.6(S)	Fowler, unpubl.
spp. (?)	Sao Paulo	200(?)	Mendes-Filho 1979

[a]Localities in Brazil are listed by city.

Since leaf-cutting ants demonstrate territorial behavior (Fowler and Stiles 1980; Jaffe, Chapter 18), the sizes of their foraging territories also can be used to estimate colony densities (Table 5). Inter- and intraspecific aggression are important regulatory mechanisms of populations; thus, estimates of colony densities based on the sizes of exclusive foraging territories are in good agreement with estimates obtained by other means (Tables 4 and 5).

Man, through his activities, also profoundly affects the population dynamics of leaf-cutting ants. Ever since the report of Soares de Sousa (1587), investigators have attributed nomadic agriculture in the tropical regions of Central and South America to rapid population increases of some species of leaf-cutting ants in clearings and a concurrent decline in food production (Wheeler 1907; Goncalves 1945). Various studies have documented that colony densities for some species are much higher in man-simplified habitats than in natural ones (Cherrett 1968; Fowler 1983a). As additional forested regions are cleared for agriculture and ranching, especially in the Amazon basin, the importance of leaf-cutting ants will undoubtedly increase.

Even on a smaller scale, the impact of man may greatly influence the densities of colonies. The most obvious intervention that man performs is that of trying to kill colonies thereby inducing year-to-year variations in colony densities (Fig. 8). Indirectly, other factors also influence colony densities. Fowler (1977a) and Claver (unpublished) have found that grazing by cattle, especially overgrazing, causes an increase in the colony density of pasture leaf-cutting ants. Fowler and Haines (1983) demonstrated that colony densities were highly dependent upon the history of disturbance and that some species may be capable of influencing plant succession. Forti (1985) was able to document the effects of disturbance on the colony densities of A. capiguara through a series of aerial photographs. In 1962, densities in natural pasture were 2.01 mature colonies/ha; but in 1972, coffee was planted and the colony density dropped to 1.3/ha. In 1975, a killing frost removed the coffee; and in 1976, an introduced grass was planted. In 1977, colony densities were 1.0/ha; and by 1982, they had increased to 1.8 colonies/ha. Jonkman (1980) also showed that A. vollenweideri may, through its nesting activities, eventually create unfavorable conditions for the survival of the species which apparently results in a change in colony distribution patterns.

COMMUNITY EFFECTS

Inter- and intraspecific aggression are among the regulatory mechanisms controlling leaf-cutting ant populations. Aggression results in the death of founding queens and in the death or emigration of smaller colonies. As leaf-cutting ants do not compete with

138

other species of ants for their food resources, their principal competition comes from other colonies of leaf-cutting ants as well as some vertebrate and invertebrate herbivores and man. In the southern subtropics of South America, species' richness (Fowler 1983b) and consequently species' packing (Fowler 1983a) are much higher than in tropical regions, and both inter- and intraspecific competition are likely to be more intense.

In the field, strong negative correlations have been found between the densities of colonies of species of <u>Atta</u> and <u>Acromyrmex</u> (Fowler 1983a and Fig. 8) suggesting that competition also occurs at the generic level. Fowler (in press) and Bucher and Montenegro (1974 and unpublished data) have established that high niche overlap values are commonly encountered between some pairs of species in any given community. When high niche overlap values are present, colony spacing between the pairs of species tends to be regular (Fowler, in press), as one would expect from maintenance of foraging territories (Fowler and Stiles 1980). Obviously, much more needs to be done to understand the factors that influence the structure of leaf-cutting ant communities and how these species coexist in species-rich communities.

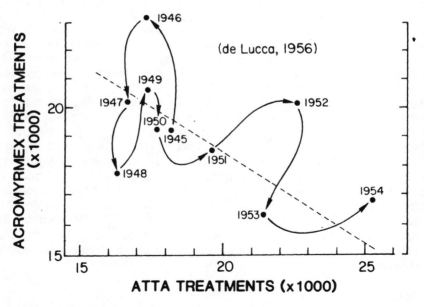

FIGURE 8. The relationship between yearly treatments for controlling colonies of <u>Acromyrmex</u> and <u>Atta</u> in Sao Paulo. The plot tends to suggest a negative relation between the abundance of colonies of these two genera, if control is only a function of abundance.

TABLE 5. Numbers and lengths of foraging trails and area of foraging territories for some leaf-cutting ants.

Species	Foraging trails				Foraging territory area (m²)[b]		Source[a]
	numbers present		mean lengths (m)				
	min.	max.	min.	max.	min.	max.	
Acromyrmex							
crassispinus	3	7	15.0	110.0	—	1300 (7.7)	1
rugosus	1	2	8.0	40.0	60 (167)	280 (36.0)	1
lundi	4	7	25.0	45.0	600 (17)	2000 (5.0)	1
heyeri	4	8	20.0	70.0	434 (23)	542 (18.5)	1
landolti	1	1	0.3	13.0	0.2 (50000)	2.0 (5000)	2
landolti	1	1	0.3	6.6	1.0 (10000)	6.0 (1667)	3
Atta							
capiguara	1	11	4.8	16.5	507 (19.7)	986 (10.2)	4
capiguara	1	23	2.4	21.1	1723 (5.8)	2352 (4.3)	4
vollenweideri	9	16	38.2	46.6	1041 (9.6)	10613 (0.9)	5
vollenweideri	5	14	55.0	152.0	3545 (2.8)	7415 (1.3)	1
mexicana	20	23	1.0	65.0	—	8000 (1.2)	6
sexdens	5	12	8.5	32.6	2125 (4.7)	6675 (1.5)	7
sexdens	1	10	3.3	7.0	1992 (5.0)	3408 (2.9)	8
sexdens	6	14	2.9	12.6	2060 (4.8)	5823 (1.7)	9

[a]References for each species according to number as follows: (1) Fowler, unpublished, (2) Fowler 1977a, (3) Everts, unpublished, (4) Forti 1985, (5) Carvalho 1976, (6) Mintzer 1979, (7) Fowler and Robinson 1979b, (8) Forti 1984, (9) Forti and Pereira-da-Silva 1979.
[b]Numbers in parentheses are estimated colony densities per ha assuming an exclusive foraging territory.

REFERENCES CITED

Amante, E. 1967a. Sauva tira boi da pastagem. Coopercotia. 23: 38-40.

Amante, E. 1967b. A sauva Atta capiguara, praga das pastagens. O Biologico. 33: 133-120.

Amante, E. 1972a. Preliminary observations on the swarming behavior of the leaf-cutting ant, Atta capiguara (Hymenoptera: Formicidae). J. Ga. Entomol. Soc. 7: 82-83.

Amante, E. 1972b. Influencias de alguns fatores microclimaticos sobre a formiga sauva Atta laevigata (F. Smith, 1858), Atta sexdens rubropilosa Forel, 1908, Atta bisphaerica Forel, 1908, e Atta capiguara Goncalves, 1944 (Hymenoptera, Formicidae) em formigueiros localizados no Estado de Sao Paulo. Ph.D. Thesis, Esc. Sup. Agric. L. Queiroz, Piracicaba, Sao Paulo.

Autuori, M. 1941. Contribuicao para o conhecimento da sauva (Atta spp.). I. Evolucao do sauveiro (Atta sexdens rubropilosa Forel, 1908). Arq. Inst. Biol. Sao Paulo. 12: 197-228.

Autuori, M. 1950. Contribuicao para o conhecimento da sauva (Atta spp.). V. Numero de formas aladas e reducao dos sauveiros inicias. Arq. Inst. Biol. Sao Paulo. 19: 325-331.

Bitancourt, A. A. 1941. Expressao matematica do crescimento de formigueiros de Atta sexdens rubropilosa representado pelo aumento do numero de olheiros. Arq. Inst. Biol. Sao Paulo. 12: 229-236.

Bucher, E. H., and R. Montenegro. 1974. Habitos forrajeros de cuatro hormigas simpatridas del genera Acromyrmex (Hymenoptera, Formicidae). Ecologia (Argentina). 11: 47-53.

Bucher, E. H., and R. B. Zuccardi. 1967. Significacion de los hormigueiros de Atta vollenweideri Forel como alteradores del suelo en la provincia de Tucuman. Acta. Zool. Lill. 23: 83-95.

Carvalho, S. 1976. Atta (Neoatta) vollenweideri Forel, 1908, no Brasil: Ocorrencia, aspectos externos e internos do sauveiro. L.D. Thesis. Univ. Fed., Santa Maria, Rio Grande do Sul.

Cherrett, J. M. 1968. Some aspects of the distribution of pest species of leaf-cutting ants in the Caribbean. Am. Soc. Hort. Sci. Trop. Reg. 12: 295-310.

Cherrett, J. M. 1969. A flight record for queens of Atta cephalotes L. (Hym., Formicidae). Entomol. Monthly Mag. 104: 255-256.

Cherrett, J. M., G. V. Pollard, and J. A. Turner. 1974. Preliminary observations on Acromyrmex landolti (For.) and Atta laevigata (Fr. Smith) as pasture pests in Guyana. Trop. Agric. 51: 66-74.

Corso, C. R., and A. Serzedello. 1981. A study of multiple mating habits in Atta laevigata based on the DNA content. Comp. Biochem. Physiol. B. 69: 901-902.

Espina, E. R., and A. Timaure. 1977. Caracteristicas de los nidos de Acromyrmex landolti (Forel) en el oeste de Venezuela. Rev. Fac. Agron. Univ. Zulia. 4: 53-62.

Forti, L. C. 1984. Relacoes entre plantas atacadas e a sauva Atta sexdens rubropilosa Forel, 1908 (Hymenoptera, Formicidae) em ecossistemas naturais e artificiais. Relatorio, Convenio IPEFI FIPEC-Formigas Cortadeiras. 99 pp.

Forti, L. C. 1985. Ecologia da atividade forageira da sauva Atta capiguara Goncalves, 1944 (Hymenoptera, Formicidae) e seu impacto em pastagens. Ph.D. Thesis, Esc. Sup. Agric. L. Queiroz, Piracicaba, Sao Paulo.

Forti, L. C., and V. Pereira-da-Silva. 1979. Distribuicao espacial dos nihos de Atta spp. (Hymenoptera, Formicidae) em povoamento de Eucalyptus spp. Acta VII Jorn. Cient. A.D.C. Botucatu.

Forti, L. C., S. Silvereiri-Neto, and V. Pereira-da-Silva. 1985. Atividade forageira de Atta sexdens rubropilosa Forel, 1908, (Hymenoptera, Formicidae): Fluxo e velocidade dos individuos na trilha, caracterizacao dos individuos forageiros e duracao e numero de jornadas de coletas de vegetais. Rev. Bras. Entomol. 28: 275-284.

Fowler, H. G. 1977a. Some factors influencing colony spacing and survival in the grass-cutting ant, Acromyrmex landolti fracticornis (Forel) (Formicidae: Attini) in Paraguay. Rev. Biol. Trop. 25: 89-99.

Fowler, H. G. 1977b. Acromyrmex (Moellerius) landolti Forel en el Paraguay: Las subespecies balzani (Emery) y fracticornis (Forel) (Insecta: Hymenoptera). Neotropica 23: 39-44.

Fowler, H. G. 1982. Male induction and function of workers excitability during swarming in leaf-cutting ants (Atta and Acromyrmex) (Hymenoptera: Formicidae). Int. J. Invert. Reprod. 4: 333-335.

Fowler, H. G. 1983a. Distribution patterns of Paraguayan leaf-cutting ants (Atta and Acromyrmex) (Hymenoptera: Formicidae: Attini). Stud. Neotrop. Fauna Environ. 18: 121-138.

Fowler, H. G. 1983b. Latitudinal gradients and diversity of the leaf-cutting ants (Atta and Acromyrmex) (Hymenoptera: Formicidae). Rev. Biol. Trop. 31: 213-216.

Fowler, H. G. 1983c. Alloethism in a leaf-cutting ant: Laboratory studies on Atta texana (Hymenoptera: Formicidae: Attini). Zool. Jb. Physiol. 87: 529-538.

Fowler, H. G. 1984. Population dynamics of the leaf-cutting ant, Atta capiguara, in Paraguay. Cienc. Cult. 36: 628-632.

Fowler, H. G. A organizacao das comunidades de formigas cortadeiras (Atta e Acromyrmex). An. IV. Sem. Reg. Ecol., Sao Carlos, Sao Paulo. (In press).

142

Fowler, H. G., and B. L. Haines. 1983. Diversidad de especies de hormigas cortadoras y termitas de tumulo en cuanto a la sucesion vegetal en praderas paraguayas, pp. 187-201. In P. Jaisson (ed.), Social insects in the tropics, Vol. II. 252 pp.

Fowler, H. G., and S. W. Robinson. 1979a. Foraging ecology of the grass-cutting ant, Acromyrmex landolti fracticornis (Formicidae: Attini) in Paraguay. Int. J. Ecol. Environ. Sci. 5: 29-37.

Fowler, H. G., and S. W. Robinson. 1979b. Foraging by Atta sexdens: Seasonal patterns, caste, and efficiency. Ecol. Entomol. 4: 239-247.

Fowler, H. G., S. W. Robinson, and J. Diehl. 1984. Effect of mature colony density on colonization and initial colony survivorship in Atta capiguara, a leaf-cutting ant. Biotropica 16: 51-54.

Fowler, H. G., and N. B. Saes. Padroes de colonizacao das formigas cortadeiras Atta bisphaerica e Acromyrmex multiconodis e a regulacao natural de suas populacoes. Cienc. Cult. (In press, a).

Fowler, H. G., and N. B. Saes. Colonization patterns of the leaf-cutting ant, Atta sexdens rubropilosa. Trop. Ecol. (In press, b).

Fowler, H. G., and E. W. Stiles. 1980. Conservative resource management by leaf-cutting ants? The role of foraging territories and trails, and environmental patchiness. Sociobiology 5: 25-41.

Geijskes, D. C. 1953. Nuptial flight time of Atta ants in Surinam. Tids. Plantz. 59: 181-184.

Goncalves, C. R. 1945. Sauvas do sul e do centro do Brasil. Bol. Fitossanitario 2: 183-218.

Goncalves, C. R. 1951. Sauvas do nordeste do Brasil. Bol. Fitossanitario 5: 1-43.

Goncalves, C. R. 1967. As formigas cortadeiras da amazonia dos generos Atta Fabr. e Acromyrmex Mayr (Hym., Formicidae). Atas Simp. Biota Amazon 5: 181-202.

Huber, I. 1907. The founding of colonies by Atta sexdens L. Ann. Rept. Smith. Inst. 1906: 355-372.

Jacoby, M. 1944. Observacoes e experiencias sobre Atta sexdens rubropilosa Forel, visando facilitar seu combate. Bol. Min. Agric., Brazil 1943. 55 pp.

Jonkman, J. C. M. 1977. Biology and ecology of the leaf-cutting ant Atta vollenweideri Forel, 1983, (Hymenoptera, Formicidae) and its impact in Paraguayan pastures. Ph.D. Thesis, Univ. Leiden, Netherlands.

Jonkman, J. C. M. 1979. Population dynamics of leaf-cutting ant nests in a Paraguayan pasture. Z. Angew. Entomol. 87: 281-293.

Jonkman, J. C. M. 1980. The external and internal structure and growth of nests of the leaf-cutting ant Atta vollenweideri Forel, (Hymenoptera: Formicidae). Z. Angew. Entomol. 89: 158-173.

Jutsum, A. R., and R. J. Quinlan. 1978. Flight and substrate utilization in laboratory reared males of Atta sexdens. J. Insect Physiol. 24: 821-826.

Kerr, W. E. 1961. Acasalamento de rainhas com varios machos em duas especies da tribu Attini. Rev. Brasil Biol. 21: 45-48.

Labrador, J. R., I. J. Martinez, and A. Mora. 1972. Acromyrmex landolti Forel, plaga del pasto guinea (Panicum maximum) en el estado de Zulia. Rev. Fac. Agron. Univ. Zulia. 2: 27-38.

Leston, D. 1978. A neotropical ant mosaic. Ann. Entomol. Soc. Am. 71: 649-653.

Lewis, T. 1975. Colony size, density and distribution of the leaf-cutting ant, Acromyrmex octospinosus (Reich) (Formicidae: Attini) in cultivated fields. Trans. R. Entomol. Soc. London. 127: 51-64.

Lewis, T., G. V. Pollard, and G. C. Dibley. 1974. Rhythmic foraging in the leaf-cutting ant Atta cephalotes (L.) (Formicidae: Attini). J. Anim. Ecol. 43: 129-142.

Lucca, F. de. 1956. O combate as formigas cortadeiras no municipio de Sao Paulo. O. Biologico 22: 28-32.

Mariconi, F. A. M. 1970. As sauvas. Ed. Agron. Ceres. Sao Paulo.

Mariconi, F. A. M. 1974. Contribuicao para o conhecimento do sauveiro inicial da sauva parda Atta capiguara Goncalves, 1944. An. Soc. Entomol. Bras. 3: 5-13.

Mariconi, F. A. M., and A. P. L. Zamith. 1963. Nova contribuicao para o conhecimento das sauvas de Piracicaba. Rev. Agric. 38: 85-93.

Martin, M. M., G. A. Carls, R. F. N. Hutchins, J. G. McConnell, J. S. Martin, and D. D. Steiner. 1967. Observations on Atta colombica tonsipes (Hymenoptera: Formicidae). Ann. Entomol. Soc. Am. 60: 1129-1130.

Mendes-Filho, J. M. 1979. Tecnicas de combate as formigas. IPEF (Piracicaba), Circular Technica No. 75: 1-14.

Mintzer, A. 1979. Foraging activity of the Mexican leaf-cutting ant, Atta mexicana (F. Smith) in a Sonoran Desert habitat. Insect. Soc. 26: 364-372.

Mintzer, A., and S. B. Vinson. 1985. Cooperative colony foundation by females of the leaf-cutting ant, Atta texana, in the laboratory. J. N. Y. Entomol. Soc. 93: 1047-1051.

Moraes, R. L. M. S. de, M. A. S. de Cunha, and C. da Cruz Landim. 1980. Modificacoes dos corpora allata de rainhas de Atta sexdens rubropilosa Forel (Hymenoptera: Formicidae) durante o primeiro ciclo reprodutivo. Rev. Bras. Entomol. 24: 267-273.

144

Moser, J. C. 1963. Contents and structure of Atta texana nest in summer. Ann. Entomol. Soc. Am. 51: 286-291.

Moser, J. C. 1967. Mating activities of Atta texana (Hymenoptera: Formicidae). Insect. Soc. 14: 295-312.

Moser, J. C., and J. R. Lewis. 1981. Multiple nest queens of Atta texana (Buckley, 1860) (Hymenoptera: Formicidae). Turrialba 31: 256-257.

Pereira-da-Silva, V. 1973. Contribuicao ao estudo das populacoes de Atta sexdens rubropilosa Forel, 1908, e Atta laevigata (Fredrick Smith, 1858) (Hymenoptera: Formicidae) no Estado de Sao Paulo. Ph.D. Thesis, Fac. Filos. Cienc. Letras, Rio Claro, Sao Paulo.

Pereira-da-Silva, V. 1975. Contribuicao ao estudo das populacoes de Atta sexdens rubropilosa Forel e Atta laevigata (Fr. Smith) no Estado de Sao Paulo (Hymenoptera: Formicidae). Studia Entomol. 18: 201-250.

Pereira-da-Silva, V. 1979. Dinamica populacional, biomasse e estrutura dos ninhos iniciais de Atta capiguara Goncalves, 1944 (Hymenoptera: Formicidae) na regiao de Botucatu, SP. L.D. Thesis. UNESP, Botucatu.

Pereira-da-Silva, V., L. C. Forti, and L. G. Cardoso. 1981. Dinamica populacional e caracterizacao dos ninhos de Acromyrmex coronatus (Fabricius, 1804) (Hymentoptera: Formicidae). Rev. Bras. Entomol. 25: 87-93.

Porter, S. D., and M. A. Bowers. 1982. Caste partitioned survivorship and route fidelity of leaf-cutting ant workers, p. 254. In M. D. Breed, C. D. Michener, and H. E. Evans (eds.), Biology of social insects. Westview Press, Boulder, Colorado. 419 pp.

Ribeiro, G. T., and R. A. Woessner. 1979. Teste de eficiencia com seis sauvacidas no controle de sauvas (Atta spp.) na Jari, Para, Brasil. Ann. Soc. Entomol. Bras. 8: 77-84.

Robinson, S. W., and H. G. Fowler. 1982. Foraging and pest potential of Paraguayan grass-cutting ants (Atta and Acromyrmex) to the cattle industry. Z. Angew. Entomol. 93: 45-52.

Rockwood, L. L. 1973. Distribution, density and dispersion of two species of Atta (Hymenoptera: Formicidae) in Guanacaste Province, Costa Rica. J. Anim. Ecol. 42: 803-817.

Rodrigues, M. G. 1966. Ecologia das formigas cortadeiras do genero Acromyrmex da mata amazonica. Ann. Rep. Min. Agric. D.P.E.A., Inst. Pesq. Agropec. Norte, Belem, Para, Brazil.

Sales, F. S. M., V. P. O. Alves, N. G. Gomes, and M. T. Alves. 1983. Vulnerabilidade de algumas culturas a sauva do Nordeste, Atta opaciceps Borgmeier, 1939. Res. VII Congr. Bras. Entomol. Brasilia, Brazil.

Schade, F. H. 1973. Ecology and control of the leaf-cutting ants of Paraguay. In R. J. Gorham (ed.), Paraguay: Ecological essays. Acad. Arts Sci. Americas, Miami, FL.

Soares de Souza, G. 1587. Tratado descritivo do Brasil. Vol. 117. Colecao Brasiliera. 5ªSerie, Sao Paulo.

Stradling, D. J. 1978. The influence of size on the foraging in the ant, Atta cephalotes, and the effect of some plant defense mechanisms. J. Anim. Ecol. 47: 173-188.

Walter, E. V., L. Seaton, and A. A. Mathewson. 1938. The Texas leaf-cutting ant and its control. U.S.D.A. Circ. No. 494: 1-18.

Warter, S. L., J. C. Moser, and M. S. Blum. 1962. Some correlations between the foraging behavior of the common nighthawk, Chordeiles minor (Forster) and the swarming behavior of two species of ants, Atta texana (Buckley) and Iridomyrmex pruinosis (Roger). Proc. La. Acad. Sci. 25: 42-46.

Weber, N. A. 1937. The biology of the fungus-growing ants. Part II. Nesting habits of the bachac (Atta cephalotes L.). Trop. Agric. 14: 223-226.

Weber, N. A. 1972. Gardening ants: The attines. Mem. Am. Phil. Soc. Vol. 92, 146 pp.

Weber, N. A. 1976a. A 10-year colony of Acromyrmex octospinosus (Hymenoptera: Formicidae).. Proc. Entomol. Soc. Wash. 79: 284-292.

Weber, N. A. 1976b. A 10-year laboratory colony of Atta cephalotes. Ann. Entomol. Soc. Am. 69: 825-829.

Weber, N. A. 1977. Recurrence of Atta colonies at a Canal Zone site (Hymenoptera: Formicidae). Entomol. News 88: 85-86.

Wheeler, W. M. 1907. The fungus-growing ants of North America. Bull. Amer. Mus. Nat. Hist. 23: 669-807.

Wilson, E. O. 1980. Caste and division of labor in leaf-cutter ants (Hymenoptera: Formicidae: Attini). I. The overall pattern in A. sexdens. Behav. Ecol. Sociobiol. 7: 143-156.

12
The Foraging Ecology
of *Atta texana* in Texas

D. A. Waller

The foraging patterns of leaf-cutting ants have been the subject of speculation dating back to Wallace (1891). More recently, this subject has been targeted for study by a number of researchers who have studied several Atta species (Cherrett 1968a; Hubbell and Wiemer 1983; Mintzer 1979; Pintera 1983; Rockwood 1976; Shepherd 1982). Most of this research was directed at describing and explaining the nature of plant selection by these polyphagous herbivores. Belt (in Wallace 1891) was probably the first to postulate a coevolutionary struggle between leaf-cutters and native plants in their habitats, suggesting that plants were under pressure to become unsuitable as forage for the ants. Current research indicates that leaf-cutter ant forage selection is influenced by repellent chemicals in the plants (Hubbell et al. 1983), by physical characteristics of the leaves (Cherrett 1972a; Stradling 1978; Waller 1982a), by an interaction between leaf palatability and leaf density (Waller 1982b), by the proximity of plants to established trails (Shepherd 1982), and by resident ant defense of plants against Atta attack (Belt, in Wallace 1891; Jutsum et al. 1981). In addition, plant choice may vary during the year in response to the dynamics of the leaf-cutter ant colony and their fungus garden. Thus, brood status in the colony, temperature, and moisture may influence the type of plants upon which they feed (Lewis et al. 1974; Rockwood 1976).

An overview of leaf-cutter ant foraging patterns must be put together piece-by-piece with detailed observations on foraging by individual species and populations. In the following study, plant selection by the northernmost leaf-cutting ant species, Atta texana, was studied in a central Texas (USA) habitat. A determination was made of plant species attacked or avoided, either totally or seasonally. A bioassay was used to investigate whether or not plant species were avoided due to chemical/nutritional unpalatability or because of physical features. Additionally, leaf density was related to ant forager size in order to evaluate the cost to foragers of

146

harvesting plants with tough leaves. Thus, this study focused on the role of plant quality in forage selection by A. texana.

METHODS

Study Site and General Procedures

All observations and experiments were conducted at the Brackenridge Field Laboratory (BFL) of the University of Texas at Austin. BFL is 7 km from the university campus and is situated along the Colorado River. The 30-ha preserve was a housing area but has been undisturbed for a number of years and is now a mixed savannah and Ulmus-Celtis-Quercus woodland. Mean monthly temperatures range from 10°C in January to 29°C in July and August. Average annual rainfall is 75 cm, with peak precipitation usually in May, September, and November. At least seven mature A. texana nests occur on BFL grounds, and others are common along the nearby river banks.

Observations on ant activity were made from 1977 to 1982. Records of plants attacked by the ants were kept from the fall of 1979 to the fall of 1981, although for some plant species records were kept for longer periods. Six mounds were observed irregularly, and three were monitored at least monthly during their active periods, as were two foraging areas with seasonally-active tunnel entrances (Area A: Melia-Celtis-Ulmus americana woodland; Area B: Quercus-Celtis-Ligustrum-Ulmus crassifolia woodland). Visits to the different areas of BFL were made approximately every other day during the five-year study.

Bioassays of Leaf Characteristics in Relation to Seasonal Foraging

A bioassay was designed to determine whether or not leaf palatability and/or cuttability could be correlated with the seasonality of ant foraging. The ants were presented with two sizes of leaf discs cut from sun leaves (i.e., those leaves exposed to the sun) of plants on which the ants were foraging. The small discs (6 mm diam.) were retrievable by the ants without cutting; however, the big discs (25 mm diam) had to be cut before the ants could carry them. It was assumed that, if ants removed both small and big discs, the leaves were cuttable and palatable. If both sizes of discs were rejected, the leaves were considered unpalatable. If only big discs were rejected, it was assumed that the leaves were palatable but uncuttable. One hundred small and 10 large discs of each plant species, and some cracked dry corn, were placed in a 15 x 100 mm petri dish. The corn served as a control to insure that the ants were foraging, since it is highly palatable to A. texana (Waller, unpublished). Dishes were placed at five active ant exit/entrance holes at the commencement of foraging activity. Discs remaining after 24 hours were removed and counted. If less than half of the discs were removed, they were considered rejected. Tests were usually made on two successive days to reduce the effects of novelty (Cherrett 1972b). In 1980, small discs were tested; in 1981 both big and small discs were tested. Cut leaves and intact leaves were also tested in 1980 and 1981, respectively, but results are not presented here since leaf size variability rendered comparisons between dates and species inappropriate. Intact leaves were always cut and/or removed when big discs were removed, so it can be assumed that hard leaf edges or cuticle did not mask attractive odors (Barrer and Cherrett 1972).

Bioassays in Relation to Leaf Density and Worker Size

Leaf density, or dry weight per unit size, is positively correlated with leaf toughness (Cherrett 1968b) and is a measure of cuttability. The effect of density was determined by punching five 6 mm diam discs from each of five fresh leaves similar to those attacked by ants. Dry weights were measured to 0.1 mg, and density was expressed as mean dry weight per disc.

Leaf-bearing ants were collected on ground or tree-trunk trails within 1 m of the nest entrance. Thirty consecutive ants that crossed a point on the trail were taken per sample. Ants were killed with HCN and stored in 80% ethanol. Headwidth (HW) was measured to 0.06 mm with an ocular micrometer and a dissecting microscope.

RESULTS

Seasonal Foraging Behavior

A. texana showed a strong seasonal pattern in its foraging activity. Beginning in late November, the ants foraged from their central mound on leaves of grass, herbs, vines, and the pulp of Melia berries except for a few cases in which they foraged from their tunnels on berry pulp and senescent leaves. Most of their foraging continued from the mounds until mid to late March when foraging tunnel entrances became very active again. At this time, foraging from the mound ceased, and plants grew on these denuded mounds. However, these plants were cut sporadically from April to June just prior to mating flights, but leaf fragments were left scattered on the mounds and apparently were not used as fungal substrate. After all the mating flights had ceased, foraging activity on the mounds stopped until November.

Foraging tunnel exits/entrances were often located at the bases of trees and became active or inactive at approximately the same times each year. For example, a tunnel ending at a Quercus fusiformis tree was opened around March 24, 20, 19, and 24 and closed around June 9, May 29, 29, and 28 in the years 1979, 1980, 1981, and 1982 respectively (first and last days that foraging ants were observed each year). While the entrance was not active the rest of the year, ants could be enticed out to forage on cracked corn and palatable leaves.

Ants foraged during the day from September through May and at night from April through October with transition periods of day and night foraging in spring and fall. The lowest air temperature recorded when ants were foraging was 10°C (January), and the highest was 31°C (March and September).

<u>Seasonal Plant Selection</u>

Many common plants near <u>A. texana</u> mounds and tunnel entrances were rarely or never attacked. These included <u>Ambrosia trifida</u> (Asteraceae), <u>Berberis trifoliolata</u> (Berberidaceae), <u>Bumelia lanuginosa</u> (Sapotaceae), <u>Celtis reticulata</u> and <u>C. laevigata</u> (Ulmaceae), <u>Condalia hookeri</u> (Rhamnaceae), <u>Cornus drummondii</u> (Cornaceae), <u>Forestiera pubescens</u> (Oleaceae), <u>Malvaviscus arboreus</u> var. <u>drummondii</u> (Malvaceae), <u>Rhus toxicodendron</u> (Anarcardiaceae), <u>Sapindus saponaria</u> var. <u>drummondii</u> (Sapindaceae), <u>Verbesina virginica</u> (Asteraceae), and <u>Vitis mustangensis</u> (Vitaceae). Other plants were uncommon in the study area, which may account for the reports of only occasional attacks. These included <u>Ilex vomitoria</u> (Aquifoliaceae), <u>Parthenocissus</u> sp. (Vitaceae), <u>Pyracantha</u> sp. (Rosaceae), <u>Rhus aromatica</u> (Anacaradiaceae), and <u>Rubus trivialis</u> (Rosaceae).

In Table 1 are listed various plants and leaf types (new, mature and senescent) that were attacked. Superimposed on this table are the months when leaf-cutting ant attacks occurred. Some of the patterns of attack are particularly intriguing. For example, leaves of <u>Lonicera japonica</u> were ignored by <u>A. texana</u> except in late November and December when new and mature leaves were attacked and cut heavily until the plants were defoliated in February. <u>Lonicera</u> produced new leaves at this time but was defoliated again. By the time new <u>Lonicera</u> leaves flushed once more in March, the <u>A. texana</u> had shifted their foraging activity once again to tunnel exit/entrances, and the <u>Lonicera</u> near their mounds was ignored.

Violet plants grew on one of the study mounds, and their leaves were ignored until mid to late December each year. Prior to attack on <u>Viola</u>, ants were observed walking over <u>Viola</u> leaves while foraging on <u>Lonicera</u>, <u>Smilax</u>, and herbs. When foragers did concentrate on <u>Viola</u>, they defoliated all of the plants on the mound. Next, the ants attacked mature <u>Nemophila phacelioides</u> leaves which had been ignored during attacks on <u>Viola</u>. New leaves that developed on

TABLE 1. Plants and month in which they were attacked by *A. texana*. (N = new leaves, M = mature leaves, S = senescent leaves, P = berry pulp, D = seeds)

Plant species (family)	Plant stage available for foraging and months (underlined) in which they were attacked												Area of BFL study site [a]
	Jan	Feb	Mar	Apr	May	Jun	Jul	Aug	Sep	Oct	Nov	Dec	
Grass (Poaceae)	<u>N</u>	M	M	M	M	M	M	M	M	M	M	<u>N</u>	mounds
Ligustrum japonicum (Oleaceae)	M	M	M	<u>N</u>	M	M	M	M	M	M	M	M	B
Lonicera japonica (Caprifoliaceae)	<u>M</u>	<u>N</u>	N	N	M	<u>M</u>	<u>M</u>	<u>M</u>	M	M	M	<u>N</u>	mounds, A
Melia azedarach leaves (Meliaceae)	-	-	-	<u>N</u>	<u>M</u>	<u>M</u>	<u>M</u>	<u>M</u>	<u>M</u>	-	-	-	A
M. azedarach berry pulp	<u>P</u>	<u>P</u>	<u>P</u>	<u>P</u>	<u>P</u>	-	-	-	-	<u>S</u>	<u>S</u>	<u>P</u>	mounds, A,B
Nemophila phacelioides (Hydrophyllaceae)	<u>M</u>	<u>S</u>	<u>N</u>	N	M	-	-	-	-	-	N	M	mounds
Quercus fusiformis (Fagaceae)	M	-	<u>N</u>	<u>N</u>	M	M	M	M	M	M	M	N	B
Salix nigra (Salicaceae)	<u>S</u>	-	-	M	M	M	M	M	<u>S</u>	<u>S</u>	-	-	B
Smilax bona-nox (Smilacaceae)	<u>S</u>	<u>S</u>	<u>N</u>	<u>N</u>	M	<u>M</u>	<u>M</u>	<u>M</u>	<u>M</u>	<u>M</u>	<u>S</u>	<u>S</u>	mounds
Ulmus americana (Ulmaceae)	-	-	<u>N</u>	<u>N</u>	M	M	M	M	M	M	-	-	mounds, A
Ulmus crassifolia leaves (Ulmaceae)	-	-	N	N	M	M	M	M	M	M	M	-	B
U. crassifolia seeds	-	-	-	-	-	-	-	-	<u>D</u>	<u>D</u>	-	-	B
Viola missouriensis (Violaceae)	<u>M</u>	<u>M</u>	M	M	-	-	-	-	N	N	M	<u>M</u>	mounds

[a] A = *Melia–Celtis–Ulmus americana* woodland.
B = *Quercus–Celtis–Ligustrum–Ulmus crassifolia* woodland.

the Viola were ignored until after the plants flowered in March and produced fruit in April. At this time, which was prior to mating flights, the Viola was defoliated once again, but the cut leaves were left on the mound.

At tunnel entrances, new leaves were cut from Quercus fusiformis and Ulmus americana while mature leaves were ignored. Leaves of all ages were cut from Ligustrum japonicum, Melia azedarach, and Salix nigra, although very young leaves of Melia were ignored the first week, even though ants foraged among these leaves for berry pulp. Only senescent leaves of Ulmus crassifolia were cut even though the ants were foraging among mature leaves as they cut seeds from the trees in September and October. Smilax bona-nox is a thorny vine that grows near almost all A. texana mounds and tunnel entrances/exits. Only senescent leaves were cut beginning in November. These were usually infected with the rust fungus Puccinia smilacis and sometimes with fungal spots of Phyllostricta sp.

Leaf Characteristics and Seasonality of Foraging

The bioassay data are presented in Tables 2 (1980) and 3 (1981). Removal of less than one-half of the small leaf discs of Lonicera, Viola, U. americana, and U. crassifolia was taken to mean that they were unpalatable at this time. For example, although 58.4% of the small discs of Lonicera were taken on July 24, 1980, only 29.6% were taken the following day, so on average less than half were removed. The ants removed 57.5% of the small discs of U. crassifolia on September 7, 1981, but on the following day I found more discs than were placed out the preceding day suggesting that previously accepted discs had been discarded. In contrast, small discs of Quercus and Smilax and big discs of the latter were taken in large numbers on all dates tested. Thus, these two species were acceptable out-of-season. Smilax remained cuttable; big discs of Quercus were rejected as the leaves became tougher with maturity. (The dynamics of A. texana attack on Quercus are reported elsewhere by Waller 1982b.) Although discs of Ligustrum japonicum were not tested out of season, 93% of whole leaves were cut in bioassays (10 leaves/dish, 5 entrances tested) on December 22, 1981. Generally, I have observed that L. japonicum leaves are cut in large numbers whenever placed near active A. texana nests, so they are palatable and cuttable throughout the year. In addition, big and small discs of Ligustrum Quihoui were taken in large numbers in December 1980 and January 1981. L. Quihoui is similar in appearance to L. japonicum and occurs on BFL grounds but is not found near A. texana tunnel entrances.

152

TABLE 2. Effect of seasonality on collection of small leaf discs by A. texana at five tunnel entrances (1980).

Plant species, test date, and percent of discs collected									
Ligustrum japonicum[a]		Ligustrum Quihoui[b]		Lonicera japonica[b]		Quercus fusiformis[b]		Smilax bona-nox[b]	
Date	%	Date	%	Date	%	Date	%	Date	%
07-15	100	12-28	80.4	07-24	58.4	07-21	98.0	09-15	94.2
07-16	100	12-30	80.6	07-25	29.6	07-22	98.4	09-19	82.6
08-07	100			08-15	15.0	08-26	66.4	09-20	91.2
08-20	100			08-23	26.8	08-29	100.0		
						09-13	100.0		
						10-04	98.2		
						10-05	97.2		
						12-06	100.0		
						12-10	83.0		

[a]Tests with new leaves that are attacked in nature.
[b]Tests with mature leaves that are not attacked in nature.

Overall, the results revealed that different plant mechanisms are involved in seasonal defense against attacks by leaf-cutting ants. Some plants were unpalatable, one was palatable but not cuttable, and others were both palatable and cuttable, yet seasonally ignored. It is possible that some plants are attacked only when they are damaged. For example, Rhus toxicondendron was not attacked by ants; but when broken stems were placed near active entrances, their leaves were readily cut. The bioassay did not address possible defense mechanisms of standing intact plants.

Effect of Leaf Density and Worker Size on Foraging

Densities of forage plants and mean HW of ants attacking these plants are listed in Table 4. A positive correlation was found between mean ant HW and plant density for ground foragers (tree foragers are larger than ground foragers, thus tree and ground samples were treated separately). A significant negative correlation was noted between mean ant HW and plant water content. This means that larger ants cut denser, drier leaves including those of Ligustrum, Quercus, Salix, and Smilax while the smaller ants cut grass, Lonicera, Nemophila, and Viola.

TABLE 3. Effect of seasonality on collection of small and large leaf discs by <u>A</u>. <u>texana</u> at five tunnel entrances (1981).

Plant species	Leaf stage	Test date	% discs removed Small	Large
Ligustrum Quihoui	new[a]	1-1	81.6	—
		1-2	100.0	100.0
	mature[a]		100.0	99.0
	new[a]	1-3	85.6	99.0
	mature[a]		100.0	100.0
Lonicera japonica	new	3-15	74.4	48.0
		3-16	99.8	68.0
Melia azedarach	mature	4-26	63.4	73.0
		4-27	79.2	44.0
		5-10	87.2	81.0
		5-11	78.8	74.0
		6-22	91.2	86.0
		6-23	99.0	66.0
Quercus fusiformis	senescent[a]	1-23	85.0	4.0
		1-24	91.1	13.0
	new	4-4	100.0	100.0
		4-5	100.0	100.0
		4-26	100.0	100.0
		4-27	100.0	100.0
		5-10	54.8	3.0
		5-11	71.8	19.0
		5-18	91.5	62.0
		5-19	100.0	20.0
Smilax bona-nox	new[a]	4-4	84.2	82.0
		4-5	95.6	88.0
	senescent[a]	7-18	80.0	61.0
		7-19	85.4	82.0
Ulmus americana	mature[a]	9-28	26.5	16.0
		9-29	23.3	5.0
Ulmus crassifolia	mature[a]	9-7	57.5	10.0
		9-8	-0.2	3.0
Viola missouriensis	mature[a]	4-11	35.4	2.0
		4-12	4.6	15.0

[a]Leaf stage or plant not attacked in nature.

TABLE 4. Mean headwidth (HW) of foragers attacking different plant species as related to leaf density.[a]

Plant species	Developmental stage	Age	HW Replications	HW mm[b]	Leaf (n=5) Density (mg)	Leaf (n=5) Water (%)
			Tree Collections			
Melia azedarach	leaves	mature	19	2.02±0.10	1.4±0.1	65.0±1.2
	berry pulp	—	17	1.92±0.14		
Quercus fusiformis	leaves	new	19	2.12±0.13	2.6±0.2	64.2±2.8
Ulmus crassifolia	seeds	—	9	2.12±0.09		
			Ground Collections[c]			
Grass	leaves	new	5	1.56±0.08	1.0±0	75.2±1.3
Ligustrum japonicum	leaves	new	7	1.94±0.10	2.5±0.3	61.8±1.3
Lonicera japonica	leaves	new	12	1.74±0.08	1.5±0.3	65.6±2.9
Nemophila phacelioides	leaves	mature	4	1.67±0.06	1.1±0.1	84.6±1.5
Salix nigra	leaves	mature	4	1.91±0.08	2.1±0.1	58.2±1.3
Smilax bona-nox	leaves	senescent	21	1.98±0.10	2.2±0.08	59.2±4.7
Viola missouriensis	leaves	mature	7	1.71±0.07	1.4±0.2	71.4±0.9

[a] Density = mean dry weight/6 mm diam. disc.
[b] Mean and standard deviation of the mean headwidth of 30 ants from each site replicate.
[c] Ground forager HW correlation with leaf-density: Kendall's tau = 0.9048, $p < 0.003$, n=7.
Ground forager HW correlation with water content: Kendall's tau = -0.7143, $p < 0.013$, n=7.

Nest or tunnel entrances, temperature, and season may influence the size of the ants foraging as well as time of day and year irrespective of leaf density. To control for these variables, I examined ants and the leaf material they carried to the nest on the same trail at similar times. On February 7, 1981 (1600 hr, 26°C), ants carrying Lonicera were significantly larger than ants carrying grass on the same trail (Lonicera mean ant HW = 1.70 mm ± 0.30 SD, n = 33; grass mean ant HW = 1.47 mm ± 0.29 SD, n = 31; Mann-Whitney U = 775, p < 0.05).

Ants were also observed trailing from another colony. In this case, the ants foraged on three plants as follows:

1. Viola leaves on December 26 at 1510 h (17°C) and December 27 at 1539 h (19°C); mean ant HW was 1.72 mm ± 0.03 SD (n = 2 samples).

2. Smilax on January 26 at 1630 h (22°C) and January 27 at 1136 h (18°C); mean ant HW was 1.92 mm ± 0.14 SD (n = 2 samples).

3. Nemophila on February 7 at 1615 h (24°C) and February 15 at 1520 h (22°C); mean ant HW was 1.64 mm ± 0.04 SD (n = 2 samples).

Again, the foragers were significantly different in size according to the plant species attacked (Kruskal-Wallis H = 26.629, p < 0.0001). Other examples of this relationship between ant size and plant species can be found in Waller (1982c).

Comparison of the data in Tables 2, 3, and 4 reveals that out-of-season leaves that were palatable in the bioassays were among the densest leaves attacked (Quercus, Smilax, and Ligustrum) and required the largest foragers to cut them. Although the ants cut leaves of the densest leaf stage of Smilax, the senescent leaves of Quercus and the mature/senescent leaves of Ligustrum japonicum are much denser (Quercus = 4.9 mg/dry disc ± 0.04 SD; Ligustrum = 5.1 mg/dry disc ± 0.4 SD). Thus, leaf density of these latter two species may deter ant attack at certain times of the year even though the leaves are palatable.

DISCUSSION

The research described illustrates the complexity of forage selection by Atta. A. texana attacks a number of plant species including monocots and dicots. Like other Atta spp., A. texana seasonally avoids many of its regular hosts. Some of these were unpalatable out of season, while others remained palatable throughout the year. These latter plants had the densest leaves of those attacked and were cut by the largest foragers. Wilson (1980) has found that the larger A. sexdens workers cut tough leaves most efficiently while small ants cut soft vegetation most efficiently. Thus, like A. sexdens, A. texana appears to be using its polymorphism to effectively exploit the resource environment. Although the relative costs of producing large and small workers have not been determined, it seems likely that using large foragers to harvest tough leaves is more ergonomically taxing on Atta colonies than cropping soft leaves. Thus increased leaf density may protect palatable plants from ant attack.

Alternatively, other unexamined factors may play a role in avoidance of bioassay-palatable species. Mudd and Bateman (1979) found that although Ligustrum ovalifolium leaves were readily cut by A. cephalotes in the laboratory, they supported no fungal growth in culture. Perhaps over a period of time, Atta responds to retarded fungal growth and ceases foraging on certain plants (Cherrett, personal communication). Barrer and Cherrett (1972) found that cut leaves were more attractive to ant foragers, so perhaps intact standing plants have some protection (similar to my observation with Rhus toxicodendron).

Some plants, including Lonicera, Ulmus spp., and Viola, are unpalatable to ants out of season. Unpalatability is a nebulous designation that requires much further analysis in order to sort out the true components. First, unpalatability may derive from changes in ant behavior alone. Ants may require different nutrients in different seasons depending on brood, temperature, or moisture conditions in the nest (Lewis et al. 1974; Rockwood 1976). Also, ants may show a spontaneous decline in preference for unchanging foods as demonstrated by Littledyke and Cherrett (1975) for cereal grains. Alternatively, preference for a given food may be enhanced by exposure (Fowler 1982). Finally, since ants cut plants that they normally avoid on the surface of the mound at the times of the mating flights, it seems likely that the decision to cut a plant is sometimes regulated by complex factors extrinsic to the condition of the plant. Second, seasonal unpalatability may derive from fluctuations in plant levels of nutrients, water, or secondary chemicals. These fluctuations may reflect natural seasonal variation or they may be induced by environmental stresses including fungal or herbivore attack (Haukioja and Niemela 1977). In addition, secondary chemicals may be produced by endophytic fungi that protect host plants from herbivores (T. Hardy and K. Clay, personal communication). These fungi represent a new avenue of investigation in Atta forage selection and may answer questions about intraspecific variability and palatability.

Probably many or all of these factors act in a dynamic flux, so that shifting levels of plant nutrients, secondary chemicals, and leaf density are in some way assessed by foragers who are, in turn, influenced by their own age and experience and the unique needs of the home colony. However, this dynamic flux yields a pattern of choice that is consistent from year to year, which suggests that the multitude of factors shaping this pattern can eventually be elucidated.

ACKNOWLEDGMENTS

I thank L. E. Gilbert, A. C. Lewis, D. Mackay, J. Moser, and M. C. Singer for valuable discussions and help during the progress of this research. I am grateful for the use of the Brackenridge Field Laboratory facilities, and I thank its staff for assistance, including J. Crutchfield, N. Eichler, and L. E. Gilbert. University of Texas subvention funds and a Sigma Xi Grant-in-Aid contributed to the support of this work.

REFERENCES CITED

Barrer, P. M., and J. M. Cherrett. 1972. Some factors affecting the site and pattern of leaf-cutting ant activity in the ant Atta cephalotes L. J. Entomol. (A) 47: 15-27.

Cherrett, J. M. 1968a. The foraging behavior of Atta cephalotes L. (Hymenoptera: Formicidae) 1. Foraging pattern and plant species attacked in tropical rain forest. J. Anim. Ecol. 27: 387-403.

Cherrett, J. M. 1968b. A simple penetrometer for measuring leaf toughness in insect feeding studies. J. Econ. Entomol. 61: 1736-1738.

Cherrett, J. M. 1972a. Some factors involved in the selection of vegetable substrate by Atta cephalotes L. (Hymenoptera: Formicidae) in tropical rain forest. J. Anim. Ecol. 41: 647-660.

Cherrett, J. M. 1972b. Chemical aspects of plant attack by leaf-cutting ants. In J. Harborne (ed.), Phytochemical ecology. Academic Press, London.

Fowler, H. G. 1982. Habitat effect on fungal substrate selection by a leaf-cutting ant. J. N. Y. Entomol. Soc. XC: 64-69.

Haukioja, E., and P. Niemela. 1977. Retarded growth of a geometrid larva after mechanical damage to leaves of its host tree. Annal. Zool. Fennici 14: 48-52.

Hubbell, S. P., and D. F. Wiemer. 1983. Host plant selection by an attine ant, pp. 133-154. In P. Jaisson (ed.), Social insects in the tropics, Vol. 2. Universite Paris-Nord.

Hubbell, S. P., D. F. Wiemer, and A. Adejare. 1983. An antifungal terpenoid defends a neotropical tree (Hymenaea) against attack by fungus growing ants (Atta). Oecologia 60: 321-327.

Jutsum, A. R., J. M. Cherrett, and M. Fisher. 1981. Interactions between the fauna of citrus trees in Trinidad and the ants Atta cephalotes and Azteca sp. J. Appl. Ecol. 18: 187-195.

Lewis, T., G. V. Pollard, and G. C. Dibley. 1974. Rhythmic foraging in the leaf-cutting ant Atta cephalotes (L.) (Formicidae: Attini). J. Anim. Ecol. 43: 129-141.

Littledyke, M., and J. M. Cherrett. 1975. Variability in the selection of substrate by the leaf-cutting ants Atta cephalotes (L.) and Acromyrmex octospinosus (Reich) (Formicidae: Attini). Bull. Entomol. Res. 65: 33-47.

Mintzer, A. 1979. Foraging activity of the Mexican leaf-cutting ant, Atta mexicana (F. Smith), in a Sonoran desert habitat (Hymenoptera: Formicidae). Insect. Soc. 26: 364-372.

Mudd, A., and G. L. Bateman. 1979. Rates of growth of the food fungus of the leaf-cutting ant Atta cephalotes (L.) (Hymenoptera: Formicidae) on different substrates gathered by the ants. Bull. Entomol. Res. 69: 141-148.

Pintera, A. 1983. Selection of plants utilized by Atta insularis in Cuba (Hymenoptera: Formicidae). Acta Ent. Bohemoslov. 80: 13-20.

Rockwood, L. L. 1976. Plant selection and foraging patterns in two species of leaf-cutting ants (Atta). Ecology 57: 48-61.

Shepherd, J. D. 1982. Trunk trails and the searching strategy of a leaf-cutter ant, Atta colombica. Behav. Ecol. Sociobiol. 11: 77-84.

Stradling, D. J. 1978. The influence of size on foraging in the ant, Atta cephalotes, and the effect of some plant defense mechanisms. J. Anim. Ecol. 47: 173-188.

Wallace, A. R. 1891. Natural selection and tropical nature. McMillan and Company, London.

Waller, D. A. 1982a. Leaf-cutting ants and avoided plants: Defenses against Atta texana attack. Oecologia 52: 400-403.

Waller, D. A. 1982b. Leaf-cutting ants and live oak: The role of leaf toughness in seasonal and intraspecific host choice. Ent. Exp. and Appl. 32: 146-150.

Waller, D. A. 1982c. Foraging ecology of the Texas leaf-cutting ant, Atta texana Buckley (Formicidae: Attini): Host choice and foraging size polymorphism. Ph.D. Dissertation, University of Texas, Austin, Texas.

Wilson, E. O. 1980. Caste and division of labor in leaf-cutter ants (Hymenoptera: Formicidae: Atta) 11. The ergonomic optimization of leaf cutting. Behav. Ecol. Sociobiol. 7: 157-165.

13
Territoriality in Leaf-Cutting Ants, *Atta* spp.

E. F. Vilela and P. E. Howse

There are few reports of aggression in relation to territory for leaf-cutting ants. In fact, very little has been learned about the basic behavioral mechanisms used by leaf-cutting ants to establish and maintain foraging territories. Holldobler (1976) suggested that the territory of an ant colony includes the areas immediately surrounding the nest and the foraging trails of the colony, which may be changed frequently.

In the present research, the possible establishment of territories by leaf-cutting ants was investigated by studying interspecific aggressive behavior. Also, the potential dominance of one species of leaf-cutting ants over another, as postulated by Jutsum (1979), was investigated. Finally, the effect of group size on the aggressiveness of two species of leaf-cutting ants, Atta laevigata and A. cephalotes was investigated to test Wilson's (1971) hypothesis that individual aggressiveness increases with the number of nearby nest mates.

MATERIALS AND METHODS

Laboratory colonies of A. cephalotes and A. laevigata were used for these experiments. The ants were maintained in a 12:12 hr light/dark cycle. Two bioassays per day per colony were performed during the light portion of the cycle. The test design involved exposing simultaneously ants from a colony of each of the two species on 12.5 cm filter paper surfaces that had been marked previously by placing them in the foraging area (near the nest jar) of one of the ant colonies for exposure times of 2 or 24 to 48 hours. The number of yellow territorial marking spots deposited by the workers on the paper was recorded. In each test, one of these marked papers, or an unmarked paper (without previous ant contact), was placed in a 12.5-cm diameter plastic dish (arena), the walls of which were painted with Fluon® (ICI, England) to prevent climbing by the ants. Also, the arena was covered with a sheet of glass to prevent ant escape. All tests were conducted in a temperature-controlled room with $27° \pm 2°C$ and 75 to 85% R.H.

In each experiment, a single alien ant and one or five "resident" ants from the colony which had marked the paper, were gently placed on the test paper. In the one versus one experiments, the control or untreated paper provided data for comparison of the results of both ant species since the two ants, one of each species, were introduced simultaneously into the arena. Duplicating the control test was therefore considered unnecessary. Care was taken to insure minimal disturbance before the start of each test. The ants selected were chosen randomly and picked up with clean forceps (washed in Teepol®) from among the workers walking near the nest.

Tests with two proportions of ants (1:1 and 1:5) and three different sizes (media, maxima, soldier) were conducted as follows:

> 1 or 5 resident media versus 1 alien media
> 1 or 5 resident maxima versus 1 alien maxima
> 1 or 5 resident maxima versus 1 alien media
> 1 or 5 resident media versus 1 alien maxima
> 1 or 5 resident soldier versus 1 alien soldier

The size categories as measured by the body length of the ants were the following: (1) for A. laevigata - media 6 to 7 mm; maxima 9 to 11 mm; (2) for A. cephalotes - media 5 to 6 mm; maxima 7 to 8 mm.

The reason that five ants were used to form a group was based on a series of trials carried out prior to the test. In these trials, we noted that when the ants were in groups of two or three, their behavior was not as clear-cut as when there were five in a group. The times for which the papers were exposed to the ant colonies (2 and 24 to 48 hrs) for territorial marking were chosen on the basis of other preliminary trials in which the discrimination of the ants for papers marked for different time periods was assayed. These trials revealed a difference in the level of responsiveness of ants to papers marked over a 5 to 12 hr period and those marked over 2 or 24 to 48 hrs (analysis by Fisher exact test). For the final analysis, the results of 24 and 48 hr test papers were considered together as they proved to be similar according to X^2 tests.

For each experimental situation (i.e., 1 resident vs. 1 alien on papers marked 2 hrs), the data were collected from 50 replications which represented a combination of the 10 replicates for each of the five categories of ants listed above.

In the case of five residents versus one alien, observations were recorded up to the end of the first fight. Any subsequent fight was not recorded. Since all resident workers were of the same size category and were chosen at random from within a few centimeters of the nest entrance, we assumed that motivational levels for aggression were equal.

The behavior of individual ants in the arena was recorded for 30 minutes, starting just after the introduction of the alien ant. The behaviors shown toward the alien ant consisted of: (1) antennation at least once followed by fighting; (2) antennation, threatening, and dragging (both considered aggressive encounters) and then fighting; or (3) no aggressive response to the alien even after successive examinations and encounters.

Continuous aggressive maintenance of the territory was measured by the following parameters: (1) total number of fights; (2) number of fights started and dominated by resident ants; (3) number of fights fatal to the resident ants; (4) time (sec) between introduction of ants and start of fighting (latency); (5) number of aggressive encounters before the fight; and (6) number of ants trying to escape from the arena.

A fight was considered fatal if either or both of the fighters were killed or they were seriously injured or mutilated. Sometimes the fighters were observed for an extra period of time up to 30 min after the original observation period to provide clearer results. In some cases the dominant ant "chopped" the subordinate into small pieces in a few minutes. Other times, the fights were non-fatal even when injuries, such as loss of a leg, occurred since the ants were still accepted in the colony after a normal "inquiry" period. When ants did not contact each other during the 30-min observation period, the trial was rejected for this study.

RESULTS

Except for an initial period of alarm, no problems were encountered in getting the ant colonies to mark the new papers as part of their territory. The ants soon resumed normal activity, and many of them explored the new surface on the table, dipping the tip of their gasters on the paper and depositing yellow secretions to mark the new territory. The spots were visible on the papers, and they were counted to give an indication of how well the papers were marked as follows:

Ant species	Mean ± S.D. of yellowish spots per cm on the papers after	
	2 hrs	24–48 hrs
A. laevigata (n=10)	0.4 ± 0.5	16.5 ± 4.8
A. cephalotes (n=10)	0.1 ± 0.3	8.9 ± 3.6

The results of the aggression tests showed that when the resident ants were on 2-hr-marked papers, the total number of fights was not significantly greater than the control (Table 1) with one exception, A. cephalotes with five residents to one alien. In all situations on the 24 to 48-hr-marked papers, the total number of fights increased significantly.

When five A. laevigata workers encountered one A. cephalotes worker, a significant increase in aggression occurred only on paper marked for 24 to 48 hrs. In the case of A. cephalotes as residents, the reverse was true and increased aggression was observed on both unmarked and 2-hr-marked papers but not 24 to 48-hr-marked papers.

The aggressive repertoire of both species analyzed in this study was identical. The initiation of aggression sometimes occurred immediately after the ants examined the paper and the other ant(s) present in the arena. Aggression was occasionally preceded by a threat behavior; i.e., mandible gaping or by a more complex display in which the ant stopped, lifted its body, raised the tip of its abdomen, opened its mandibles, shook its body and then initiated the fight by seizing its opponent.

All castes showed aggressive responses toward an opponent of about the same size. Larger workers were more successful in fighting smaller workers. Thus, the defense of the colony is not the responsibility of any one size of workers.

According to data in Table 2, A. laevigata is more aggressive since in the experiments with blank papers at a ratio of one to one, they started 69% of the fights compared to 12% by A. cephalotes; 19% showed simultaneous aggression (ratio 11:2, X^2: $p < 0.02$). Aggression by A. cephalotes increased significantly on the 2-hr-marked papers in tests where they were the residents at a ratio of 1 to 1. When groups of 5 residents encountered 1 alien, the results indicated that residency was the most important factor.

In Table 3, there are indications that the aggressive action of A. laevigata against A. cephalotes was always successful, as they dominated most of the fights regardless of marking time with one exception, A. cephalotes as residents in groups of five. In this case the results indicated a dominance of A. cephalotes also unrelated to the time the papers were marked (X^2: $p > 0.05$). For these species, it seems that aggressiveness is directly related to the number of nestmates.

The number of alien ants killed did not increase significantly as the marking time increased, and the number of fatal fights was independent of the number of nestmates for both species (Table 4).

The data in Table 5 show that the number of aggressive encounters (Index of aggression or Ia) for A. laevigata residents was greater than the control for both the 2- and 24 to 48-hr-marked papers. For A. cephalotes, an increase in the Ia value was evident when there were five residents vs one alien on the 24 to 48-hr-marked papers.

TABLE 1. Influence of territorial marking and worker ratio on number of fights between workers of A. laevigata and A. cephalotes. (Avg. of 50 replicates)

Alien species	Resident species	Marking time (h)[a]	No. of fights between 1 or 5 ants of resident species and 1 alien ant		
			1 ant[b]	5 ants[b]	(X^2)[c]
A. cephalotes	A. laevigata	0	16a	25a	N.S.
		2	25a	35a	N.S.
		24–48	38b	48b	6.7*
A. laevigata	A. cephalotes	0	16a	32a	9.0*
		2	26a	43b	11.9**
		24–48	39b	45b	N.S.

[a]Length of time papers were exposed to territory-marking species. Terms resident and alien do not apply at "0" time since worker ants were exposed simultaneously to unmarked papers.
[b]Numbers followed by same letter in each column not significantly different (P=0.05); X^2 (2X2) contingency test (with Yates' correction), 1 d.f.
[c]X^2 (2X2) contingency test (with Yates' correction), 1 d.f. Significance: *P < 0.01, **P < 0.001; N.S. P=0.05.

TABLE 2. Influence of territorial marking on number of fights started by A. laevigata or A. cephalotes when they were the resident (R) or alien (A) ant(s).

Alien species	Resident species	Territorial marking time	Number of fights[a] started by either or both species at ratios of					
			1-R to 1-A			5-R to 1-A		
			R	A	Both	R	A	Both
A. cephalotes	A. laevigata	0	11	2	3	19	2	4
		2	14	4	7	26	5	4
		24-48	29	2	7	39	6	3
			X^2 (3X3):N.S.			X^2 (3X3):N.S.		
A. laevigata	A. cephalotes	0	2	11	3	27	4	1
		2	20	4	2	31	8	4
		24-48	29	5	5	36	4	5
			X^2 (3X3):P < 0.001			X^2 (3X3):N.S.		

[a]For comparison of proportions of ants that started fights: a X^2 (3X3) test with 4 d.f. used. N.S. (not significant) at P=0.05.

TABLE 3. Influence of territorial marking on number of fights dominated by A. laevigata or A. cephalotes when they were the resident (R) or alien (A) ant(s).

Alien species	Resident species	Territorial marking time	Number of fights dominated[a] by either or both species at ratios of					
			1-R to 1-A			5-R to 1-A		
			R	A	Both	R	A	Both
A. cephalotes	A. laevigata	0	8	1	7	18	3	4
		2	14	4	7	20	7	8
		24-48	23	7	8	28	9	11
A. laevigata	A. cephalotes	0	1	8	7	19	6	7
		2	10	9	7	21	14	8
		24-48	13	15	11	22	10	13

[a]No significant difference in proportion of ants dominating fights in any of four test series; X^2 (3X3) test, 4 d.f., P=0.05.

TABLE 4. Influence of territorial marking on number of fatal fights between A. laevigata and A. cephalotes when they were the resident (R) or alien (A) ant(s).

Alien species	Resident species	Territorial marking time	Number of fatal fights[a] to either or both species at ratios of					
			1-R to 1-A			5-R to 1-A		
			R	A	Both	R	A	Both
A. cephalotes	A. laevigata	0	1	4	0	2	7	0
		2	4	9	0	5	9	1
		24–48	6	12	2	3	13	1
A. laevigata	A. cephalotes	0	4	1	0	4	7	1
		2	7	7	1	5	8	0
		24–48	8	8	0	4	11	2

[a]Comparison of proportion of fights fatal to resident and alien ants calculated by Fisher's exact probability test. No significant differences noted at $P=0.05$.

TABLE 5. Influence of territorial marking on mean number of aggressive encounters during latent period (start of encounter to fight) with A. laevigata and A. cephalotes.

Alien species	Resident species	No. of residents to aliens	Territorial marking time (h)	Latent period (sec)	Avg. no. of aggressive encounters	Index of Aggression (Ia)[a]
A. cephalotes	A. laevigata	1:1	0	141	3.0	1.8a
			2	75	2.2	4.0b
			24–48	107	4.3	3.5b
						(H=16.1)***
		5:1	0	72	1.7	2.6a
			2	69	3.0	4.0b
			24–48	70	3.8	4.3b
						(H=15.7)***
A. laevigata	A. cephalotes	1:1	0	77	3.0	3.1ab
			2	128	2.3	2.2a
			24–48	65	2.9	4.2b
						(H=13.2)**
		5:1	0	48	1.8	3.7a
			2	78	3.8	4.2a
			24–48	46	3.0	5.7b
						(H=19.5)***

[a]The Index of Aggression (Ia) represents the average frequency of aggressive encounters per minute made by the resident ant(s) up to the start of the fight, considering the start of the fight as an aggressive encounter. Data analyzed with Kruskal-Wallis one-way analysis of variance with individual Ia data (not means): **P < 0.01; ***P < 0.001.

In Table 6 the data reveal that resident ants were more likely to remain in the arena when they were on their own marked papers, especially when the papers had been marked for the longer period. They probably recognized the latter as their territory and became more confident in fighting and thus were less likely to attempt to escape from the arena. It was noticed also that they moved around the arena more slowly than ants on the control papers or when they were the aliens. When A. cephalotes workers were residents, there was a significant difference between the number of ants trying to escape from the control papers compared to the 2-hr-marked papers. This showed that 2 hrs was sufficient to initiate territorial behavior. The duration of fights varied considerably without any specific pattern and did not appear worthy of analysis.

A solitary media-sized resident ant attacked a maxima-sized alien only when the resident was on its own territory and usually when it was on paper marked over a 24-hr period. Even a group of five media-sized A. laevigata did not attack a maxima A. cephalotes on blank paper, but they were aggressive on the 24 to 48-hr-marked paper. Nevertheless, five media A. cephalotes attacked a maxima A. laevigata alien even on blank paper, which suggests that the association of nestmates may play an important role in the territorial behavior of this species.

TABLE 6. Influence of territorial marking on escape attempts by a resident ant in the presence of alien ant. (Avg. of fifty 30-minute observation periods.)

Alien species	Resident species	Territorial marking time (h)	Escape attempts[a]
A. cephalotes	A. laevigata	0	30a
		2	22ab
		24-48	14b
A. laevigata	A. cephalotes	0	38a
		2	26b
		24-48	16b

[a]Proportion of ants trying, or not trying, to escape analyzed by X^2 (2X2) contingency test (with Yates' correction). Numbers followed by the same letter are not significantly different at P=0.05, 1 d.f.

DISCUSSION AND CONCLUSIONS

All attacks observed between the two species of Atta studied centered around the use of their mandibles, sometimes with the help of their forelegs. Involvement of defensive secretions that could paralyze or even cause damage to the opponent was not detected. The first body parts attacked were the legs, which were obviously the easiest part on the body to grasp. Subsequently, the fighters attacked the petiole which could cause the death of the opponent, although that was not a very common outcome of the fight. Only about one-third of the total number of combatants were killed in fights. According to Jutsum (1979) ants that win fights seem to learn to attack specific parts of the body, especially the petiole.

In the present experiments, single ants of two different Atta species placed together on unmarked paper did not exhibit intense aggressive behavior toward the other. They fought in only a small percentage of the encounters, and then the fights were dominated by the most aggressive species (A. laevigata, according to Jutsum 1979). Weber (1972) said that large nests of leaf-cutting ants tend to be more aggressive than small ones, but this is not applicable in this case as the A. laevigata colony was only half the size of the A. cephalotes colony. On the other hand, in all size categories, A. laevigata workers were slightly larger than A. cephalotes workers.

Ants on marked papers appeared to respond more aggressively to alien intruders with an increased level of response on territories marked for over 24 hrs. For A. cephalotes, there was a change in the aggressive behavior compared with the controls even on papers marked for two hours (Table 2). It is interesting to note that being on marked territory causes an increased willingness of A. cephalotes residents to fight although it does not determine the outcome of the fight with either species.

The threat response of the ants was more frequent when they were residents. They also spent more time in the arena when on their own marked paper.

The number of nestmates associating together also seems to play an important role in the territorial and aggressive behavior of leaf-cutting ants. The results indicate that the resident ants behaved as dominants at the 5 to 1 ratio, and this was unrelated to the marking time for the papers. According to Sale (1972), aggression increases in certain animals as a function of group size independent of density. We may conclude that territorial marking influences the initiation of a fight, whereas the number of ants in each group may determine the direction of dominance.

When a small ant was alone on a neutral or alien territory, it did not fight a larger ant, but it did fight if it was on its own territory or in a group of five. The conclusion of Jutsum (1979) that when opponents get larger, the dominant species takes longer to attack unless opponents are much larger in number, in which case the usually dominant species is defeated, seems valid only when the ants meet each other on an unmarked area.

The territory in these Atta species is defended physically as in Pogonomyrmex (Holldobler 1976). Wilson (1980) and Fowler (1977) believed this could be the case during interspecific encounters in leaf-cutting ants. The interspecific aggression may be a consequence of specific territoriality and thus would serve for the defense of the nest.

Based on our experiments, we might expect that in the field leaf-cutting ant aggressiveness should decrease as the distance from the nest increases. The existence of territorial behavior and the number of nestmates present in a given area are factors to be taken into account when the ants are away from the nest. The territory of leaf-cutting ants under natural conditions may be thought of as an area in the vicinity of the nest only where normally a large number of nestmates are found together. Thus, we might expect experiments conducted outside the nest area in the field to show that the ants are less aggressive than ants from laboratory colonies. This could explain some confusion in the literature on aggressive behavior. Possibly, aggression of leaf-cutting ants can be understood only when correlated with territorial factors.

The leaf-cutting ant territory, according to Jaffe et al. (1979), is marked with a pheromone which has at least two components: colony-specific and species- or genus-specific chemicals. The authors suggest that the valves gland secretion is used by A. cephalotes to mark their territories. The active component of the pheromone is a volatile substance, and a period of one hour seems to be sufficient to allow the substance to volatilize near or below the threshold concentration. The number of territorial marks per unit time must, therefore, be important. Marking should be done economically; e.g., by placing valves gland secretions in the immediate vicinity of the nest and possibly, as Jaffe (1979) postulated, in areas where the ants are cutting leaves. The trails are already marked with the trail pheromone of the poison gland. The trails are also defended but are best regarded as chemically separate entities.

Agonistic behavior is usually preceded by antennal inspection of the ants' surroundings. The ant moves from place to place and immediately tries to identify surrounding objects such as another ant or the ground on which it is walking. The odors involved in mutual recognition are surface factors which generate very thin active layers in the air and must be perceived by very close or direct contact chemoreception (Jutsum 1979). This means that the so-called "colony odor" may be recognized at the same time as the territorial odor, giving the ant information about its location.

Scent-marking around the nest and the foraging areas could

orient the resident ants within their territory, serving to maintain the ants' familiarity with its environment. Further behavioral studies on field colonies of leaf-cutting ants are required to test this statement. Recent reviews (for instance, see Gosling 1982) have emphasized the complexity and variety of signals that might be transmitted by scent marks.

REFERENCES CITED

Fowler, H. G. 1977. Some factors influencing colony spacing and survival in the grass–cutting ant Acromyrmex fracticornis (Forel) (Formicidae: Attini) in Paraguay. Rev. Biol. Trop. 25: 89-99.
Gosling, L. M. 1982. A reassessment of the function of scent marking in territories. Z. Tierpsychol. 60: 89-118.
Holldobler, B. 1976. Recruitment behavior, home range orientation and territoriality in harvester ants, Pogonomyrmex. Behav. Ecol. Sociobiol. 1: 3-44.
Jaffe, K. 1979. Chemical communication among workers of leaf-cutting ants (Hymenoptera: Formicidae). Ph.D. Thesis, University of Southampton, Southampton, England.
Jaffe, K., M. Bazire-Benazet, and P. E. Howse. 1979. An integumentary pheromone–secreting gland in Atta sp.: Territorial marking with a colony specific pheromone in Atta cephalotes. J. Insect Physiol. 25: 833-839.
Jutsum, A. R. 1979. Interspecific aggression in leaf-cutting ants. Anim. Behav. 27: 833-838.
Sale, P. F. 1972. Effect of cover on agonistic behavior of a reef fish: A possible spacing mechanism. Ecology 53: 753-758.
Weber, N. A. 1972. Gardening ants, the attines. Mem. Am. Phil. Soc. Vol. 92, 146 pp.
Wilson, E. O. 1971. The insect societies. Belknap Press of Harvard Univ. Press. Cambridge, Mass. 548 pp.
Wilson, E. O. 1980. Caste and division of labor in leaf-cutting ants (Hymenoptera: Formicidae: Atta). II. The ergonomic optimization of leaf-cutting. Behav. Ecol. Sociobiol. 7: 157-165.

14

Ecological Studies
of the Leaf-Cutting Ant,
Acromyrmex octospinosus, in Guadeloupe

P. Therrien, J. N. McNeil,
W. G. Wellington, and G. Febvay

The leaf-cutting ant, Acromyrmex octospinosus Reich (Hymenoptera: Formicidae), was first found in Guadeloupe in 1954 on Grande-Terre island near Morne-a-l'Eau and is now found throughout most of Grande-Terre as well as in the northeast section of Basse-Terre (Fig. la). This insect is considered an agricultural pest, particularly in home gardens, and has been the object of an active chemical control programme since 1956 (Malato et al. 1977). While the foraging of Ac. octospinosus has been studied in other countries (e.g., Weber 1972), an ecological study in Guadeloupe was of interest given that this introduced ant species occurs in habitats on the island that have very different climates, particularly in the amount and frequency of rainfall. The project was initiated in June 1983 and continued until September 1984 to compare the daily and seasonal foraging patterns, the length and duration of foraging trails, and the choice of plants collected by foragers in the two regions. Here we present an overview of our results, as all the analyses are not yet finished.

Active nests were located at Lemesle (A on Fig. la), near the site where the ant was first observed following its introduction, and at Bois de Lomard (B on Fig. la). We had hoped to observe nests on Basse-Terre, in a more westernly region of the ant's distribution where the annual rainfall is considerably higher than Lemesle (Fig. lb), but the active destruction of nests in this area by the Crop Protection Service made it impossible to have a site that would remain undisturbed during the entire study period.

The activity of five nests in both regions was evaluated during four 12-hour periods (two from 0600 to 1800 and two from 1800 to 0600) each month. Each hour the total number of ants entering and leaving the nest during a 3-minute period was counted. It is readily apparent that the seasonal activity patterns, as estimated by the mean number of ants/min. for all sampling periods during each month, vary considerably at both sites (Fig. 2a) with a general

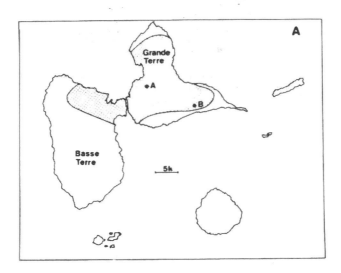

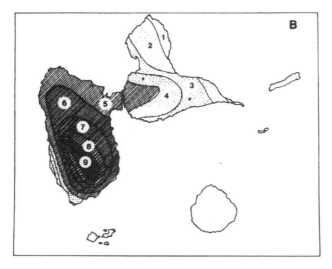

FIGURE 1. (a) The 1982 distribution of <u>Ac. octospinosus</u> in Guadeloupe. Our study sites were at Lemesle (A) and Bois de Lomard (B) on Grande-Terre. (b) The annual rainfall in Guadeloupe for zones 1 to 9 are respectively: <1.0, 1.0-1.25, 1.25-1.5, 1.5-1.75, 1.75-2.0, 2.0-4.0, 4.0-6.0, 6.0-8.0, and >8 m per year. (Maps obtained at I.N.R.A., Guadeloupe.)

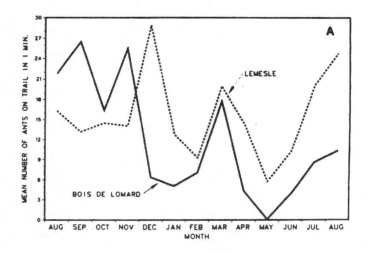

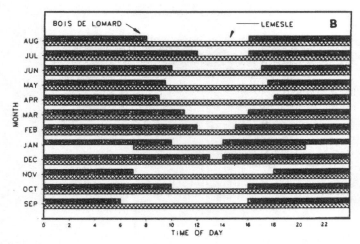

FIGURE 2. The activity of Ac. octospinosus over a 12-month period in 1983-1984 at Lemesle and Bois de Lomard: (a) Seasonal activity based on the mean number of ants/min observed (n=4). (b) Seasonal changes in the daily activity based on the presence or absence of ants during the four monthly sampling periods.

tendency for lower activity during the dry season from December through May. The presence of a temporary rise in activity during March may be the result of favorable abiotic conditions (e.g., a temporary increase in humidity due to rainfall), but may also reflect an increase in foraging associated with the production of the sexuals which left the nests in early May. Daily activity patterns, determined by the presence or absence of foragers on the trails over a 12-month period, are markedly different in the two regions (Fig. 2b). Periodicity is more evident at Bois de Lomard than at Lemesle, and the absence of activity throughout most of the day at Bois de Lomard, especially during the months of low rainfall, suggests that abiotic factors may be of considerable importance, indirectly through induced changes in host plant quality or directly on the foraging ants themselves. The absence of activity during the night at Lemesle in January resulted from very intense rainfall during the sampling periods, a condition which is known to inhibit foraging by attines (Hodgson 1955; Weber 1972; Lewis et al. 1974b).

Although all activity and weather data have not been analysed, in this symposium we present daily activity patterns, trail temperatures, and vapor density deficits for one nest at each site for September 1983 and March 1984 (Fig. 3). The resulting patterns clearly show that foraging is not only markedly different between the two habitats but also that it changes throughout the year at the same site. In both months activity at Bois de Lomard was practically nonexistent from sunrise to sunset, the period when both the temperature and the vapor density deficit were highest, even though temperature and relative humidity conditions were less extreme during September than March (Fig. 3). At Lemesle the difference in foraging activity between the two months was more pronounced. During the rainy season, activity was greatest between sunrise and sunset, although a temporary lull in activity occurred in the middle of the day when the temperature and vapor density deficit were the highest (Fig. 3). In March, activity increased during the night (scotophase); and while activity during the day (photophase) did not stop entirely, the drop associated with maximum temperature and vapor density deficit conditions was very pronounced. These results support the hypothesis that meteorological conditions modify the foraging activity of Ac. octospinosus, as shown for this (Weber 1972) and other attine species (Lewis et al. 1974a; Gamboa 1976; Fowler and Robinson 1979a, b; Mintzer 1979). However, other factors not included in this study, such as host plant chemistry, natural enemies and competitors, are undoubtedly involved. Lewis et al. (1974b) have suggested that the switching of foraging activity may result from a complex relationship between the colony's nutritive needs and the availability of foragers. This could explain the situation where in two adjacent Ac. octospinosus nests, one was mainly diurnal while the other was nocturnal (Cherrett 1968).

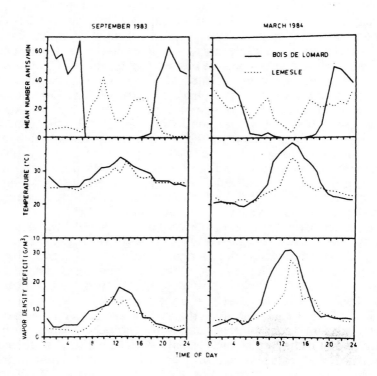

SEPTEMBER 1983   MARCH 1984

BOIS DE LOMARD

LEMESLE

FIGURE 3. The daily activity pattern for two <u>Ac.</u> <u>octospinosus</u> nests, one at Lemesle and the other at Bois de Lomard, together with daily changes in temperature and vapor density deficit in September 1983 and March 1984.

The food sources used in the two study areas were evaluated every 2 hours during each 12-hour study period by collecting the load of every third returning forager, until a total of 30 samples was obtained. In order to minimize the impact of sampling on foraging activity, no leaf samples were taken when trail activity was less than 15 ants/3 min. Thus, for each 12-hour period, a maximum of 180 leaf fragments could be collected. Samples were preserved and subsequently identified using herbarium collections and a guide to the flora of the French Antilles (Fournet 1978). The total number of plant species brought back to the nests was very similar in the two habitats, with approximately 75% being common to both (Table 1). However, as found with this and other attines (e.g., Cherrett 1968; Cherrett and Seaforth 1970; Littledyke and Cherrett 1975), very few plant species were heavily exploited (Table 1). Furthermore, the relative importance of certain plant species collected may vary between the two sites (Table 2). These differences may be the result of actual preferences of individual nests and/or the relative availability of the different plants in each habitat. A map of the distribution and abundance of vegetation was prepared for both sites and these data support the preliminary observations that the apparent differences in plant utilisation result from differential availability rather than different nest preferences.

TABLE 1. Comparison of plant species brought back to the nest by Ac. octospinosus workers in two different areas of Guadeloupe.

Parameters	Lemesle (5 nests)	Bois de Lomard (3 nests)
Total number of species	50	53
Species that represent < 5% of total samples collected	46	50
Species that represent >10% of total samples collected	4	2
Number of species common to both areas	38	

Each month the numbers and lengths of foraging trails associated with all the nests were recorded in each habitat to assess any changes over time. The varying lengths and the changes in the number of trails during a 3-month period for four nests at each site underline the dynamics of foraging activity of Ac. octospinosus (Table 3). The most marked changes were observed at nest 2 at Lemesle and nest 3 at Bois de Lomard, where the number of active trails varied due to foragers responding to the temporary availability of certain fruits.

Given the dynamic nature of foraging trails, we were interested in determining the fade-out time of the trail pheromone that is laid down by Ac. octospinosus foragers (Cross et al. 1982) using the technique described by Jander and Jander (1979). A series of sticks was placed end to end on an active trail in such a way that all foragers using that trail traveled along them when leaving and entering the nest (Fig. 4a). After a 24-hour period, the sticks were removed and care was taken not to touch the marked surfaces. The sticks were then arranged in such a manner that at one point foragers returning to the nest had a binary choice between one marked and one unmarked stick (Fig. 4b). The choice made by 25 foragers was recorded before the position of the two sticks was reversed to avoid possible bias and the choice of the next 25 individuals was noted. The choice of the first and second group of 25 ants was compared using a G test of independence. Since the position of the marked stick had no significant effect, the results were pooled. This procedure was repeated every 12 hours until foragers did not differentiate between the marked and unmarked sticks. Between tests, which took ≤ 2 min., both marked and control sticks were suspended in a clearing within the habitat, in such a way that no ants could come in contact with them. The experiment was repeated twice and three marked sticks tested in each experiment. It is very clear from the results (Fig. 5a) that an Ac. octospinosus trail does not remain active for much more than 24 hours without repeated deposition of the pheromone. In a habitat where many of the plant species exploited are only available for specific periods of the year, the rapid degradation of the pheromone would be important in the maintenance and survival of colonies, as persistent trails leading to depleted resources would waste both foraging time and energy.

TABLE 2. Relative importance of certain plant species, expressed as a percentage of all leaf fragments collected by Ac. octospinosus workers in two different areas of Guadeloupe.

Species	Lemesle	Bois de Lomard
Acacia spp.	0.7	6.5
Artocarpus altilis	12.1	0.05
Bidens pilosa	2.9	1.3
Borreria laevis	1.3	0.9
Cassia obtusifolia	1.6	3.0
Euphorbia hirta	4.9	15.7
Euphorbia hypericifolia	1.4	0.7
Mangifera indica	10.3	6.0
Phyllanthus debilis	0.4	2.6
Priva lappulacea	0.1	0.8
Senecoides cinera	11.7	12.2
Sida rhombifolia	1.9	9.9
Teramnus labialis[a] Cracca caribaea[a]	25.8	3.4

[a]These two species can only be separated when flowers are present and, as most samples collected were leaf fragments, they have been grouped together.

TABLE 3. Turnover in trails from Ac. octospinosus nests in two different areas of Guadeloupe between June 15 and September 17, 1983.

Location	Nest	Trails Total	Formed	Disappeared	Length (m)
Lemesle	N2[a]	7	5	3	1.5,2.0,2.0, 3.3,5.0,5.0, 11.0
	N3	3	1	1	3.6,8.7,8.7
	N4	4	1	0	2.0,15.0, 15.0,21.0
	N5	2	0	0	3.0,5.6
Bois de Lomard	N1	3	1	2	6.0,10.0, 25.0
	N2	1	0	0	5.0
	N3[a]	6	4	3	3.5,4.0,4.0, 10.5,23.0, 24.0
	N5	3	1	1	2.0,3.0,4.0

[a]Trails formed to temporary fruit sources.

It is of interest to note that even at the beginning of the tests not all foragers chose the marked stick, perhaps because of the somewhat artificial substrate they encountered. However, it is possible that certain individuals do not necessarily follow previously established trails but instead forage independently as scouts, thereby locating previously undiscovered resources in the habitat.

Our trail pheromone experiments were always started at sundown and it is apparent that the rate of dissipation during the first 12 hours, which coincides with the scotophase, is much less pronounced than in the subsequent 12-hour period. Higher temperature and vapor density deficit conditions (see Fig. 3), together with possible light-catalysed decomposition, could explain the increase in pheromone degradation rate during the daylight hours. To test the possibility that U.V. may affect pheromone decomposition, we used the same binary choice bioassay, but in this case the marked stick was exposed to an artificial U.V. light source at a distance of 1 m for 30 min. every 12 hours. The U.V. source was a Mazda TG 15 germicide lamp with >90% of the radiant flux emitted at 253.7 nm and an irradiance of 37 microwatts/cm^2 at 1 m from the source. The experiment was repeated twice, with two replicates in each, and the results compared with those obtained from a control where the marked stick was not exposed to the U.V. source. Throughout each experiment, all sticks were held in the laboratory, except for the period required to execute the bioassays, to minimize exposure to the natural U.V. It is evident that exposure to the artificial U.V. source significantly ($F_{1,3}=250.74$, $P < 0.001$) accelerated pheromone degradation when compared with the control (Fig. 5b). These results do not prove that U.V. plays a role in the degradation of the pheromone trail under field conditions, as the artificial source emits a shorter wavelength of U.V. than observed in nature. However, the fact that no step-like degradation, as seen in the field (Fig. 5a), was observed in the control of the U.V. experiment (Fig. 5b) strongly suggests that the increased decomposition noted during the photophase is related to the action of one or several prevailing abiotic conditions, including U.V. light.

FIGURE 4. The arrangement of sticks (a) along an active Ac. octospinosus trail that was used to obtain a substrate marked with the trail pheromone, and (b) where, at the junction, foraging ants had a choice between a marked and an unmarked stick.

Our preliminary results demonstrate that the foraging activity of Ac. octospinosus varies not only on a seasonal basis within a given habitat, but also between sites where the abiotic conditions differ. When all analyses have been completed, further comparisons between nests, and individual trails of the same nest, will help clarify the relative importance of food availability and prevailing abiotic factors on the foraging of Ac. octospinosus.

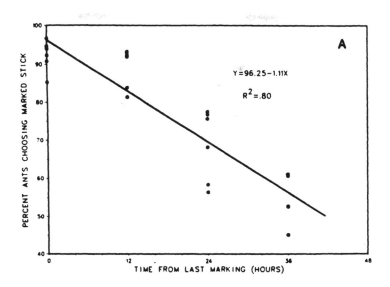

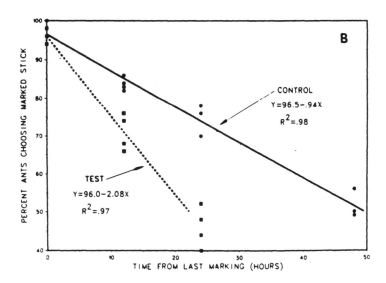

FIGURE 5. (a) The fade-out time of the pheromone trail of
<u>Ac. octospinosus</u> estimated by a binary choice test. (b) A
comparison of the fade-out times of two pheromone trails
when one (test) was exposed to an artificial U.V. source for
30 min. every 12 hours (see text for details).

182

ACKNOWLEDGMENTS

We would like to thank Dr. Alain Kermarrec for the support provided by La Station de Zoologie et Lutte Biologique at the INRA station at Petit Bourg, Guadeloupe, throughout the study, as well as A. Jean Bart, V. Erho, and L. Mills for their valuable technical assistance. This work was supported in part by graduate scholarships to P. Therrien from NSERC and FCAC, and by NSERC research grants to J. N. McNeil and W. G. Wellington.

REFERENCES CITED

Cherrett, J. M. 1968. The foraging behaviour of Atta cephalotes L. (Hymenoptera: Formicidae) 1. Foraging pattern and plant species attacked in tropical rain forest. J. Anim. Ecol. 37: 387-403.

Cherrett, J. M., and C. E. Seaforth. 1970. Phytochemical arrestants for the leaf-cutting ants, Atta cephalotes (L). and Acromyrmex octospinosus (Reich), with some notes on the ants' response. Bull. Entomol. Res. 59: 615-625.

Cross, J. H., J. R. West, R. M. Silverstein, A. R. Jutsum, and J. M. Cherrett. 1982. Trail pheromone for the leaf-cutting ant, Acromyrmex octospinosus (Reich), (Formicidae: Myrmicinae). J. Chem. Ecol. 8: 1119-1124.

Fournet, J. 1978. Flore illustree des phanerogames de Guadeloupe et de Martinique. I.N.R.A., Paris. 1654 pp.

Fowler, H. G., and S. W. Robinson. 1979a. Foraging ecology of the grass-cutting ant, Acromyrmex landolti fracticornis (Formicidae: Attini) in Paraguay. Int. J. Ecol. Environ. Sci. 5: 29-37.

Fowler, H. G., and S. W. Robinson. 1979b. Foraging by Atta sexdens (Formicidae: Attini): Seasonal patterns, caste and efficiency. Ecol. Entomol. 4: 239-247.

Gamboa, G. J. 1976. Effects of temperature on the surface activity of the desert leaf-cutter ant, Acromyrmex versicolor versicolor (Pergande) (Hymenoptera: Formicidae). Amer. Mid. Nat. 95: 485-491.

Hodgson, E. S. 1955. An ecological study of the behavior of the leaf-cutting ant Atta cephalotes. Ecology 36: 293-304.

Jander, R., and U. Jander. 1979. An exact field test for the fade-out time of the odor trails of the Asian weaver ant Oecophylla smaragdina. Insect. Soc. 26: 165-169.

Lewis, T., G. V. Pollard, and G. C. Dibley. 1974a. Rhythmic foraging in the leaf-cutting ant Atta cephalotes (L.) (Formicidae: Attini). J. Anim. Ecol. 43: 129-141.

Lewis, T., G. V. Pollard, and G. C. Dibley. 1974b. Micro-environmental factors affecting diel patterns of foraging in the leaf-cutting ant Atta cephalotes (L.) (Formicidae: Attini). J. Anim. Ecol. 43: 143-153.

Littledyke, M., and J. M. Cherrett. 1975. Variability in the selection of substrate by the leaf-cutting ants Atta cephalotes (L.) and Acromyrmex octospinosus (Reich) (Formicidae: Attini). Bull. Entomol. Res. 65: 33-47.

Malato, G., A. Kermarrec, and J. M. Troup. 1977. Donnees nouvelles sur l'invasion de la Guadeloupe par Acromyrmex octospinosus Reich (Formicidae: Attini). 1. Aspects biogeographiques et perspectives de controle chimique. Nouv. agron. Antilles-Guyane 3: 473-484.

Mintzer, A. 1979. Foraging activity of the Mexican leaf-cutting ant Atta mexicana (F. Smith) in a sonoran desert habitat (Hymenoptera: Formicidae). Insect. Soc. 26: 364-372.

Weber, N. A. 1972. Gardening ants, the attines. Mem. Am. Phil. Soc. Vol. 92, 146 pp.

15
Perspectives on Some Queen Pheromones of Social Insects with Special Reference to the Fire Ant, *Solenopsis invicta*

D.J.C. Fletcher

Progress in the analysis of the regulatory functions and chemical identity of queen pheromones in the social Hymenoptera has been extraordinarily slow during the 30 years that have passed since Butler (1954) demonstrated the existence of a "queen substance" in the honey bee, Apis mellifera. This queen pheromone system has become better known than any other, yet our understanding of it remains far from complete. Early on, it was shown (Butler and Simpson 1958) that the mandibular glands of a queen are the source of an inhibitory pheromone that prevents worker bees from building queen cells, and the principal inhibitory component of the secretion was identified as (E)-9-oxo-2-decenoic acid (9OD) (Barbier and Lederer 1960; Butler et al. 1961). Later, a second constituent, (E)-9-hydroxy-2-decenoic acid, was also found to have an inhibitory effect (Butler and Callow 1968). However, queens from which the mandibular glands were removed retained an appreciable inhibitory capability (Gary and Morse 1962); and additional inhibitory substances (as yet unidentified) were shown to originate in glands located in the abdomen (Velthuis 1970) and in tarsal glands (Lensky and Slabezki 1981). With similar diversity, 9OD was found to have more than one function. It has activity in inhibiting ovarian development in worker bees (e.g., Butler and Fairey 1963), it is the sex pheromone of the queen honey bee (Gary 1962; Boch et al. 1975), and it attracts worker bees during swarming (Morse 1963; Avitable et al. 1975). The chemical composition of the odor that attracts workers to their queen within the nest remains unknown.

This summary of the progress that has been made in understanding the most intensively studied of all queen pheromone systems among social insects suggests that those who attempt to analyze the pheromonal systems of queen ants can expect to have a difficult time. There is remarkable complexity. The products of several exocrine glands may contribute to a single function, yet the secretion of a single gland may serve more than one function. Moreover, a particular glandular secretion is likely to contain a blend of active components. There is also another source of complexity, not evident in the honey bee paradigm, to which I shall draw attention later. These difficulties notwithstanding, our knowledge of the queen pheromones of ants is gradually advancing.

INHIBITION OF WORKER OVIPOSITION

No inhibitory pheromones of ants have yet been identified chemically, but there is ample anecdotal and experimental evidence of their existence (see review by Fletcher and Ross 1985). For example, Holldobler and Wilson (1983) showed that in laboratory nests, queenless fragments of colonies of the African weaver ant, Oecophylla longinoda, will rear males from the eggs of laying workers. They reported, however, that in the field a single queen is capable of inhibiting colonies of up to 50,000 workers from laying reproductive eggs, even though these workers occupy numerous nests in up to 17 trees. In this ant, as in many species, the queen inhibits workers from laying eggs that will hatch, but stimulates them to produce nonviable trophic eggs that she eats and that are fed to larvae.

In a number of ant genera such as Tetramorium, Pheidole, and Solenopsis, workers lack functional ovaries, so that an inhibitory queen influence over them is not necessary.

INHIBITION OF OVIPOSITION BY VIRGIN QUEENS

Several years ago, we discovered that queens of the fire ant, Solenopsis invicta, produce an inhibitory pheromone that prevents virgin queens (female alates) from shedding their wings (dealating) and laying eggs in the parental nest. In queenless laboratory colonies, a number of female alates dealate and one or more of them becomes a replacement queen, laying unfertilized eggs that develop into males (Fletcher and Blum 1981a). This is a parallel situation to the development of laying workers in queenless honey bee or weaver ant colonies. I and my colleagues have continued to investigate this and other pheromonal systems of the fire ant, and I will summarize part of the work here.

First, consider the question of whether the pheromone acts directly on female alates or whether it acts on the workers and influences their behavior towards female alates. In ants of the genus Myrmica, queen influence inhibits female larvae from developing into sexuals, but it apparently does not act on the larvae directly. Instead, it alters the behavior of the workers towards them. Specifically, it causes workers to feed large larvae less, and even to bite them, thereby accelerating their metamorphosis into workers (Brian and Hibble 1963; Brian 1973). We, therefore, considered it possible that in S. invicta the pheromone prevented workers from actively causing female alates to dealate.

We found that if we kept single sexually mature female alates with small groups of workers, about 89% dealated within 72 hours, whereas, if we isolated them without workers only 22% dealated during the same period. This indicated that workers do facilitate dealation in some way (Fletcher et al. 1983). However, when we isolated female alates for longer periods, they dealated, their wing muscles underwent histolysis, they laid numerous eggs, and behaved like colony-founding queens after a nuptial flight (Fletcher and Blum 1983a). Evidently, the inhibitory pheromone acts directly on the female alates. In the presence of workers, the dealates do not behave like colony-founding queens but as virgin replacement queens. Under these circumstances there is a clear advantage to be gained by rapid dealation, since replacement queens acquire reproductive dominance over their nestmates by being first to produce the inhibitory pheromone (which is associated with dealation), thereby preventing other virgin queens from dealating and ovipositing (Fletcher et al. 1983). The stimulus provided by the presence of workers is probably nonspecific, although it seems likely that the food female alates receive from them is an important factor in promoting more rapid dealation and oviposition compared with virgin queens in isolation.

It is also of interest that although isolated female alates were sexually mature, as judged by their individual body weights (see Fletcher and Blum 1983a) and could therefore have participated in a nuptial flight at any time, they did not dealate and oviposit as rapidly as newly mated queens, which normally dealate almost as soon as they have landed and often lay their first eggs less than 24 hours later. Typically, isolated female alates took 1 to 7 days to both dealate and to lay eggs. This suggests that the effects of disinhibition are reinforced either by flight, or insemination, or both, after a female alate has left the parental colony.

We developed an assay for the inhibitory pheromone using the prevention of dealation by virgin queens as an indicator of the presence of the pheromone. Assay units consisted of 2 cm^3 of worker ants, 0.5 cm^3 of brood, and two sexually mature female alates. The pheromone-bearing corpse of a mated queen was added to each experimental unit, and the control units were without a queen corpse. The mean length of time required for the first live female alate to dealate in a number of experimental and control units was then compared. The presence of the inhibitory pheromone in the experimental units delayed dealation to a significant extent (Fletcher and Blum 1981b).

We have used this assay and a modification of it for a variety of purposes. First, we compared the inhibitory capability of the heads, thoraces, and gasters of mated queens to determine where the glandular source is located. The results (Table 1) indicated that the pheromone is produced in the gaster. The procedure has been criticized by Vander Meer (1983) on the grounds that the poison sac in the abdomen is the source of a queen attractant pheromone (see Glancey, Chapter 19), so that the retention of queen gasters in assay units for longer periods than heads and thoraces "may simply be indicative of the duration of tending activity (mediated by the queen attractant-recognition pheromone) and may not demonstrate the probable site of inhibitory pheromone production or storage." It is true that if the glandular source is located in the head or thorax and the attractant is necessary to cause workers to collect it (which may well be the case), workers would discard the heads or thoraces while these body parts still contained inhibitory pheromone. However, the gaster remained inhibitory for a much longer time than it could have if it were not a source of the pheromone. Some gasters have remained inhibitory for more than three months (unpublished data).

TABLE 1. Effectiveness of the corpses of queens, or parts of corpses, in preventing dealation among sexually mature virgin queens.

Part of queen's body	Mean no. days to dealation		SE of difference	DF	P (t-test)
	Queen	Control			
Whole body	9.9	1.4	1.39	9	0.001
Head & thorax	2.6	1.2	0.37	9	0.01
Gaster	8.0	1.2	0.93	9	0.001

We have also shown by means of the assay that the inhibitory capability of mated queens is positively correlated with their fecundity (Fletcher and Blum 1983b). Recently, we compared many additional categories of queens (D. Willer and D. Fletcher, unpublished data) and found that queens of monogynous colonies are more potent than queens of polygynous colonies. This is in agreement with the above result, since the queens of monogynous colonies are more fecund than those of polygynous colonies (Fletcher et al. 1980). We also confirmed that while sexually mature alate queens in queen-right colonies do not produce the pheromone, dealate virgin replacement queens assayed only one week after disinhibition produce it in substantial amounts.

The assay can be used effectively to detect the presence of inhibitory pheromone in solvent extracts of whole queens. For this purpose, we dispersed the extracts on the corpses of freshly killed female alates after removing their wings and added untreated corpses to the control units (K. Ross and D. Fletcher, unpublished data). We are using this procedure to assay purified extracts for chemical identification of the pheromone.

EXECUTION OF QUEENS BY WORKERS

The workers of queenright colonies of social insects almost invariably attack and kill conspecific alien queens introduced to them experimentally. When queenless, however, they will frequently accept an unfamiliar queen. This, of course, is the principle employed by beekeepers when they requeen their colonies. Wilson (1966) introduced multiple queens to queenless fire ant colonies by the stratagem of chilling all the occupants of a nest. Initially, all the queens were accepted, but the workers then executed all but one in each case. Why did the workers kill the supernumerary queens, how did they select which queens to kill, and why did they avoid the mistake of attacking all the queens? These are questions to which we believe we have found plausible answers.

When virgin queens dealate in queenless S. invicta colonies, the workers soon begin to execute most of them. Usually, 1 to 3 survive as virgin replacement queens. We have shown, by means of allozyme markers, that queens are singly inseminated (Ross and Fletcher 1985), so that all the female alates and workers in a colony are full siblings, having a mean coefficient of relatedness of 0.75, the theoretical maximum. Under these circumstances, it seems unlikely that female alates would possess individually distinguishable qualitative pheromonal differences that would cause workers to discriminate among them during a sequence of executions. Therefore, we hypothesized (Fletcher and Blum 1983b) that workers respond in two ways to the quantity of pheromones produced by these queens. First, when the total quantity of queen pheromones circulating in the colony exceeds a certain threshold, it causes workers to reduce the amount present by killing off most of the producers. Second, the workers kill the poorest pheromone producers first, leaving only the best of them alive at the end of the execution sequence. We further hypothesized that a queen's pheromone production is positively related to her fecundity, so that the workers automatically retain the best egg producer(s) and leave her (them) unharmed.

These hypotheses could be more easily tested by using mated queens, since we could adjust their egg production by regulating their food supply. Moreover, by using only queens that were unfamiliar to the workers of the experimental colonies, we could control qualitative pheromonal differences among them. If an excess of queen pheromones in a colony causes workers to execute supernumerary queens, queenless workers should kill one of two unfamiliar queens introduced simultaneously to them. In 11 of 12 trials, they indeed killed one queen, and in the twelfth they killed both queens. In essence this was a repeat of Wilson's (1966) experiment. To test whether the workers would discriminate among queens on the basis of their egg production, we simultaneously introduced to them a pair of queens, one of high and the other of low fecundity. They executed the queen of low fecundity in 22 of 24 trials. In one of the remaining two trials, they killed the highly fecund queen, and in the other they killed both queens. These results support the quantitative pheromonal hypothesis.

We do not know which queen pheromones the workers are responding to when they execute supernumerary queens. It is possible, however, to determine whether queens of high and low fecundity produce different amounts of the inhibitory pheromone by using the bioassay described previously. We found that queens of low fecundity indeed produce significantly less of this pheromone as measured by our bioassay (Fletcher and Blum 1983b).

POTENTIAL OF PHEROMONES IN PEST MANAGEMENT

At this stage we still know too little about the queen pheromones of ants to claim that it will be possible to use them in management programs. However, they play a major role in the regulation of reproduction in at least three ways (Fletcher and Ross 1985), so that in principle they have considerable potential. The three ways to which I refer are: (1) inhibition of oviposition by workers and/or virgin queens, (2) regulation of queen number; i.e., the execution of supernumerary queens by workers, and (3) regulation of the production of sexuals; i.e., caste determination.

I have reviewed very briefly the research we are doing on the first two of the above, but we are also working on the third. Our primary objective is to develop an assay for a pheromone that appears to inhibit the sexualization of female larvae; and currently, our results look very promising. We are also making good progress with the development of an assay for the pheromone(s) involved in queen execution. Other investigators (Rocca et al. 1983a, 1983b) have identified components of an attractant pheromone of fire ant queens, these being (E)-6-(1-pentenyl)-2-H-pyran-2-one and tetra-hydro-3,5-dimethyl-6-(1-methylbutyl)-2-H-pyran-2-one (also, see Glancey, Chapter 19).

Given the complexity of queen pheromonal systems in the social Hymenoptera as revealed by research on honey bees and fire ants, it seems improbable that any particular queen pheromone of a pest species would be of much practical value on its own. To achieve success, it will be necessary, in my estimation, to use a combination of interacting pheromones that represent less of a departure from the natural situation. This goal will undoubtedly entail considerable research effort; but if artificially raising the level of queen pheromones in colonies (via food baits, perhaps) is found to have adverse effects on reproduction, that effort would be worth the cost. It is also worth noting that once the pheromonal mechanisms are better understood, they could provide leads to other avenues of control. For example, since pheromones serve vital regulatory functions, it may be sufficient to merely block the action of one or more of them rather than to use the pheromones themselves.

REFERENCES CITED

Avitabile, A., R. A. Morse, and R. Boch. 1975. Swarming honey bees guided by pheromones. Ann. Entomol. Soc. Am. 68: 1079-1082.

Barbier, M., and E. Lederer. 1960. Structure chimique de la 'substance royale' de la reine d'abeille (Apis mellifica L.). C. R. Acad. Sci. Paris Ser. D. 250: 4467-4469.

Boch, R., D. A. Shearer, and J. C. Young. 1975. Honey bee pheromones: Field tests of natural and artificial queen substance. J. Chem. Ecol. 1: 133-148.

Brian, M. V. 1973. Caste control through worker attack in the ant Myrmica. Insect. Soc. 20: 87-102.

Brian, M. V., and J. V. Hibble. 1963. Larval size and the influence of the queen on growth in Myrmica. Insect. Soc. 10: 71-81.

Butler, C. G. 1954. The method and importance of the recognition by a colony of honeybees of the presence of its queen. Trans. R. Entomol. Soc. London 105: 11-29.

Butler, C. G., and R. K. Callow. 1968. Pheromones of the honeybee (Apis mellifera L.): the "inhibitory scent" of the queen. Proc. R. Entomol. Soc. London Ser. A. 43: 62-65.

Butler, C. G., R. K. Callow, and N. C. Johnston. 1961. The isolation and synthesis of queen substance, 9-oxodec-trans-2-enoic acid, a honeybee pheromone. Proc. R. Soc. London Ser. B. 155: 417-432.

Butler, C. G., and E. M. Fairey. 1963. The role of the queen in preventing oogenesis in worker honeybees. J. Apic. Res. 2: 14-18.

Butler, C. G., and J. Simpson. 1958. The source of the queen substance of the honeybee (Apis mellifera L.). Proc. R. Entomol. Soc. London Ser. A. 33: 120-122.

Fletcher, D. J. C., and M. S. Blum. 1981a. Pheromonal control of dealation and oogenesis in virgin queen fire ants. Science 212: 73-75.

Fletcher, D. J. C., and M. S. Blum. 1981b. A bioassay technique for an inhibitory primer pheromone of the fire ant, Solenopsis invicta Buren. J. Ga. Entomol. Soc. 16: 352-356.

Fletcher, D. J. C., and M. S. Blum. 1983a. The inhibitory pheromone of queen fire ants: effects of disinhibition on dealation and oviposition by virgin queens. J. Comp. Physiol. 153: 467-475.

Fletcher, D. J. C., and M. S. Blum. 1983b. Regulation of queen number by workers in colonies of social insects. Science 219: 312-314.

Fletcher, D. J. C., M. S. Blum, T. V. Whitt, and N. Temple. 1980. Monogyny and polygyny in the fire ant, Solenopsis invicta. Ann. Entomol. Soc. Am. 73: 658-661.

Fletcher, D. J. C., D. Cherix, and M. S. Blum. 1983. Some factors influencing dealation by virgin queen fire ants. Insect. Soc. 30: 443-454.

Fletcher, D. J. C., and K. G. Ross. 1985. Regulation of reproduction in eusocial Hymenoptera. Ann. Rev. Entomol. 30: 319-343.

Gary, N. E. 1962. Chemical mating attractants in the queen honeybee. Science 136: 773-774.

Gary, N. E., and R. A. Morse. 1962. The events following queen cell construction in honeybee colonies. J. Apic. Res. 1: 3-5.

Holldobler, B., and E. O. Wilson. 1983. Queen control in colonies of weaver ants (Hymenoptera: Formicidae). Ann. Entomol. Soc. Am. 76: 235-238.

Lensky, Y., and Y. Slabezki. 1981. The inhibiting effect of the queen bee (Apis mellifera L.) foot-print pheromone on the construction of swarming queen cups. J. Insect Physiol. 27: 313-323.

Morse, R. A. 1963. Swarm orientation in honeybees. Science 141: 357-358.

Rocca, J. R., J. H. Tumlinson, B. M. Glancey, and C. S. Lofgren. 1983a. The queen recognition pheromone of Solenopsis invicta: Preparation of (E)-6-(1-pentenyl)-2-H-pyran-2-one. Tetrahedron Lett. 24: 1889-1892.

Rocca, J. R., J. H. Tumlinson, B. M. Glancey, and C. S. Lofgren. 1983b. Synthesis and stereochemistry of tetrahydro-3,5-dimethyl-6-(1-methylbutyl)-2-H-pyran-2-one, a component of the queen recognition pheromone of Solenopsis invicta. Tetrahedron Lett. 24: 1893-1896.

Ross, K. G., and D. J. C. Fletcher. 1985. Comparative study of genetic and social structure in two forms of the fire ant, Solenopsis invicta (Hymenoptera: Formicidae). Behav. Ecol. Sociobiol. 17: 349-356.

Vander Meer, R. K. 1983. Semiochemicals and the red imported fire ant (Solenopsis invicta Buren) (Hymenoptera: Formicidae). Fla. Entomol. 66: 139-161.

Velthuis, H. H. W. 1970. Queen substance from the abdomen of the honey bee queen. Z. Vgl. Physiol. 70: 210-222.

Wilson, E. O. 1966. Behaviour of social insects. Symp. R. Entomol. Soc. London 3: 81-96.

16
Chemical Communication
in Leaf-Cutting Ants

P. E. Howse

INTRODUCTION

The last fifteen years have seen an enormous increase in the volume of publications on chemical secretions of ants and on the role of such secretions in controlling their behaviour (for example, Bradshaw and Howse 1984). It has been tempting to assume that chemical control of behaviour is achieved by 'trail pheromones,' 'alarm pheromones,' 'brood pheromones,' and so forth. This has suffered from the discovery that pheromones in general are multi-component, and that minor components may have major roles in controlling behaviour. Thus 'alarm' in ants is not a unitary behaviour pattern, but may, as in the example of the African weaver ant, Oecophylla longinoda, consist of alerting, reduction in behavioural thresholds, orientation up an odour gradient, and prolonged biting and pulling of an alien object (Bradshaw et al. 1979a). These separate actions appear to be controlled by different components of the mandibular gland 'alarm pheromone.' In addition to this, spraying of formic acid and mass attack are partly under the control of other pheromones from the gaster (Bradshaw et al. 1979b). Clearly, what we call 'alarm' in this case is by no means unitary and neither is it controlled by a single chemical stimulus.

Particularly in social insects, the effect of a chemical secretion may be influenced by the presence of nestmates, by other chemical signals, signals in other sensory modalities, and by after-effects of preceding actions. In the following, I will describe the results of experiments on laboratory colonies of leaf-cutting ants in which it is clear that the context of a chemical signal is important in deciding the ensuing behaviour and that the simple concept of, for example a trail pheromone, is inadequate to explain the complexities of ant behaviour (see Vander Meer, Chapter 17).

TRAIL FOLLOWING

Atta cephalotes workers that have found a food source recruit nestmates with the aid of a trail laid by dipping the tip of the gaster on the ground. The quantity of pheromone on the trail determines the degree of recruitment. Individual ants appear to be able to regulate the amount of pheromone on the trail and so control recruitment directly (Jaffe and Howse 1979). As the strength increases, the ants are more likely to follow the trail to the end without diverting to other sources of food. Trail-following can also be induced by a poison gland extract, or with the synthetic major component of the trail pheromone, methyl-4-methyl-pyrrole-2-carboxylate (M4MP2C) (Tumlinson et al. 1972). In either case, recruitment to a food source slows down markedly after 5 to 10 minutes unless the ants find food, in which case they reinforce the trail (Jaffe and Howse 1979). This contrasts with the behaviour of ants following a natural trail in which they appear to reinforce it by touching the tip of the gaster to the substrate. A trail of poison gland extract or of M4MP2C lacks an autostimulatory element that would help to maintain its existence.

Studies on A. cephalotes, A. laevigata, and Acromyrmex octospinosus (Vilela et al., in press) have shown that the trail pheromone is only one of several cues to which the ants may respond in orientation back to the nest. If leaf-carrying ants following a trail across a bridge are confronted with the bridge in a reversed position, they are not impeded in daylight. In the dark, however, Ac. octospinosus workers take twice as long to cross the bridge as ants encountering the bridge simply turned through 360°. In this latter control situation, ants also take longer in the dark than in the light, indicating that visual cues are used (Table 1). Further, if the bridge is turned around while the ants are traversing it, they then find their way back to the nest if they are in the light, but in the dark, A. cephalotes appears to make random choices while Ac. octospinosus moves away from the nest. This suggests that in the case of A. cephalotes, visual cues dominate over odour cues, but that Ac. octospinosus relies on the trail pheromone, which can apparently impart directional information in the absence of visual cues. When these experiments were repeated with a bridge of two parallel runways, joined at each end where they were in contact with the nest tables, the ants established trails down one of the runways, thus taking a left- or right-hand deviation. Rotation of the bridge through 180° in the light significantly delayed Ac. octospinosus, and rotation in the dark delayed both species. This appears to be due to the previous association of the trail with a left or right deviation and the confusion when the two are no longer associated. A memory of turns must therefore be involved.

TABLE 1. The mean time(s) required for the first leaf-carrying ant to cross a bridge when its natural trail over the bridge has been rotated either 360° (control) or 180°. (n = 24).

Species	Rotation of trail	One-runway bridge		Two-runway bridge	
		Light	Dark	Light	Dark
A. cephalotes	360°	21±5	29±7	29±9	45±20
	180°	25±11	34±17	28±13	65±29
		N.S.	N.S.	N.S.	P < 0.01
Ac. octospinosus	360°	10±3	18±5	16±4	33±17
	180°	11±2	43±27	28±12	68±30
		N.S.	P < 0.001	P < 0.001	P < 0.001

Mean time (±SD) for condition indicated[a]

[a]Mann-Whitney U-test (n > 20); N.S. = Not significant at P = 0.05.

It was further possible to show that both A. cephalotes and Ac. octospinosus could orientate towards the nest in a tri-radiate tubular glass maze (one end of which was connected to the nest-table) equally well in the light or in the dark. However, when the maze was rotated through 120°, A. cephalotes could find the new nest-end equally well in the light, but took significantly longer in the dark. Ac. octospinosus took over twice as long whether in the light or in the dark. Both species can therefore detect odour asymmetries in the tri-radiate maze, even when they have laid trails in each arm. In addition, Vilela et al. (in press) were able to show that, in the absence of chemical cues, A. cephalotes would respond to gravitational cues in the form of an inclined maze when it had previously been associated with the homeward direction of the nest.

It can therefore be concluded that leaf-cutting ants have a variety of cues available for orientation to and from the nest, including the pheromone trail, visual cues, the spatial layout of the trail, odour differences on the trail, and gravitational cues. The ability of ants to return repeatedly to specific cutting sites may well depend on their ability to integrate these various kinds of information, including patterning of chemical trail signals.

TERRITORY

One way in which chemical trail signals may be modulated is through marking of areas around the nest or around cutting sites. Such marking has been shown to occur in laboratory colonies of A.

cephalotes (Jaffe et al. 1979). Ants that encounter a clean substrate advance little by little, dipping their gaster tips onto the substrate from time to time. This results in faintly visible yellow spots. Such marked "territory" (a paper surface, for example) changes the behaviour of the ants on it, although it must be continually renewed, since the ants apparently cannot detect the secretion after about an hour. In fact, the secretion is autostimulatory, and ants of the same colony renew it continuously, while those of other colonies adopt a stance with the abdomen raised. Similar behaviour occurs in A. sexdens rubropilosa (Vilela 1983). Ants on their own marked territory show a much reduced tendency to leave it (Vilela and Howse, in press, a), and a greater tendency to initiate fights with aliens (Table 2) particularly when in a group.

TABLE 2. Initiation of fights during encounters between A. cephalotes (residents) and A. laevigata (aliens) on paper substrates marked over different periods with the territorial pheromone of the former (Vilela and Howse, in press, a).

	Test Conditions					
	1 resident vs. 1 alien			5 residents vs. 1 alien		
	Fights started by			Fights started by		
	Resident	Alien	Both	Resident	Alien	Both
Control	2	11	3	27	4	1
Marked (2 hrs.)	20	4	2	31	8	4
Marked (24-48 hrs.)	29	5	5	36	4	5
	X^2(3x3): P < 0.001			X^2(3x3): N.S.		

On new substrate, or on one marked with the secretions of other attine species, A. sexdens curls the abdomen ventrally, which, like abdomen raising behaviour in A. cephalotes, has been used in a bioassay for recognition of territorial markers. The source of the territorial pheromone in A. cephalotes is the valves gland (Jaffe et al. 1979), and experiments with naturally-marked substrates suggest that the secretion has both species-specific and colony-specific components. In A. sexdens rubropilosa, the Dufour's gland may be implicated, as (Z)-9-nonadecene, which is a major component of the Dufour's gland secretion, reduces the level of abdomen curling on test substrates compared to that on substrates marked by nestmates (Vilela 1983).

TRANSPORT

The collection of food by leaf-cutting ants is clearly not a process that can be understood simply in terms of a trail of pheromone. Ants recruited to the cutting sites show a sequence of behaviour which includes pivoting the abdomen ventrally so that the gaster brushes the leaf surface (Bradshaw et al., in press). When the leaf is detached it is held vertically in the mandibles and the tip of the gaster is again applied several times to the cut edge. Cuttings in laboratory colonies housed under glass jars on open foraging arenas are commonly taken to a pile outside the nest, and the marking behaviour is carried out again if the leaf particle is moved again. Freshly-marked leaves placed close to a trail are picked up sooner than unmarked leaves, and attract ants over a distance of about 2 cm. Filter-paper discs with Dufour's gland extract and poison gland extract were picked up sooner than rectal sac or fat body extracts, or solvent controls, but the Dufour's gland extract significantly increased the total time for which the discs were transported and also stimulated further marking of the discs. This gland therefore appears to be the source of the marking pheromone which increases the probability of pick-up, increases transport time, and, like the natural trail pheromone and the territorial pheromone, is autostimulatory. The two principal components have been shown to be n-tridecane and (Z)-9-nonadecene, both of which stimulate a high frequency of marking, but the tridecane, in particular, is effective only within a narrow concentration range. The poison gland secretion, or M4MP2C, is known to be attractive to ants (Robinson et al. 1982) and may contaminate the Dufour's gland secretion.

The Dufour's gland pheromone provides chemical signals which increase the probability of cut-leaf fragments reaching the nest. This may be important particularly for chain transport of the leaf fragment under field conditions when the leaf-cutting workers let their fragments fall to the ground where nestmates retrieve and carry them to the nest.

There is no evidence that the marking secretion increases the probability of acceptance of the leaf fragments within the nest. Material that is taken to the nest may be incorporated into the fungus garden, or rejected and consigned to rubbish piles. The trails that lead to such sites where rubbish is dumped are distinct from the foraging trails and not frequented by leaf-carrying ants. In A. cephalotes, the decision about whether or not the collected material is to be treated as rubbish is made after it has been transported to the vicinity of the nest, where all trails start (Jaffe 1983).

Orientation along refuse trails may depend upon visual cues in A. cephalotes (Jaffe 1983) but in A. sexdens rubropilosa, rotating the substrate on which rubbish trails have been established confuses the ants, suggesting that chemical cues are also of importance (unpublished observations). Refuse particles presented on the food trails are normally picked up by ants going away from the nest, and are taken back to the nest entrance before being transported to the rubbish dumps. There are, however, certain exceptions (Table 3). Fresh litter washed in methylene chloride or water, as well as fresh fungus-garden material dried for 2 hours or more, or left to dry in air for 72 hours or more, is picked up from food trails and taken back to the nest. It is then left there except for litter extracted in distilled water which is accepted immediately into the nest, sometimes after marking. It appears, therefore, that such material has lost volatile chemical signals which mark it as non-food. Conversely, pieces of freshly removed garden fungus are taken to the refuse pile, evidently because they retain volatile signals, but these have a different meaning outside the context of the nest. The importance of the context of the chemical signal is also shown by the fact that fungus garden litter 120 hours old is taken to the nest if picked up on a food trail and to the rubbish dump if picked up on a refuse trail.

DISCUSSION

The work discussed here shows that the response to a given chemical signal can be modulated by a variety of other factors, and that worker ants may continually renew their own pheromone markers. It may, therefore, be relatively simple to elicit trail following, or bait pick-up, for example, in controlled laboratory conditions in the short term. In field conditions, on the other hand, an insect following a trail may encounter a wide variety of stimuli which give directional information and which, at certain times, may be more reliable than a pheromone trail. Laboratory studies suggest that some of these may be visual, spatial, and gravitational cues. Learnt features of a route may thus supplement the purely directional information provided by a pheromone trail, and may replace them where a trail has weakened by volatilisation or disruption on a loose substrate. Likewise, the ability of ants to defend areas on which they are active and recruit nestmates will be influenced by the marking of the substrate with their own secretions. Leaf transport depends to an extent on secretions that (a) make particles attractive to other nestmates, (b) induce pick-up, (c) induce transport, and (d) stimulate others to mark again. Even if all that occurs, volatile stimuli may be present which may result in the particle being rejected at the nest entrance or elsewhere and treated as refuse.

TABLE 3. The final destination of bait particles offered to A. sexdens rubropilosa workers on trails in laboratory tests (n = 10) (Howse, unpublished).

| Material | Trail | Taken to | | Number transported and marked |
		Nest	Rubbish dump	
Vermiculite	Food	–	–	–
Oven-dried litter after 2 hours	Food	10	–	–
Oven-dried litter after 24 hours	Food	10	–	–
Air-dried litter after 120 hours	Food	7	–	3
Litter washed in water	Food	7	1	5
Litter washed in dichloromethane	Food	10	–	–
Fresh fungus garden	Food	1	8	–
Water extract on vermiculite	Food	1	6	–
Dichloromethane extract on vermiculite	Food	–	5	–
Air-dried litter after 120 hours	Refuse	1	9	–
Dichloromethane extract on vermiculite	Refuse	–	6	–

In view of the likelihood that ants respond to a multiplicity of cues in orientation on the trail and in transporting leaf fragments from cutting sites to the nest, it is not surprising that attempts to enhance the efficiency of baits in field applications by addition of synthetic trail pheromone for A. cephalotes (Robinson et al. 1982) and for A. sexdens rubropilosa (Cross et al. 1979) have not been successful. In our own experiments (Vilela and Howse, in press, b), vermiculite particles impregnated with abdominal extracts of A. sexdens rubropilosa were picked up more times than blank controls by the same species, but only a small percentage was taken back to the nest, the number being not significantly different from blank controls. The same effect was achieved with synthetic 3-ethyl-2,5-dimethylpyrazine, the major component of the trail pheromone of this species (Cross et al. 1979), but in field trials with citrus pulp bait there was no significant increase in pick-up. It may therefore be necessary to incorporate a number of different pheromone signals into a bait before enhanced pick-up in the field can be obtained.

REFERENCES CITED

Bradshaw, J. W. S., and P. E. Howse. 1984. Sociochemicals of ants, pp. 429-473. In W. J. Bell and R. T. Carde (eds.), Chemical ecology of insects. Chapman and Hall, London. 524 pp.

Bradshaw, J. W. S., R. Baker, and P. E. Howse. 1979a. Multi-component alarm pheromones in the mandibular glands of major workers of the African weaver ant, Oecophylla longinoda. Physiol. Entomol. 4: 15-25.

Bradshaw, J. W. S., R. Baker, and P. E. Howse. 1979b. Chemical composition of the poison apparatus secretions of the African weaver ant, Oecophylla longinoda and their role in behaviour. Physiol. Entomol. 4: 39-46.

Bradshaw, J. W. S., P. E. Howse, and R. Baker. A novel auto-stimulatory pheromone regulating transport of leaves in Atta cephalotes. Anim. Behav. (In press).

Cross, J. H., R. C. Byler, U. Ravid, R. M. Silverstein, S. W. Robinson, P. M. Baker, J. Sabino de Oliveira, A. R. Jutsum, and J. M. Cherrett. 1979. The major component of the trail pheromone of the leaf-cutting ant, Atta sexdens rubropilosa Forel: 3-Ethyl-2,5-dimethylpyrazine. J. Chem. Ecol. 5: 187-203.

Jaffe, K. 1983. Chemical communication systems in the ant Atta cephalotes, pp. 165-180. In P. Jaisson (ed.), Social insects in the tropics, vol. 2. Universite Paris-Nord, 252 pp.

Jaffe, K., and P. E. Howse. 1979. The mass recruitment system of the leaf-cutting ant Atta cephalotes. Anim. Behav. 27: 930-939.

Jaffe, K., M. Bazire-Benazet, and P. E. Howse. 1979. An integumentary pheromone-secreting gland in Atta sp.: Territorial marking with a colony-specific pheromone in Atta cephalotes. J. Insect Physiol. 25: 833-839.

Robinson, S. W., A. R. Jutsum, J. M. Cherrett, and R. J. Quinlan. 1982. Field evaluation of methyl-4-methylpyrrole-2-carboxylate, an ant trail pheromone, as a component of baits for leaf-cutting ant (Hymenoptera: Formicidae) control. Bull. Entomol. Res. 72: 345-356.

Tumlinson, J. H., J. C. Moser, R. M. Silverstein, R. G. Brownlee, and J. M. Ruth. 1972. A volatile trail pheromone of the ant Atta texana. J. Insect Physiol. 18: 809-814.

Vilela, E. F. 1983. Behaviour and control of leaf-cutting ants (Hymenoptera: Attini). Ph.D. Thesis, University of Southampton, Great Britain.

Vilela, E. F., and P. E. Howse. Aggressiveness and territory in leaf-cutting ants (Hymenoptera: Attini). (In press, a).

Vilela, E. F., and P. E. Howse. Laboratory and field evaluations of a new aldrin bait for the control of the leaf-cutting ant, Atta sexdens rubropilosa, and pheromone performance as an attractive component in baits. (In press, b).

Vilela, E. F., K. Jaffe, and P. E. Howse. Orientation in leaf-cutting ants (Hymenoptera: Formicidae). (In press).

17
The Trail Pheromone Complex
of *Solenopsis invicta*
and *Solenopsis richteri*

R. K. Vander Meer

Research on the trail pheromone complex of Solenopsis invicta
Buren and S. richteri Forel began over 25 years ago, when Wilson
(1959) reported on its source and possible chemical nature. Since
then, we have found that the trail pheromone system is part of a
highly integrated symphony of pheromone and context related interactions, whose behavior releaser effects are not wholly independent.
As will be discussed later, the trail pheromone can be divided into
several sub-categories, although again we must not lose sight of the
whole. With this thought in mind, I will present a brief background
discussion on fire ant recognition and orientation mechanisms,
followed by a review of the status of research on the trail
pheromone.

TERRITORIALITY AND RECOGNITION

Solenopsis species are territorial by nature and tend to
monopolize a food source and foraging area due to their aggressive
behavior and ability to utilize trail pheromones that recruit and
orient workers to specific locations (Wilson 1962). Fire ant workers,
given a choice between soil from their own nest, soil from another
colony's nest, or unnested soil, invariably choose to move into soil
from their home nest (Hubbard 1974). To maintain a territory,
worker ants must be able to distinguish nestmates from non-
nestmates on at least two levels: (1) interspecific; i.e., related
species such as S. invicta and S. geminata display aggressive
behavior toward each other both in and out of the context of their
home colony, and (2) intraspecific nestmate recognition; i.e.,
workers from one colony discriminate workers from another colony
of the same species and defend their nest and territory. Intraspecific recognition is subtle compared to interspecies recognition,
as we have observed that non-nestmate conspecifics out of their
home territory context do not fight.

Conspecifics reared under identical laboratory conditions are not as aggressive toward each other as they are toward individuals from field colonies (Obin and Vander Meer, unpublished results). This suggests that context and overlying environmental odors play important roles in Solenopsis nestmate recognition (see Jaffe, Chapter 18).

Qualitatively, pheromone components are identical, but they vary quantitatively, as evidenced by significant overall differences in cuticular hydrocarbon patterns of several S. invicta colonies (Vander Meer, unpublished). An argument can be made that the natural variation of all chemicals, environmental and innate, form a constantly changing "gestalt" or nest odor that is integrated over time and iteratively learned by workers throughout the continuum of a colony.

ORIENTATION MECHANISMS

Ants have evolved a wide variety of orientation mechanisms. Although trail pheromone systems receive the most notoriety, they represent only one possible mechanism and even when utilized may not always be the dominant stimulus. The carpenter ant, Camponotus pennsylvanicus produces trail pheromones from its hindgut; but when a strong direct light is available, worker attention to trails diminishes and the ants eventually use light as their main orientation cue to reach food (Hartwick et al. 1977). The army ant, Neivamyrmex nigrescens, utilizes a combination of tactile and chemical stimuli to follow a trail (Topoff and Lawson 1979; Topoff and Mirenda 1978). Pheidole militicida uses a combination of trail pheromones (originating in the poison gland) and visual cues (Holldobler and Moglich 1980). The wood ant, Formica rufa, imprints visual cues that are maintained into the next spring by overwintering workers (Rosengren 1977).

All of the above examples represent after-the-fact orientation mechanisms where visual and/or tactile cues have already been imprinted or a chemical trail has already been laid down providing the orientation mechanism. This leaves the question of how ants initially orientate to a trail. There are several systems described in the literature. The desert ant, Cataglyphis bicolor, uses sun compass orientation to find its way back to its nest (Wehner and Lanfranconi 1981). F. pratensis workers use visual cues determined from the asymmetry of their surroundings to locate their home after finding a food source (Kaul 1983). The arboreal ant, Paltothyreus tarsatus, keeps track of its location by orientating to forest canopy patterns (Holldobler 1980). From these studies, it is evident that there are a number of orientation cues that can be used, singly or in combination.

Fire ants are very efficient foragers and search an area by

walking in a looping pattern until they find a food source (Wilson 1962). Then they lay a chemical trail directly back to their nest. Since they do not retrace their looping search path, they must utilize a different homing mechanism while laying down the initial chemical trail. Marak and Wolken (1965) found that trail-following S. invicta workers reversed direction if the light source was reversed. More recently, we investigated the effects of light on initial trail formation and showed conclusively that a light source is the dominant visual cue (not in the sun compass sense) for successful trail formation. If the light source is continuously rotated, the ants are unable to form a trail. However, once the chemical trail is established, rotation of the light has no effect on trail activity (Vander Meer and Lofgren, unpublished results). In addition to light, fire ants have been reported to exhibit context-specific positive geotaxis, the context being those workers involved as sanitary engineers (Howard and Tschinkel 1976). Stratten and Coleman (1973) found that fire ants were capable of learning the position of a food source by using distal-visual cues. So, all in all, the more we learn about fire ants, the more sophisticated and versatile their behavior appears.

The above review illustrates that multiple "correct" answers to broad behavioral questions are the rule rather than the exception when studying fire ants or any other social insect. This is further illustrated by the following detailed discussion of the trail pheromone complex.

TRAIL PHEROMONE

Wilson (1959) established that the Dufour's gland is the source of the fire ant trail pheromone. This gland is located at the base of the sting apparatus, and its contents are emitted through the stinger. A foraging worker coming back from a food source lays a chemical trail by periodically touching its stinger to the substrate on which it is walking. Other workers are recruited to the trail and, depending on the nature of the food source, reinforce the trail with pheromone. Soon a line of ants following the trail is evident. This is in fact the end product of the trail pheromone response. An understanding of the fire ant trail pheromone system, however, is dependent on a knowledge of its chemistry and the behavioral hierarchy of the trail pheromone responses. For the ants to exploit a new food source, the trail pheromone must first attract or recruit workers to the trail. One might expect that orientation along the trail would be an automatic next step, but remarkably this is not the case. It is evident now that, in S. invicta, the Dufour's gland contents contain an orientation primer pheromone that is required to release the orientation response. Thus, successful foraging for S. invicta involves a hierarchy of behavior starting with recruitment, followed by orientation priming, followed by orientation. I will discuss our work with S. invicta, followed by a discussion of the less complicated S. richteri system, and then present an explanation of some of these results based on fire ant trail pheromone chemistry.

Orientation Sub-category

Over the many years since the discovery of S. invicta's trail pheromone (Wilson 1959), a very reliable trail pheromone bioassay has been developed that utilizes a natural worker trail leading up a ramp and over a platform to a food source. A section of the platform can be removed and replaced with a section treated with the test material (Barlin et al. 1976; Jouvenaz et al. 1978). This bioassay was used to isolate and identify two alpha-farnesene and two homofarnesene components (Fig. 1) of the trail pheromone (Vander Meer et al. 1981; Alvarez and Vander Meer 1983). Z,E-alpha-farnesene was quantitatively the major component (ca 6 ng/worker) and elicited the most sensitive response in trail orientation bioassays (0.4 pg/cm). Z,E-alpha-farnesene and an equivalent amount of Dufour's gland extract had identical activities in an orientation bioassay, which indicated that Z,E-alpha-farnesene was solely responsible for orientation activity. Williams et al. (1981a) used the same type of bioassay to isolate and identify Z,Z,Z-allofarnesene (Fig. 1) as the trail pheromone of S. invicta. However, in our laboratory we found the following: (1) the major active component in Dufour's gland extracts is not thermally labile, whereas Z,Z,Z-allofarnesene is thermally labile; (2) Z,Z,Z-allofarnesene synthesized by the method of Williams et al. (1981b) had a different retention time than the major trail pheromone component identified from Dufour's gland extracts as Z,E-alpha-farnesene; and (3) Z,Z,Z-allofarnesene gave positive results only at high concentrations compared to Dufour's gland extracts and synthetic Z,E-alpha-farnesene. Therefore, the rest of this discussion will be based on Z,E-alpha-farnesene as the major and most active trail pheromone component.

S. richteri's Dufour's gland profile is dominated by a single peak composed of two isomeric tricyclic homosesquiterpenes designated C-1/C-2. These two compounds are as active in the orientation bioassay as an equivalent amount of Dufour's gland extract. Therefore, they elicit 100% of the orientation response. C-1 and C-2 are found in S. invicta Dufour's glands at only about 75 pg/worker compared with 4000 pg/worker in S. richteri Dufour's glands. No alpha-farnesenes or homofarnesenes have been found in S. richteri's Dufour's gland extracts. Barlin et al. (1976) and Jouvenaz et al. (1978) used the ramp/platform food trail bioassay to show that the two imported fire ant species are capable of orienting to each other's Dufour's gland extracts. We now know that this bioassay specifically measures the orientation sub-category of the trail pheromone, and we can provide an explanation of these results based on trail pheromone chemistry. No species specificity was observed in this bioassay because S. invicta is moderately sensitive to C-1/C-2 (>4 pg); and at the bioassay concentration of 0.01 worker equivalents (WE)/cm (Barlin et al. 1976), there is >10 times enough C-1/C-2 in S. richteri's Dufour's gland extract to elicit a response from S. invicta. S. richteri is not sensitive to Z,E-alpha-farnesene (>60 pg/cm) and would elicit a marginal response to 0.01 WE/cm (equivalent to 60 pg/cm Z,E-alpha-farnesene) of S. invicta's Dufour's gland. However, the amount of C-1/C-2 in S. invicta's Dufour's gland is about 10X that needed (>50 fg/cm) to elicit an orientation response from highly sensitive S. richteri.

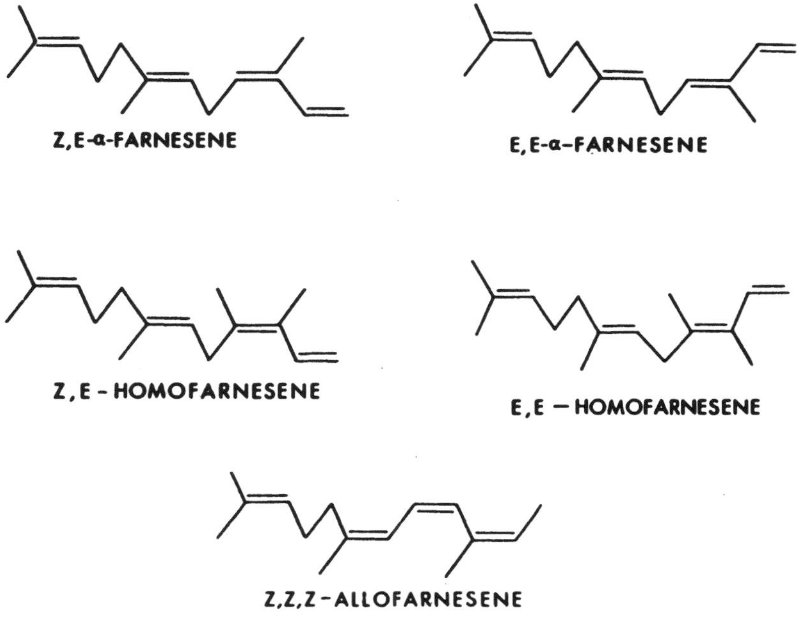

Z,E-α-FARNESENE

E,E-α-FARNESENE

Z,E - HOMOFARNESENE

E,E − HOMOFARNESENE

Z,Z,Z-ALLOFARNESENE

FIGURE 1. Structures of S. invicta trail pheromones isolated using a trail orientation bioassay (Vander Meer et al. 1981; Alvarez and Vander Meer 1983; and Williams et al. 1981a, b).

Recruitment Sub-category

Two bioassays were used to measure the recruitment sub-category. One was based on the response of workers to a point source of Dufour's gland extract. In this test workers are not only attracted to the spot, but they also aggregate and bite at the substrate. The bioassay was structured so that multiple samples and controls could be run simultaneously. It was quantified by counting the number of ants responding at 5-minute intervals for a total time of 30 minutes. Unexpectedly, neither Z,E-alpha-farnesene nor the other compounds isolated using the orientation bioassay elicited recruitment activity when presented to S. invicta workers as a point source. However, we determined that 85% of Dufour's gland recruitment activity could be obtained with a mixture of Z,E-alpha-farnesene and the two isomeric tricyclic homofarnesenes, C1/C2. These compounds occur at only 75 pg of C-1/C-2 per S. invicta worker and have a profound behavioral effect in combination with the major orientation pheromone. However, the fact that we did not obtain a 100% response indicates that there are still some components missing. A similar situation exists with S. richteri, where C-1/C-2 gave 85% of Dufour's gland activity in the point source bioassay (Vander Meer et al. 1985).

Cross species tests using the point source bioassay showed clear species specificity (Vander Meer et al. 1985). This can be rationalized chemically by the fact that S. invicta requires a combination of C-1/C-2 and Z,E-alpha-farnesene. That combination is not present in S. richteri Dufour's gland extracts. S. richteri, however, requires C-1/C-2, which is over 50 times less concentrated in S. invicta Dufour's glands. Concentration/activity studies have shown that S. richteri workers do not respond in the point source bioassay to the quantities of C-1/C-2 found in S. invicta Dufour's glands (Vander Meer et al. 1985).

The second recruitment bioassay measured the attraction of workers to volatile components of Dufour's gland extracts using a Y-tube olfactometer. S. invicta responded poorly to Z,E-alpha-farnesene or C-1/C-2 alone when compared to Dufour's gland extracts; however, their response to a combination of the compounds was statistically indistinguishable from those of an equivalent concentration of Dufour's gland extract. Similarly, the response of S. richteri workers in the olfactometer to C-1/C-2 and an equivalent amount of S. richteri Dufour's glands was identical (Vander Meer, unpublished data). Olfactometer species-specificity tests comparing the responses of S. richteri and S. invicta to each others Dufour's gland extracts did not show clear differentiation. Both species showed significant attraction to each others trail pheromones, although S. invicta was most capable of responding both to its own and S. richteri's Dufour's gland extract.

Orientation Primer Sub-category

The hypothesis for an orientation primer pheromone originated with the observation that the orientation pheromone, Z,E-alpha-farnesene, did not induce orientation in randomly foraging workers.

A positive response was obtained only when test workers were already following a trail. Therefore, a bioassay was devised that measured the response of non-trailing ants to streaked test samples. We found that the addition of an equivalent amount of Dufour's gland extract at the beginning of a Z,E-alpha-farnesene trail increased the orientation activity of Z,E-alpha-farnesene greater than 4X (Vander Meer et al. 1984). In contrast, the recruitment mixture Z,E-alpha-farnesene and C-1/C-2 had no orientation priming effects. Based on the orientation and orientation primer bioassay results, we concluded that the components required to initiate orientation are not themselves required for orientation. These statements define a primer pheromone (Nordlund 1981).

The orientation primer pheromone alters a worker's physiological state such that it maximizes the behavioral releaser effects of the trail orientation pheromone. There are precedents for this concept in the literature. Ants have evolved a wide variety of mechanisms to recruit and orientate workers to food sources, ranging from primitive tandem calling in Leptothorax acervorum (Moglich et al. 1974) to the complex multiple recruitment systems of the African weaver ant, Oecophylla longinoda (Holldobler and Wilson 1978). A successful L. acervorum forager attracts a nestmate with chemicals and then physically guides them to the food source; consequently, trail orientation pheromones are not utilized. This recruitment mechanism may represent one of the first evolutionary steps toward trail pheromones and mass communication in myrmecine ants (Holldobler 1978). In other ant species, foraging workers lay down chemical trails that have orientation effects but do not release recruitment and orientation behavior. In these examples, motor displays and mechanical signals are important to induce or stimulate actual trail following (Holldobler 1978). For instance, F. fusca scouts must perform a waggle display to excite nestmates into following their trail pheromone (Moglich and Holldobler 1975). The orientation primer pheromone is the chemical analog to the physical waggle display.

Further up the evolutionary ladder, recruitment strategies involve trail pheromones that elicit recruitment, priming or modulation, and/or orientation. S. invicta falls into this category because a single glandular source is responsible for all mass-foraging activities. The orientation primer pheromone has not been chemically defined; however, species-specificity tests showed that S. invicta was primed by its own Dufour's gland extract. Surprisingly, their response to the S. richteri Dufour's gland extract increased by a factor of two. In contrast, S. richteri only responded to its own Dufour's gland extract, ignoring the material from S. invicta (Vander Meer et al. 1985). This is an example of one-way species specificity. At the present time, we do not have enough information to explain these results in terms of Dufour's gland chemistry.

SUMMARY

When species specificity bioassay results are collated, we find an almost complete range of responses (Table 1). There is (1) total lack of specificity in the orientation bioassay, (2) total specificity with the recruitment or point-source bioassay, (3) one-sided specificity using the orientation primer bioassay, and (4) ambiguous results using the olfactometer. Although this appears to be about as confusing a situation as possible, we are developing an understanding of these results based on our knowledge of the chemistry of S. invicta's and S. richteri's Dufour's gland contents. Besides the alpha-farnesenes and homofarnesenes, n-heptadecane and n-nonadecane had been reported previously in S. invicta's Dufour's gland extract but had no associated behavioral releaser effects (Barlin et al. 1976). Also present in Dufour's gland extracts (but inactive) are the five normal methyl and dimethyl branched hydrocarbons that are ubiquitous to S. invicta and are characteristic of the species (Nelson et al. 1980; Vander Meer and Wojcik 1982). Preliminary experiments indicate that these compounds are important in synergizing orientation primer effects (Vander Meer, unpublished results).

Mass-foraging in both species is moderated by chemicals produced by a single gland, and there are at least three broad behavioral categories (recruitment, orientation primer, and orientation) released by these chemicals. The simplest situation is that a single chemical or group of chemicals is responsible for all three trail pheromone sub-categories. The most complicated is if a different chemical or set of chemicals is responsible for each sub-category. S. invicta comes close to the latter situation except Z,E-alpha-farnesene is a common component for all elements of the trail pheromone. In contrast, S. richteri utilizes a single mixture of components (C-1/C-2) to release 100% of the orientation and recruitment sub-categories, and these same components release 85% of the orientation priming activity. Future research will further define trail pheromone sub-categories and their interrelationships, as well as their potential as an adjunct to control methods.

TABLE 1. Species specificity of S. invicta and S. richteri to four sub-categories of their respective trail pheromones.

Sub-category	Species	
	S. invicta	S. richteri
Recruitment (olfactometer)	±	±
Recruitment (point source)	+	+
Orientation primer	−	+
Orientation	−	−

REFERENCES CITED

Alvarez, F. M., and R. K. Vander Meer. 1983. Synthesis of homo-farnesenes: Trail pheromone components of the red imported fire ant. Nat. Am. Chem. Soc. Mtg., Washington, DC, Abstract ORGN 72.

Barlin, M. R., M. S. Blum, and J. M. Brand. 1976. Fire ant trail pheromones: Analysis of species specificity after gas chromatographic fractionation. J. Insect Physiol. 22: 839-844.

Hartwick, E. B., W. G. Friend, and C. E. Atwood. 1977. Trail laying behaviour of the carpenter ant, Camponotus pennsylvanicus (Hymenoptera: Formicidae). The Can. Entomol. 109: 129-136.

Holldobler, B. 1978. Ethological aspects of chemical communication in ants. Adv. Study Behav. 8: 75-115.

Holldobler, B. 1980. Canopy orientation: A new kind of orientation in ants. Science 210: 86-88.

Holldobler, B., and M. Moglich. 1980. The foraging system of Pheidole militicida (Hymenoptera: Formicidae). Insect. Soc. 27: 237-264.

Holldobler, B., and E. O. Wilson. 1978. The multiple recruitment systems of the African weaver ant Oecophylla longinoda (Latreille) (Hymenoptera: Formicidae). Behav. Ecol. Sociobiol. 3: 19-60.

Howard, D. F., and W. R. Tschinkel. 1976. Aspects of necrophoric behavior in the red imported fire ant, Solenopsis invicta. Behavior 56: 157-180.

Hubbard, M. D. 1974. Influence of nest material and colony odor on digging in the ant, Solenopsis invicta (Hymenoptera: Formicidae). J. Ga. Entomol. Soc. 9: 127-132.

Jouvenaz, D. P., C. S. Lofgren, D. A. Carlson, and W. A. Banks. 1978. Specificity of the trail pheromones of four species of fire ants, Solenopsis spp. Fla. Entomol. 61: 244.

Kaul, R. M. 1983. Orientation of Formica pratensis (Hymenoptera, Formcidae). Zoologicheskii Zhurnal. 62: 240-244.

Marak, G. E., and J. J. Wolken. 1965. An action spectrum for the fire ant (Solenopsis saevissima). Nature (London). 285: 1328-1329.

Moglich, M., and B. Holldobler. 1975. Communication and orientation during foraging and emigration in the ant Formica fusca. J. Comp. Physiol. 101: 275-288.

Moglich, M., U. Maschwitz, and B. Holldobler. 1974. Tandem calling: A new kind of signal in ant communication. Science 186: 1046-1047.

Nelson, D. R., C. L. Fatland, R. W. Howard, C. A. McDaniel, and G. J. Blomquist. 1980. Re-analysis of the cuticular methyl-alkanes of Solenopsis invicta and S. richteri. Insect Biochem. 10: 409-418.

Nordlund, D. A. 1981. Semiochemicals: A review of terminology, pp. 13-28. In D. A. Nordlund, R. L. Jones, and W. J. Lewis (eds.), Semiochemicals: Their role in pest control. John Wiley and Sons, N.Y., NY. 306 pp.

Rosengren, R. 1977. Foraging strategy of wood ants (Formica rufa group) II. Nocturnal orientation and diel periodicity. Acta Zool. Fennica. 150: 1-31.

Stratten, L. O., and W. P. Coleman. 1973. Maze learning and orientation in the fire ant (Solenopsis saevissima). J. Comp. Physiol. Psych. 83: 7-12.

Topoff, H., and K. Lawson, 1979. Orientation of the army ant Neivamyrmex nigrescens: Integration of chemical and tactile information. Anim. Behav. 27: 429-433.

Topoff, H., and J. Mirenda. 1978. Precocial behaviour of callow workers of the army ant Neivamyrmex nigrescens: Importance of stimulation by adults during mass recruitment. Anim. Behav. 26: 698-706.

Vander Meer, R. K., F. M. Alvarez, and C. S. Lofgren. 1984. Fire ant recruitment and orientation trail pheromones: A unique species isolation mechanism. Presented at a Conference on Molecular Messengers in Nature. Nat. Inst. Health, Bethesda, MD.

Vander Meer, R. K., F. M. Alvarez, and C. S. Lofgren. 1985. Fire ant species specificity: A chemical explanation. Presented at the Sec. Annu. Mtg. Int. Soc. Chem. Ecol. Madison, WI.

Vander Meer, R. K., F. D. Williams, and C. S. Lofgren. 1981. Hydrocarbon components of the trail pheromone of the red imported fire ant, Solenopsis invicta. Tetrahedron Lett. 22: 1651-1654.

Vander Meer, R. K., and D. P. Wojcik. 1982. Chemical mimicry in the myrmecophilous beetle Myrmecaphodius excavaticolli. Science 218: 806-808.

Wehner, R., and B. Lanfranconi. 1981. What do the ants know about the rotation of the sky? Nature 293: 731-733.

Williams, H. J., M. R. Strand, and S. B. Vinson. 1981a. Trail pheromone of the red imported fire ant Solenopsis invicta (Buren). Experientia 37: 1159-1160.

Williams, H. J., M. R. Strand, and S. B. Vinson. 1981b. Synthesis and purification of the allofarnesenes. Tetrahedron. 37: 2763-2767.

Wilson, E. O. 1959. Source and possible nature of the odour trail of fire ants. Science 129: 643-654.

Wilson, E. O. 1962. Chemical communication amoung workers of the fire ant Solenopsis saevissima (Fr. Smith). 1. The organization of mass-foraging. 2. An information analysis of the odour trail. 3. The experimental induction of social responses. Anim. Behav. 10: 134-164.

18
Nestmate Recognition and Territorial Marking in *Solenopsis geminata* and in some Attini

K. Jaffe

NESTMATE RECOGNITION

It has been known for a long time that ants recognize and differentiate their nestmates from alien ants of conspecific or heterospecific colonies mainly on the basis of chemical cues. Species-specific recognition of queens has been reported for Pogonomyrmex badius (Holldobler 1976), Solenopsis invicta (Glancey et al. 1981 and Chapter 20), Solenopsis spp. (Jouvenaz et al. 1974), and Neivamyrmex spp. and Labidus spp. (Watkins and Cole 1966). In these cases, a queen pheromone was reported to induce the recognition signal.

Contact pheromones appear to be responsible for brood recognition. Examples can be found in S. invicta (Walsh and Tschinkel 1974), Myrmica spp. (Brian 1975), Atta cephalotes (Robinson and Cherrett 1974), and Formica spp. (Jaisson 1975; Jaisson and Fresneau 1978; Le Moli and Passetti 1977, 1978; Le Moli and Mori 1982). Ants learn the chemical signal shortly after eclosion with a type of imprinting mechanism (see references for Formica spp. above; Isingrini and Lenoir 1984). Thus, ants that eclose to adults among foreign brood may recognize the foreign brood as nestmates and treat brood from their original colony as aliens.

Another example of chemically-based recognition is found in the necrophoric behaviour of ants. Examples include P. badius (Wilson et al. 1958), S. saevissima and S. invicta (Blum 1970; Howard and Tschinkel 1976), Myrmecia vindex (Haskins and Haskins 1974), and Monomorium pharaonis (Legakis 1979). In the case of Solenopsis, myristic acid, palmitic acid, oleic acid, and linoleic acid seem to be the recognition cues. Triethanolamine is active as a recognition signal in dead Myrmecia, and oleic acid is active in Pogonomyrmex. Oleic acid induces waste recollection but not necrophoric behaviour in Monomorium.

Nestmate recognition among adult workers has been studied less. Odours have always been assumed to be responsible for this recognition (Forel 1874; Fielde 1904; Howse 1975; Holldobler and Michener 1980). However, experimental evidence for this appeared only recently. In Megaponera foetens (Longhurst 1977), it was shown that dummies with cuticular waxes from nestmates were accepted into the colony or attacked much less than dummies with waxes from conspecific aliens. In the Ponerinae Odontomachus bauri (Jaffe and Marcuse 1983), and Ectatomma ruidum (Marquez 1984), it was shown that volatile substances produced by workers in the different body parts are used as recognition signals. These volatiles are absorbed on the cuticle and thus can be active in nestmate recognition as both contact and volatile pheromones. The Formicinae Camponotus rufipes (Jaffe and Sanchez 1984) uses the cephalic alarm pheromone as a nestmate recognition signal, the same as the Myrmicinae A. cephalotes (Jaffe 1983). Crematogaster sumichrasti (Pino 1984) utilizes cephalic and abdominal alarm pheromones as nestmate recognition signals as do some of the Dolichoderinae (Colmenares 1982; Alvarez 1984; Jaffe et al., unpublished). In the Pseudomyrmicinae, we find examples of both; e.g., Pseudomyrmex triplarinus uses the alarm pheromone as a recognition signal (Aragort 1984), whereas P. termitarius (Lopez 1984) uses non-specific, intrinsically produced odours, the same as the Ponerinae, for nestmate recognition.

As with brood recognition, adult nestmate recognition appears to be achieved through an imprinting process (Le Moli and Passetti 1977; Jaisson 1980; Le Moli and Mori 1984); i.e., ants that eclose in an environment of adult workers will recognize them as nestmates, even if they are from a different species. Although this relation has been studied most in slave ants and their slave makers, this seems to be a more or less general phenomenon in the Formicidae (Le Masne 1952; Provost 1979; Jaisson 1980).

The Nestmate Recognition System of Solenopsis geminata and Some Attini

Two types of bioassays have been used to study nestmate recognition. In the first, two workers were drawn at random, one from the nest where the test was to be performed (control), and the other from a different conspecific colony (experimental). The workers were picked up with clean plastic forceps and placed at random on the foraging area of the control colony. The behaviour of the resident workers towards the intruders was described verbally on a tape recorder for later quantification. If statistical differences could be detected in the response of ants of the resident colony towards either the experimental or control ants for any particular behaviour, colony-specific recognition was presumed to exist.

The data from these tests (Table 1) revealed that most species differentiated between their nestmates and the intruder, although aggression towards the latter was not very strong in Acromyrmex landolti and Ac. octospinosus. The lower Attini did not exhibit nestmate recognition, but this may be because disturbance of the nests is always too intense and thus no appropriate behavioural variables can be used to measure recognition.

In the second bioassay, dead workers (killed by freezing in dry ice) or parts of them were placed near the entrance to the foraging area of the colonies. Two workers or worker parts were always presented simultaneously to the colonies. One of them was a nestmate from the test colony, whereas, the other came from a different conspecific colony. The time resident workers took before removing each of the dead workers or worker parts was measured.

In other tests, dead workers were freeze-dried by exposing them at pressures of 0.001 mm Hg at -20°C using a vacuum pump. The freeze-dried workers were impregnated with odours from different body parts of other workers by placing them together in a 2 ml glass vial with freshly crushed body parts. Care was taken in order to avoid direct contact between the freeze-dried ants and the crushed body parts. The vial was sealed and left at room temperature for 30 minutes. After this period, the freeze-dried ants were taken out and used for tests. Statistically significant differences in any behaviour of the resident workers towards live ants, bodies or body parts of dead ants, and/or impregnated freeze-dried ants were taken as evidence for colony-specifc recognition. As shown in Table 2, there are a number of possibilities in Attini nestmate recognition systems. The primitive Attini, Trachymyrmex urichi, use non-specific odours absorbed on the cuticle for nestmate recognition, Acromyrmex use both alarm pheromones and environmental odours (Jutsum et al. 1979), and Atta use only the alarm pheromone. Solenopsis seems to be located just under Atta in this gradient, using various cues, but all are alarm pheromones (Puche 1982).

TERRITORIAL BEHAVIOR

A territory, as defined by Holldobler and Wilson (1977), is "an area occupied more or less exclusively by an animal by means of overt defense or advertisement." Four operational criteria for defining territorial marking pheromones (Jaffe and Puche 1984) are as follows:

1. The organism must secrete a chemical(s) onto the substrate of a portion of the home range.
2. The organism has to recognize its own mark and has to differentiate it from marks of conspecifics. Therefore, the chemical(s) must have intraspecific differences detectable by the organism.

TABLE 1.   Results of nestmate recognition tests with conspecific colonies of several species of Atta and S. geminata.[a]

Species	Intraspecific aggression	Colony specific behaviour indicated	Reference
Atta			
cephalotes	++	x	Jaffe 1983; Vilela 1983
laevigata	++	x	Naccarata 1983; Vilela 1983
Acromyrmex			
octospinosus	+	x	Jaffe 1980; Jutsum et al. 1979
landolti	+	x	Navarro 1983; Jaffe and Navarro, in press
Trachymyrmex urichi	-	x	Jaffe and Villegas, in press
Mycocepurus sp.	-	-	Villegas 1982
Myrmicocrypta sp.	-	-	Villegas 1982
Mycetophylax sp.	-	-	Villegas 1982
Solenopsis geminata	++	x	Puche 1982; Jaffe and Puche 1984

[a]++: very aggressive; +: occasionally aggressive; x: behaviour present.

TABLE 2. Results of bioassays for detection of nestmate recognition.

Species	Source of pheromone inducing colony-specific behaviour			Reference
	Head	Thorax	Gaster	
Atta				
cephalotes	+	-	-	Jaffe 1983
laevigata	+	-	-	Naccarata 1983
Acromyrmex				
octospinosus	+	-	-	Jaffe 1980
landolti	+	-	-	Navarro 1983; Jaffe and Navarro, in press
Trachymyrmex				
urichi	+	+	+	Jaffe and Villegas, in press
Solenopsis				
geminata	+	-	+	Puche 1982

+ = behaviour present

3. The presence of a territorial signal has to give some advantage to the organism producing the signal.
4. An area marked with a territorial pheromone should be respected by conspecifics. This is a sufficient, but not necessary, condition for territoriality. Criterion 3, for example, could be fulfilled without criterion 4.

The end result of territorial marking should be a reduction in aggressive interactions between the organisms because colonies are separated more than would be expected from a random occupation of suitable habitats.

The subject of territorial behaviour in ants has been partially reviewed by Baroni-Urbani (1979) and Holldobler (1979). Evidence for territoriality has been reported for a number of species, including P. badius (Holldobler 1976), Formica rufa (Skinner 1980), and S. saevissima (Wilson et al. 1971). The gland secreting the territorial pheromone varies for each species. Thus, the rectal sac secretes the territorial mark in Oecophylla longinoda (Holldobler and Wilson 1977) and the Dufour's gland in Myrmica ants (Cammaerts et al. 1977, 1978, 1981). An abdominal gland, other than the rectal sac, Dufour's gland, or poison gland, is suspected as the pheromone source in the Formicinae, C. rufipes (Jaffe and Sanchez 1984). The metapleural gland is used by two species of Pseudomyrmicinae (Lopez 1984; Aragort 1984), as well as the Dolichoderinae Azteca foreli (Alvarez 1984) and the Myrmicinae Crematogaster sumichrasti (Pino 1984). An abdominal gland is suspected as the source of a territorial marker in a Conomyrma sp. (Colmenares 1982) of the subfamily Dolichoderinae.

Studies of Territorial Behaviour

Two approaches have been used to demonstrate the presence of territorial behaviour. In the first, petri dishes lined with 9-cm diameter filter papers were placed for 48 hours on the wooden foraging table of the colonies. For the test, the dish and filter paper were removed from the foraging table and a test was conducted by placing two workers, chosen at random from two different colonies, on the contaminated or a clean piece of filter paper. Observations were made as to which ant initiated an attack and the time each ant took in leaving the dish. Any statistically significant colony-specific differences were attributed to the existence of a colony-specific mark(s) on the paper. If the same results were obtained with a specific extract of workers, it was taken as evidence of territorial marking.

In case no evidence of territorial recognition was noted, a second bioassay was performed. In this case, every effort was made to keep topographical features of the test areas identical. For example, objects, such as glass nests and petri dishes with water or food, as well as the placement of the wooden tables in the laboratory, were the same for the different test colonies (at least from the experimenter's point of view). Two fixed visual designs were employed. Workers were taken at random from colonies with each design and were released in their colony of origin (control) or in conspecific colonies with either design (experimental). The number of ants going to the nest, and the time taken for it was measured. If no statistically significant difference could be detected between the control and the experimental colonies with the same design, but differences were detected when the control and the experimental had different designs, then visual signals were suspected as having an influence on territorial recognition.

The data (Table 3) show that Atta and Acromyrmex mark their territory with a pheromone, but other Attini do not. S. geminata also marks its territory. The source of the territorial pheromone in Atta is the valves gland (Bazire-Benazet and Zylberberg 1979; Jaffe et al. 1979); whereas in Solenopsis it is produced by the metathoracic gland (Jaffe and Puche 1984). T. urichi recognizes its territory only through visual cues.

TABLE 3. Results of bioassays for detection of territorial marking and recognition.

Species	Colony specific territorial recognition	marking	Reference
Atta			
cephalotes	++	x	Jaffe et al. 1979
laevigata	++	x	Naccarata 1983; Vilela 1983
Acromyrmex			
octospinosus	++	x	Jaffe 1980
landolti	++	x	Navarro 1983
Trachymyrmex urichi	#	–	Villegas 1982
Myrmicocrypta sp.	–	–	Villegas 1982
Mycocepurus sp.	–	–	Villegas 1982
Mycetophylax sp.	–	–	Villegas 1982
Solenopsis geminata	++	x	Jaffe and Puche 1984

Symbols under recognition column indicate that either odour (++) or visual (#) cues are the main sources of territorial recognition; x indicates the presence of marking behaviour.

CONCLUSION

Nestmate recognition systems and territorial behaviour seem to be very similar among the different Formicidae. In the advanced Attini (Atta and Acromyrmex) and in S. geminata, nestmate recognition is based on alarm pheromones (abdominal and cephalic in the case of S. geminata and cephalic only in the case of the higher Attini). All of these species mark their territory chemically. Species not marking their territory, such as T. urichi, have a nestmate recognition system based on non-specific odours and territorial behaviour based on visual cues. The very primitive species of Attini, Mycetophylax, Myrmicocrypta, and Mycocepurus, lack both.

Evolution of nestmate recognition seems to be linked to territorial behaviour (Jaffe and Marcuse 1983; Jaffe and Sanchez 1984). Three distinct steps are recognized in this evolution:

1. Non-specific odours, intrinsic and environmental, are used for nestmate recognition and no territorial recognition through odours exists.
2. Non-specific intrinsic ant odours are used as nestmate recognition signals; and, in some cases, the same odours, secreted on the substrate, are used as territorial recognition signals (Fresneau 1980).
3. Alarm pheromones serve as nestmate recognition signals. Territorial pheromones are secreted onto the substrate.

Environmental odours present on the cuticle and used for nestmate recognition are possible in any of the above steps.

Thus, evolution tends to simplify the recognition signals; i.e., reduce the number of signals involved and at the same time give them a more specific meaning or function. This would indicate that the more evolved species, such as Atta and Solenopsis must have simple nestmate recognition signals. The relative concentrations of a few compounds of the alarm pheromone complex should be enough to give a colony specific signal (Crozier and Dix 1979), such as described for A. cephalotes (Jaffe 1983).

The territorial mark is time dependent. This means that areas which are not intensively used lose the chemical mark on it (Jaffe et al. 1979; Jaffe and Puche 1984). This suggests that the territories of these ants are variable in space and time, depending on the exploitation of the resources. A permanent partition of space can occur in stable habitats, such as those reported for S. saevissima (Wilson et al. 1971), but are not likely to occur for the Attini, as the resources vary in space and time (Cherrett 1980).

REFERENCES CITED

Alvarez, M. 1984. Comunicacion quimica entre obreras de la hormiga Azteca foreli. Tesis de Licenciatura, Universidad Simon Bolivar, Caracas, Venezuela. 153 pp.

Aragort, W. 1984. Comunicacion quimica entre obreras de la hormiga Pseudomyrmex triplarinus. Tesis de Licenciatura, Universidad Simon Bolivar, Caracas, Venezuela. 110 pp.

Baroni-Urbani, C. 1979. Territoriality in social insects, pp. 91-121. In H. R. Hermann (ed.), Social insects. Academic Press, N.Y., NY.

Bazire-Benazet, M., and L. Zylberberg. 1979. An integumentary gland secreting a territorial marking pheromone in Atta spp.: Morphology, histochemistry and ultrastructure. J. Insect Physiol. 25: 751-756.

Blum, M. 1970. The chemical basis of insect sociality, pp. 61-94. In M. Beroza (ed.), Chemicals controlling insect behaviour. Academic Press, N.Y., NY. 170 pp.

Brian, M. V. 1975. Larval recognition by workers of the ant Myrmecia. Anim. Behav. 23: 745-756.

Cammaerts, M. C., R. P. Evershed, and E. D. Morgan. 1981. Comparative study of the Dufour's gland secretion of workers of four species of Myrmica ants. J. Insect Physiol. 27: 59-65.

Cammaerts, M. C., M. R. Inwood, E. D. Morgan, and K. Parry. 1978. Comparative study of the pheromones emitted by workers of the ants Myrmica rubra and Myrmica scabrinodis. J. Insect Physiol. 24: 207-214.

Cammaerts, M. C., E. D. Morgan, and R. Tyler. 1977. Territorial marking in the ant Myrmica rubra. Biol. Behav. 2: 263-275.

Cherrett, J. M. 1980. Possible reasons for the mutualism between leaf-cutting ants and their fungus. Rev. Biol. Ecol. Med. 8: 113-122.

Colmenares, O. 1982. Estudio del comportamiento social en dos especies del genero Conomyrma. Tesis de Licenciatura, Universidad Simon Bolivar, Caracas, Venezuela. 95 pp.

Crozier, R. H., and M. W. Dix. 1979. Analysis of two genetic models for the innate components of colony odor in social Hymenoptera. Behav. Ecol. Sociobiol. 4: 217-224.

Fielde, A. M. 1904. Power of recognition among ants. Biol. Bull. 7: 227-250.

Forel, A. 1874. Les Fourmis de la Suisse. Nouv. Mem. Soc. Helv. Sc. Nat. Zurich, vol. 26, 447 pp.

Fresneau, D. 1980. Fermeture des societes et marquage territorial chez des fourmis ponerines du genre Neoponera. Biol. Ecol. Med. 7: 205-206.

Glancey, B. M., A. Glover, and C. S. Lofgren. 1981. Pheromone production by virgin queens of Solenopsis invicta. Sociobiology 6: 119-127.

Haskins, C. P., and E. F. Haskins. 1974. Notes on necrophoric behaviour in the archaic ant Myrmecia vindex. Psyche 81: 258-267.

Holldobler, B. 1976. Recruitment behaviour, home range orientation and territoriality in harvester ants Pogonomyrmex. Behav. Ecol. Sociobiol. 1: 3-44.

Holldobler, B. 1979. Territoriality in ants. Proc. Am. Phil. Soc. 123: 211-218.

Holldobler, B., and C. D. Michener. 1980. Mechanisms of identification and discrimination in social Hymenoptera, pp. 35-58. In Markl (ed.), Evolution of social behaviour. Weinheim, Verlag Chemie GmbH.

220

Holldobler, B., and E. O. Wilson. 1977. Colony specific territorial pheromone in the African weaver ant. Proc. Nat. Acad. Sci. USA 74: 2072-2075.

Howard, D. F., and W. R. Tschinkel. 1976. Aspects of necrophoric behaviour in the red imported fire ant Solenopsis invicta. Behaviour 56: 157-180.

Howse, P. E. 1975. Pheromones and defensive secretions in social insects. Proc. VII Congr. IUSSI, Dijon, France.

Isingrini, M., and A. Lenoir. 1984. Le developpement de la reconnaissance coloniale dans les soins au couvain chez la fourmi Cataglyphis cursor. Cologuio Int. Etologia Universidad Autonoma de Barcelona, Espana.

Jaffe, K. 1980. Chemical communication among workers of the leaf-cutting ant Atta cephalotes. Ph.D. Thesis, University of Southampton, England. 183 pp.

Jaffe, K. 1983. Chemical communication among workers of leaf-cutting ants, pp. 165-180. In P. Jaisson (ed.), Social insects in the tropics, vol. 2. Univ. Paris-Nord, Paris, France. 252 pp.

Jaffe, K., and M. Marcuse. 1983. Individual recognition and territorial behaviour in the ant Odontomachus bauri (Emery). Insect. Soc. 30: 466-481.

Jaffe, K., and J. G. Navarro. Comunicacion quimica entre obreras de la hormiga cortadora de grama Acromyrmex landolti. Rev. Bras. Entomol. (In press).

Jaffe, K., and H. Puche. 1984. Territorial marking with metapleural gland secretion in the ant Solenopsis geminata. J. Insect Physiol. 30: 265-270.

Jaffe, K., and C. Sanchez. 1984. Nestmate recognition and territorial behaviour in the ant Camponotus rufipes. Insect. Soc. 31: 302-315.

Jaffe, K., and G. Villegas. On the communication system of the fungus growing ant Trachymyrmex urichi. Insect. Soc. (In press).

Jaffe, K., M. Bazire-Benazet, and P. E. Howse. 1979. Territorial marking with a colony specific pheromone from an integumentary gland in leaf-cutting ants. J. Insect Physiol. 25: 833-839.

Jaisson, P. 1975. L'impregnation dans l'ontogenese des comportements de soins aux cocon chez la jeune fourmi rousse Formica polyctena. Behaviour 52: 1-37.

Jaisson, P. 1980. Les colonies mixtes plurispecifiques: un modele pour l'etude des fourmis. Biol. Ecol. Med. 7: 163-166.

Jaisson, P., and D. Fresneau. 1978. The sensitivity and responsiveness of ants to their cocoons in relation to age and method of measurement. Anim. Behav. 26: 1064-1071.

Jouvenaz, D. P., W. A. Banks, and C. S. Lofgren. 1974. Fire ants: Attraction of workers to queen secretions. Ann. Entomol. Soc. Am. 67: 442-444.

Jutsum, A. R., T. S. Saunders, and J. M. Cherrett. 1979. Intraspecific aggression in the leaf-cutting ant Acromyrmex octospinosus. Anim. Behav. 27: 839-844.

Legakis, A. 1979. Aspects of chemical communication in the pharaoh ant, Monomorium pharaonis. Ph.D. Thesis, University of Southampton, England. 199 pp.

Le Masne, G. 1952. Classification et caracteristiques des principaux types de groupements sociaux realises chez les invertebres. Rapport au Coloque International sur la Structure et la Physiologie des Societes Animales. Paris, 1950. Coll. Inter. C.N.R.S. XXXIV.

Le Moli, F., and A. Mori. 1982. Early learning and cocoon nursing behaviour in the red wood ant Formica lugubris. Boll. Zool. 49: 93-97.

Le Moli, F., and A. Mori. 1984. The effect of early experience on the development of aggressive behaviour in Formica lugubris. Z. Tierpsychol. 65: 241-249.

Le Moli, F., and M. Passetti. 1977. The effect of early learning on cocoon recognition, acceptance and care of cocoons in the ant Formica rufa. Atta Soc. Ital. Sci. Nat. Museo Civ. Stor. Nat. Milano 118: 49-64.

Le Moli, F., and M. Passetti. 1978. Olfactory learning phenomena and cocoon nursing behaviour in the ant Formica rufa. Boll. Zool. 45: 389-397.

Longhurst, C. 1977. Behavioural, chemical and ecological interactions between west African ants and termites. Ph.D. Thesis, Southampton University, England.

Lopez, M. E. 1984. Comunicacion quimica entre obreras de la hormiga Pseudomyrmex termitarius. Tesis de Licenciatura, Universidad Simon Bolivar, Caracas, Venezuela. 99 pp.

Marquez, M. 1984. Comunicacion quimica entre obreras de la hormiga Ectatomma ruidum. Tesis de Licenciatura, Universidad Simon Bolivar, Caracas, Venezuela. 99 pp.

Naccarata, V. 1983. Estudio sobre Atta laevigata, plaga de las plantaciones de pino caribe en el sur del Estado Monagas. Tesis de Licenciatura, Universidad Simon Bolivar, Caracas, Venezuela. 123 pp.

Navarro, J. G. 1983. Estudio de algunos aspectos ecologicos y sistemas de comunicacion quimica en Acromyrmex landolti. Tesis de Licenciatura, Universidad Simon Bolivar, Caracas, Venezuela. 103 pp.

Pino, E. 1984. Comunicacion quimica entre obreras de Crematogaster sumichrasti. Tesis de Licenciatura, Universidad Simon Bolivar, Caracas, Venezuela. 98 pp.

Provost, E. 1979. Etude de la fermeture de la societe de fourmis chez diverses especes de Leptothorax et chez Camponotus lateralis. C. R. Acad. Sci. Paris 288: 429–432.

Puche, H. 1982. Estudio del comportamiento social de Solenopsis geminata. Tesis de Licenciatura, Universidad Simon Bolivar, Caracas, Venezuela. 83 pp.

Robinson, S. W., and J. M. Cherrett. 1974. Laboratory investigations to evaluate the possible use of brood pheromones of the leaf-cutting ant Atta cephalotes as a component in an attractive bait. Bull. Entomol. Res. 63: 519–529.

Skinner, G. J. 1980. Territory, trail structure and activity patterns in the wood ant Formica rufa in limestone woodland in Northwest England. J. Anim. Ecol. 49: 381–394.

Vilela, E. 1983. Behaviour and control of leaf-cutting ants. Ph.D. Thesis, Southampton University, England. 209 pp.

Villegas, G. 1982. Los sistemas de comunicacion quimica en la hormiga Trachymyrmex urichi y de otras Attini. Tesis de Licenciatura, Universidad Simon Bolivar, Caracas, Venezuela. 110 pp.

Walsh, J. P., and W. R. Tschinkel. 1974. Brood recognition by contact pheromone in the red imported fire ant Solenopsis invicta. Anim. Behav. 22: 695–704.

Watkins, J. F., and T. W. Cole. 1966. The attraction of army ant workers to secretions of their queens. Texas J. Sci. 18: 254–265.

Wilson, E. O., N. I. Durlach, and I. M. Roth. 1958. Chemical releasers of necrophoric behaviour in ants. Psyche 65: 108–114.

Wilson, N. L., J. H. Dillier, and G. P. Markin. 1971. Foraging territories of imported fire ants. Ann. Entomol. Soc. Am. 64: 660–665.

19
The Queen Recognition Pheromone of *Solenopsis invicta*

B. M. Glancey

Wheeler (1910) was the first to suggest that ant queens elicited unique behavior from their workers. Later, other researchers demonstrated the existence of queen pheromones in particular ant species. For example, Stumper (1956) reported that Lasius alienus and Pheidole pallidula queens produced attractant pheromones that originated from glands in their thorax or abdomen. The pheromones induced specific licking or grooming behaviors. Solvent extracts of these queens elicited queen-like responses from workers when placed on "surrogate queens" made of paper, sponge, wood, or even bodies of dead worker ants. Schneirla (1957, 1971) reported that army ant queens produced worker attractants that were important in maintaining colony cohesiveness. Watkins and Cole (1966), working with six species of Neivamyrmex, found that workers were attracted to portions of filter paper on which their queen had been confined. Brian (1973) described tests in which he found evidence for what he called a "queen recognition pheromone" in Myrmica rubra. The compound was non-volatile and was probably produced in the thorax or abdomen.

The first observations of a worker attractant produced by queens of Solenopsis invicta were made in 1971, when co-workers and I noticed that an open jar containing field queens was highly attractive to fire ant workers (Glancey 1980). Subsequent experiments showed that: (1) workers responded to air drawn over a mated queen confined in a syringe; (2) pentane extracts of mated queens applied to filter paper were attractive to workers; (3) virgin alates were not attractive; and (4) the queen abdomen was more attractive than the thorax or head. In other tests, S. invicta queen extracts painted on worker ants of S. richteri, S. geminata, Camponotus caryae discolor, and C. pennsylvanicus offered temporary protection from attack when placed in S. invicta colonies (Glancey 1980).

Jouvenaz et al. (1974) found that S. invicta and S. geminata workers aggregated within squares on absorbent paper in which their queens had been confined. S. invicta workers were more attracted to squares previously occupied by their mother queen than to areas occupied by a conspecific queen. In addition, one-way species specificity was noted, since S. invicta workers were attracted to areas previously occupied by their own or S. geminata queens, but S. geminata workers were attracted only to areas occupied by queens of their own species.

Over the past several years, we have concentrated on isolating and identifying the chemicals in S. invicta queens that mediate the behavioral responses of workers towards their queen. The following is a review of the status of this research.

LABORATORY TESTS

Over a period of several years, laboratory bioassays for the S. invicta "queen pheromone" evolved into those described briefly below and in detail by Lofgren et al. (1983). They referred to the pheromone as the "queen recognition pheromone" even though this terminology was not completely descriptive of all the associated behaviors. The same terminology will be used in this paper.

Olfactometer Bioassay

The olfactometer consisted of a Wilson cell (9 cm diam) with three of the four entrance/exit ports closed and the sides dusted with talc to prevent escape of the ants. Live queens or queen extracts were tested by placing them in a Pasteur pipet (extracts were absorbed onto a piece of filter paper). The narrow end of the pipet was inserted into a hollow plastic tube (ca 2 mm ID; 8 cm in length). The other end was rolled in cotton to form a tight swab. The swab end was inserted into the open port of the olfactometer and the space around the swab filled with clay. The volatiles were conveyed through the pipet and swab into the test arena with an air stream (0.5 L/min). The efficacy of the test materials was measured by introducing either a single worker or groups of 20 workers into the olfactometer. In the case of one worker, the total time that it spent at the swab over a 10-min period was determined. With 20 workers, we counted the number of ants attracted to the swab at 1-min intervals for 5 minutes. The sample results were compared to solvent controls in each set of experiments.

Surrogate Queen Bioassay

Surrogate queens were made of rubber septa cut to the size of a physogastric queen (8 by 3 mm; ca 20 mg); quartered lengthwise; extracted with methanol, methylene chloride, and hexane; and oven-dried at ca 80°C. Pretreatment of the septa was necessary because it increased its absorptive capacity and changed, for the better, its pheromone release properties.

The test arena was a Wilson cell with all the exit/entrance ports closed. The treated septum was placed in a 2-cm observation square in the center of the cell. Twenty worker ants with a small amount of brood were then placed in the cell. The data were quantified by counting the number of ants clustered in the square at 1-min intervals for 5 minutes. Responses to a control were determined for each series of tests.

Results

Our results demonstrated conclusively that S. invicta queens produce a worker attractant. Release of the attractant ceased when the queen was separated from her colony for more than 30 min; however, activity returned immediately after she again came in contact with her workers. Extracts of live queens were attractive in the olfactometer and gave a queen-like response when applied in the surrogate queen bioassay. The maximum response occurred at concentrations of 0.5 to 5.0 queen equivalents. The responses of workers to extracts of sexual immatures and adults were significantly less than to queen extracts (Lofgren et al. 1983).

WHOLE COLONY AND FIELD BIOASSAYS

Evaluation of test materials against queenright laboratory colonies and field colonies required a different protocol than that used in bioassays with groups of isolated workers. A three-sided tray (81 x 122 x 7.6 cm) with Fluon®-painted sides was used as a field observation arena (Glancey et al. 1983). In the summer season, it was necessary to shade the box to prevent heat kill of the ants. In each test, a shovelful of nest tumulus was scattered on the floor of the box. Treated septa were then placed on the soil and observations made of the responses of the ants. Assays were run with up to 8 septa at one time.

Workers from a colony disrupted as described above immediately collected and hid their brood beneath pieces of soil. If the queen was thrown out with the soil, the workers would immediately cluster about her and place brood near her. In about 10 min, the workers established several trails along which they carried the brood back to the nest and, if the queen was present, they also guided her back to the nest along one of the trails.

Similar trays were also used in laboratory assays after closing their open ends. A laboratory nest cell with brood, workers, and queen was placed at one end of the tray, and a live caged queen or surrogate queen was placed at the opposite end. Workers from the nest cell were then scattered over the tray floor and their behavior

·

towards the queen or test samples noted. Our observations of the responses of ants in both laboratory and field tests were categorized into five behaviors. These were used to evaluate the effectiveness of the natural queen extracts and synthetic compounds. These behaviors were as follows:

1. Intense initial attraction toward the queen;
2. Formation of a dense cluster of workers about the queen;
3. Transport of brood to the queen and touching of the queen with the brood and/or depositing the brood next to her;
4. Formation of trails to the nest, often with several of the trails coalescing into a very wide trail terminating at the nest;
5. Guidance of the queen along one of the trails (if the queen did not follow the trail by herself, the workers dragged her into the nest).

In 17 field trials, workers responded to the pheromone-treated surrogate queens by attraction and clustering in 100% of the trials, touching brood to the septum 100%, forming trails 95%, and returning the septa to the nest in 95% of the trials. The ants ignored septa that were treated with extracts of ant larvae or pupae or soybean oil. In a few cases the workers did pick up and carry septa treated with extracts of female alates and sexual brood. However, they did not carry them along an established trail (Glancey et al. 1983).

In the laboratory, the responses of attraction and clustering occurred in 100% of the trials, touching brood 91%, forming trails 94%, and returning the surrogate to the nest in 89% of the trials. Septa treated with extracts of major workers, male alates, sexual pupae, and solvent controls failed to produce any responses.

TESTING OF SYNTHETIC COMPONENTS OF THE PHEROMONE

Chemical fractionation of extracts of S. invicta queens coupled with the bioassays described above led to the isolation of three compounds (Fig. 1) that were identified through spectral data and tests of synthesized materials (Rocca et al. 1983a, b). Blends of compounds A + B and A + B + C were as active as extracts of live queens in eliciting responses from workers at four of five geographical locations (Florida, Georgia, Mississippi). At one site in Georgia, the two blends were equal in the response they elicited but were statistically less effective than the queen extract. At the Mississippi site, a combination of component C with A was also equal in attraction to the queen extract. None of the components were active by themselves. The role of component C is not clear at this time. The synthetic pheromone blends of A + B and A + B + C were most active between 5 and 10 ng per surrogate queen. In field bioassays, S. geminata, S. xyloni, and S. richteri did not respond to either the extracts of live queens or to the synthetic compounds (Glancey et al. 1984).

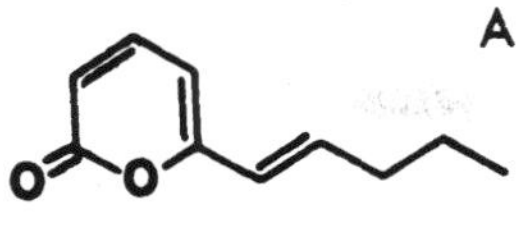

(E̲)-6-(1-pentenyl)-2H-pyran-2-one

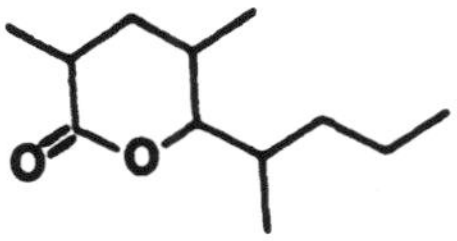

tetrahydro-3,5-dimethyl-6-(1-methylbutyl)2H-pyran-2-one

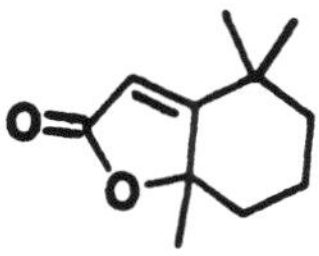

dihydroactinidiolide

FIGURE 1. Chemical structure of three of the components
of the red imported fire ant queen recognition pheromone as
determined by Rocca et al. 1983a, b.

DISCUSSION

The queen recognition pheromone of S̲. invicta is the first ant
pheromone system of its kind to be described. The pheromone was
reported by Vander Meer et al. (1980) to originate in the poison
sac. However, the three compounds isolated were obtained from
whole body extracts, thus we cannot at this time affirm that all
three originated in the poison sac or that other components might
not be discovered. Jones and Fales (1983) reported the isolation of
several compounds from the mandibular gland of the carpenter ant,
C̲. pennsylvanicus, one of which was identical to compound A from
S̲. invicta queens. They do not report any behavioral responses
controlled by this chemical.

Glancey et al. (1981) found that virgin dealated queens also
produce the pheromone as part of the biochemical and physiological
changes that occur after dealation. Maximum queen pheromone

production was determined by bioassays to occur 9 to 12 days after dealation.

The pheromone stored in the poison sac of the queen appears to remain active after her death. Glancey (unpublished results) observed a dead queen's abdomen that was maintained and tended for up to 8 months. Williams et al. (1981) reported that the gasters of queens killed with Amdro® were retained by surviving workers for up to 9 weeks. The head and thorax were never kept but discarded onto the refuse pile. Recently Glancey (unpublished results) has shown that once a colony has adopted a mated or a dealated virgin queen for its colony queen, the workers will tend the gaster of that queen after death if the poison sac remains inside the gaster. Removal of the poison sac results in rejection of the gaster. Also, in seven trials, colonies that were tending a gaster that contained a poison sac were offered a live mated physogastric queen. Only two of the colonies accepted the live queen.

Fletcher and co-workers (see Chapter 15) discovered a primer pheromone produced by the queen that prevents female alates from competing for resources with the queen. This primer pheromone inhibits dealation and oviposition of virgin queens and appears to be produced in the queen's abdomen. It is not known if the pheromone is the same as the recognition pheromone. In any event, the grooming and licking induced by the recognition pheromone dictates that it may be involved in the distribution of the primer pheromone (Vander Meer 1983).

Fowler and Roberts (1982) concluded that queens of C. pennsylvanicus release an "entourage" pheromone. They based their conclusion on worker attraction to marked squares on a paper on which a queen had been confined. This technique is similar to that used by Jouvenaz et al. (1973) with S. invicta. These investigators referred to chemicals released on the paper as "queen tending pheromones." They noted, however, that the worker response might be simply attraction rather than tending.

Another subject that needs investigation is the relationship between the queen's age and the quality and quantity of pheromones she produces. Brian (1980) found that the influence of M. rubra queens over their workers varied with queen age and season of the year. The queens communicated their presence to the workers by both chemical and topographical stimuli. The abdomen was postulated as the most likely source of the chemical stimulus.

Passera (1980) found that queens of Plagiolepis pygmaea produce an epicuticular pheromone that inhibits egg-laying by workers. He found that the pheromone could be removed by dipping the queen in acetone daily for 15 seconds.

A different case of inhibition exists in the weaver ants, Oecophylla longinoda and O. smaragdina (Holldobler and Wilson 1983). The queens apparently produce a material in a cephalic organ(s) which induces her attendant workers, at frequent intervals, to regurgitate ingested food and produce trophic eggs for her consumption. The workers are prevented from laying viable eggs, evidently as a result of pheromones produced by the queen. These pheromones persist in the corpses of dead queens for as long as 4 months. The origin of the queen pheromones may be the cephalic organs acting in conjunction with abdominal intersegmental glands.

In conclusion, it is obvious that the repertoire of pheromones produced by ant queens, and S. invicta queens in particular, has barely been scratched. The queen recognition pheromone and the inhibitory primer pheromone (Fletcher, Chapter 15) are the only queen pheromones described thus far for S. invicta. However, integration and regulation of the complex social behavior of the colony dictates the presence of other queen-produced pheromones. Perhaps the key to effective safe control techniques awaits the elucidation of one or more of these pheromones. It is to this end that part of our current research program involves the isolation and identification of other queen pheromones that control specific behaviors or physiological processes.

REFERENCES CITED

Brian, M. V. 1973. Queen recognition by brood-rearing workers of the ant, Myrmica rubra L. Anim. Behav. 21: 691-698.

Brian, M. V. 1980. Social control over sex and caste in bees, wasps and ants. Biol. Rev. 55: 379-415.

Fowler, H. G., and R. B. Roberts. 1982. Entourage pheromones in carpenter ants (Camponotus pennsylvanicus) (Hymenoptera: Formicidae) queens. J. Kans. Entomol. Soc. 55: 568-570.

Glancey, B. M. 1980. Biological studies of the queen pheromone of the red imported fire ant. Proc. Tall Timbers Conf. Ecol. Anim. Control Habitat Manage. 7: 149-154.

Glancey, B. M., A. Glover, and C. S. Lofgren. 1981. Pheromone production by virgin queens of Solenopsis invicta Buren. Sociobiology. 6: 119-127.

Glancey, B. M., C. S. Lofgren, J. R. Rocca, and J. H. Tumlinson. 1983. Behavior of disrupted colonies of Solenopsis invicta towards queens and pheromone-treated surrogate queens placed outside the nest. Sociobiology. 7: 283-288.

Glancey, B. M., J. Rocca, C. S. Lofgren, and J. H. Tumlinson. 1984. Field tests with synthetic components of the queen recognition pheromone of the red imported fire ant, Solenopsis invicta Buren. Sociobiology. 9: 19-30.

Holldobler, B., and E. O. Wilson. 1983. Queen control in colonies of weaver ants (Hymenoptera: Formicidae). Ann. Entomol. Soc. Am. 76: 235-238.

Jones, T. H., and H. M. Fales. 1983. E-6-(1-pentenyl)-2H-pyran-2-one from carpenter ants (Camponotus spp.). Tetrahedron Lett. 24: 5439-5440.

Jouvenaz, D. P., W. A. Banks, and C. S. Lofgren. 1974. Fire ants: Attraction of workers to queen secretions. Ann. Entomol. Soc. Am. 67: 442-444.

Lofgren, C. S., B. M. Glancey, A. Glover, J. Rocca, and J. Tumlinson. 1983. Behavior of workers of Solenopsis invicta (Hymenoptera: Formicidae) to the queen recognition pheromone: Laboratory studies with an olfactometer and surrogate queens. Ann. Entomol. Soc. Am. 76: 44-50.

Passera, L. 1980. La fonction inhibitrice des reines de la fourmi Plagiolepis pygmaea Latr.: Role des pheromones. Insect. Soc. 27: 212-225.

Rocca, J. R., J. H. Tumlinson, B. M. Glancey, and C. S. Lofgren. 1983a. The queen recognition pheromone of Solenopsis invicta, preparation of (E)-6-(1-pentenyl)-2H-pyran-2-one. Tetrahedron Lett. 24: 1889-1892.

Rocca, J. R., J. H. Tumlinson, B. M. Glancey, and C. S. Lofgren. 1983b. Synthesis and stereochemistry of tetrahydro-3,5-dimethyl-6-(1-methylbutyl)2H-pyran-2-one, a component of the queen recognition pheromone of Solenopsis invicta. Tetrahedron Lett. 24: 1893-1896.

Schneirla, T. C. 1957. Theoretical consideration of cyclic processes in doryline ants. Proc. Am. Philos. Soc. 101: 106-133.

Schneirla, T. C. 1971. Army ants. W. N. Freeman and Co. San Francisco, CA. 349 pp.

Stumper, R. 1956. Etudes myrmecologiques. LXXVII. Les secretions attractives des reines de fourmis. Mitt. Schweiz. Entomol. Ges. 29: 373-380.

Vander Meer, R. K. 1983. Semiochemicals and the red imported fire ant Solenopsis invicta Buren (Hymenoptera: Formicidae). Fla. Entomol. 66: 139-161.

Vander Meer, R. K., B. M. Glancey, C. S. Lofgren, A. Glover, J. Tumlinson, and J. Rocca. 1980. The poison sac of red imported fire ant queens: Source of a pheromone attractant. Ann. Entomol. Soc. Am. 73: 609-612.

Watkins, J. F. II, and T. W. Cole. 1966. The attraction of army ant workers to secretions of their queen. Texas Sci. 18: 254-265.

Wheeler, W. M. 1910. Ants: Their structure, development and behavior. Columbia Univ. Press, N.Y., NY. 663 pp.

Williams, D. F., C. S. Lofgren, and R. K. Vander Meer. 1981. Tending of dead queens by workers of Solenopsis invicta Buren (Hymenoptera: Formicidae). Fla. Entomol. 64: 545-547.

20
Leaf-Cutting Ant Symbiotic Fungi: A Synthesis of Recent Research

A. Kermarrec, M. Decharme, and G. Febvay

Attine ants comprise 13 genera and sub-genera, and about 200 species, all of which are characterized by the habit of culturing fungi.   In this synthesis, we will limit our study to <u>Atta</u> and <u>Acromyrmex</u>, the two most evolved genera (Weber 1982) which, by their defoliation of cultivated vegetation, are responsible for the damage identified by Cherrett (Chapter 2) and Fowler (Chapter 3). The fungi cultivated by these ants with much care and success has not been the subject of much research, and we believe that the paucity of publications describing these fungi is due to the lack of cooperation between mycologists and entomologists.

Many questions concerning the symbionts of <u>Atta</u> and <u>Acromyrmex</u> spp. have not been answered despite a number of biological and biochemical investigations during the last decade; therefore, we will review recent publications and personal communications with reference to the following basic questions: (1) What is the taxonomic position of the symbionts of higher attines? (2) What are the possible origins of the symbiosis? (3) What microbial flora are associated with the fungi?   (4) What is known about their morphology and cytology? (5) What is known about antibiosis and cooperative management of their environment? (6) What are the reasons for a symbiotic process and how similar is it to termite-fungus symbiosis?   (7) What are the possibilities of controlling the ants by eliminating their symbionts?   A short glossary of mycological terms is included at the end of the text.

TAXONOMIC POSITION AND ORIGIN OF THE SYMBIOTIC FUNGI

Little is known with certainty about the taxonomic position and origin of the symbionts of the two most evolved genera of leaf-cutting ants.   Weber (1938) str ssed that if only hyphae and gongylidia are considered, a specific identification is questionable. The formal proof of physical relationships between the mycelium of the comb and the sporophores (rarely seen) in ant nests cannot be established as easily as for termites (Heim 1977). Weber (1972) stated that the real goal of producing cultures of these symbionts on synthetic media is to produce the perfect sexual forms for the taxonomists to use in establishing definitive species. It was not until 1957 that the first experimentally-produced carpophores were obtained from the less evolved symbiont of the attine, Cyphomyrmex costatus Mann (Weber 1957a). It is remarkable that while the sporophore of this species has been classified in the Ascomycetes (Table 1), its organic relationship with the comb has never been formally established. This is not the case for the fungi grouped under Agaricaceae, but the classification of these closely-related genera is disputed by taxonomists: Rozites (Moeller 1893), Pholiota (Saccardo 1895 and Singer 1922 in Weber 1938), Agaricus (Singer 1951 in Heim 1957), Lentinus (Linder and Stevenson 1935 in Weber 1938), Lepiota (Smith and Locquin in Weber 1957a), and Leucocoprinus (Heim 1957).

Heim (1957) concludes (Table 1) from the respective descriptions of Moeller (1893) and Weber (1957a) the likely identity of these species even though the carpophores belong to the symbionts of very differently evolved attines. Heim suggested the name Leucocoprinus gongylophorus and he emphasized that, as with termites, it is unlikely that attines have domesticated several species of leucocoprines. On the basis of morphological and cytological descriptions of Decharme (1978, 1981), Decharme and Issaly (1980), and Angeli-Papa and Eyme (1979), it is reasonable to retain the name given by Heim (1957), at least for Atta and Acromyrmex.

Kreisel (1972) described the symbiont of Atta insularis, an endemic species of Cuba, as Attamyces bromificatus n.g., n.sp., pendant to the Afro-Asian Termitomyces. This new genus is classified among mycelia sterilia, Fungi Imperfecti (Deuteromycetes), because of the absence of a fructifying form and doubt as to its definitive generic position. This cautious opinion does not preclude its eventual classification in the Basidiomycetes. Whether or not it is identical (morphologically, biochemically, ...) with Heim's Leucocoprinus remains to be established.

Lehmann (1974, 1975, 1976, 1981) asserted that the symbiotic fungi of attines and of African termites belong to the group Aspergillus (Ascomycetes). This conclusion, surprising for its revolutionary implications for the biosystematics of several classes of fungi, merits evaluation by specialists.

Hervey et al. (1977) and Weber (1979) cautiously kept the genus Lepiota for the lower attines. The last author concluded that this genus is cultivated by four primitive attines (Weber 1957a, 1972) and that the choice between the genera Rozites, Attamyces, and Leucocoprinus, at least for advanced attines, remains a problem for expert mycologists. A large number of fungus cultures is maintained by the New York Botanical Garden (Weber 1979).

TABLE 1. Sporophores of the symbiotic fungi of Attine ants (partly from Weber 1979). If the relationship between vegetative form and sporophore has been proven, it is noted as a + after reference date.

Name	Ant	Type of Nest and Locality	Reference
I-BASIDIOMYCETES			
- Rozites gongylophora = Pholiota sp. = Agaricus sp. = Leucocoprinus gongylophorus	Atta discigera A. cephalotes	abandoned; Brasil abandoned; Surinam	Moller 1893+ Weber 1957a+ Weber 1938+ Heim 1957+
- Lepiota sp. = Leucoagaricus sp. = Leucocoprinus gongylophorus	Cyphomyrmex costatus Myrmicocrypta buenzlii M. auriculatum	live, in vitro; Panama	Weber 1957a+ Weber 1966+ Hervey et al. 1977+ Singer 1951+ (in Heim 1957) Heim 1957+
- Lentinus atticolus	A. cephalotes	abandoned; Guyana	Weber 1938
- Locellinia mazzuchi	A. vollenweideri	abandoned; Argentina	Weber 1938

234

TABLE 1 continued.

II-ASCOMYCETES

- Xylaria micrura - Bargillinia belti - Rhizomorpha formicarum (=Discoxylaria myrmecophila)	Acromyrmex lundi Ac. lundi	abandoned; Argentina abandoned; Argentina	Weber 1938 Weber 1966
- Poroniopsis bruchi - Hypocreodendron sanguineum	Ac. heyeri	abandoned; Argentina	Weber 1966
- Auricularia sp. - Xylaria sp. (=Daldinia sp.)	Apterostigma mayri C. rimosus	live, in vitro; Panama	Weber 1957a

The absence of fossil traces restricts us to speculations about the co-evolutionary dynamics of species which emanated from an Amazonian center and took place over some 60 to 80 million years. Wilson (1975) postulates that fungus-growing originated only once in a single ancestral attine living in this region during its long period of continental isolation from the late Mesozoic times to approximately 4 million years ago. Weber (1972) thought that the accumulation of organic waste by the ant societies remains the essential causative factor for the generation of the symbiotic process. Garling (1979) uses, with a certain intellectual audacity, the probabilistic opinions of Wheeler (1907), Weber (1958, 1972) and Hervey et al. (1977) to develop an elegant mycorrhizal hypothesis. She suggests that since most ectomycorrhyzal fungi are Basidiomycetes (Agaricales), like the ant fungi (including some Lepiota), the possible sequence of events leading to obligatory fungus-growing could be interpreted as an accidental encounter between ordinary mycorrhizae and ants in underground nests.

ASSOCIATED MICROBIAL FLORA

The microbiological content of attine nests appears much more complex than described previously. Craven et al. (1970), Scheld et al. (1971), Serzedello and Tauk (1974), Decharme (1981), and Papa and Papa (1982a, b) emphasize that the vegetative substrate of the symbiont harbors yeasts and bacteria. Bacteria seem to be less important than yeasts in the general symbiotic process. However, bacteria might play a role in the lysis of cellulose polyglycanes into sugars assimilable by the fungus. At least six species of Bacillus have been identified inside nests of A. laevigata, and the ants themselves seem to inoculate fresh vegetation with bacterial flora from their infrabuccal pocket during preparation of the substrate. However, a positive role for the bacteria is not recognized by all authors. Papa and Papa (1982a, b), for example, propose that the bacterial flora, although important, are in a state of equilibrium that is regulated by the attines through substrate acidification with glandular secretions. According to these authors, the preparation of substrate for the fungus does not necessitate the same microbial sequence necessary for the precomposting of the manure used to produce man's edible mushroom, Agaricus bisporus. Cellulolytic yeasts and bacteria must, therefore, be contemporaneous with colonization and growth of the symbiotic fungi.

Angeli-Papa and Eyme (1979) also related the inhibition of growth of hyphal apices in the fungus of Acromyrmex octospinosus to the presence of virus-like particles. This leads one to suspect that the virus acts as a supplementary partner in the equilibrium of the biosystem.

RECENT MORPHOLOGICAL AND CYTOLOGICAL OBSERVATIONS

Decharme (1981), in his studies of the symbionts of A. cephalotes, A. sexdens and Ac. octospinosus, confirmed the observations of several other researchers on the morphology and cytology of the symbionts (Moeller 1893; Weber 1972; Angeli-Papa and Eyme 1979). Within a staphyla, the gongylidia are formed by progressive swelling of the apices of the hyphae and, more rarely, of the subapical regions. The cytoplasm in these spherical heads, 45 um in diameter, is quiescent, whereas the merismatic cytoplasm of the club-shaped mycelia in the fungus garden is active and rich in ribonucleic acid. This leads one to assume that nutrients are produced in these structures. The basic energy reserve chemical is glycogen which is produced in massive quantities from the earliest stages of gongylidia formation. Thus, the polysaccarides concentrated within the staphylae (Quinlan and Cherrett 1978a, 1979) are in a highly assimilable form for the insect (Febvay and Kermarrec 1983). This observation is contrary to the conclusions of Martin et al. (1969a) who reported that only simple sugars were stored.

It appears also that the formation of gongylidia, both in the garden and in vitro, follows a seven-day rhythm which may be linked to the accumulation of a critical mass of fungal material. A similar rhythm (Kermarrec, unpublished) was shown by the foraging activity of several laboratory nests. Therefore, the dynamics of fungal growth in the ant colony may follow the law of supply and demand.

Differentiation in gonglylidia is accompanied by extensive architectural modifications of the cellular wall and content. The swelling is simultaneous with the formation and growth of a unique central vacuole supported by important membrane systems (plasmalemmasomes and endoplasmic reticulum sequestration aisles) described by Angeli-Papa and Eyme (1979). The islets of cytoplasm surrounded by these aisles undergo an autolytic breakdown caused by acid phosphatases. This initiates and leads to a continuous thinning of the peripheral cytoplasm and to an increase in the volume of the central vacuole. The contents of the vacuole are mixed with the thin peripheral cytoplasmic layer after destruction of the tonoplast. At this time, the gongylidia can be compared to a goat-skin bottle which has a thick wall covered with mucilage and is filled with a finely-granulated mictoplasm that maintains its turgidity (Eyme and Angeli-Papa 1978). The gongylidia then are tanks filled with glycogen, hydrolases, and viral particles.

Decharme (1981) showed that the walls of the gongylidia (as reported for A. bisporus by Michalenko et al. 1976) become scalloped during their development and may increase ten-fold in thickness. The presence of pectic compounds, cellulose, and chitin has been shown chemically. These walls, therefore, contribute substantially to the nutrition of the larvae (Febvay et al. 1984). Anastomose loops are common in Basidiomycetes but were not observed in these strains, although Weber (1972) did observe them in abundance in the symbiont of the minor attine, Apterostigma, and less frequently in other species. On the other hand, anastomoses between parallel hyphae are frequent and do not show any septum. Weber (1972) also noticed the presence of four chromocenters in all the strains studied, a finding that corresponds to a chromosome number of 2n=8 which is found in the genus Lepiota (Heim 1971). The ultrastructure of the septal apparatus of the symbiont of A. sexdens is, because of the presence of dolipores, characteristic of Hymenomycetes.

Scanning electron microscope observations and chemical analysis of the vegetative substrate show no correlation between its most active decomposition stage and the maximum development of the fungus. This high rate of degradation leads us to think that there are more complex microbial interactions in the superficial layers of the fungus garden.

ANTIBIOSIS AND COOPERATIVE MANAGEMENT OF THE MICROBIAL ENVIRONMENT: A COMPLEX REALITY

Garret (1956) identified the following five basic factors that contribute to the high degree of competitiveness of the mycelia in the nest environment: (1) a high degree of inoculation of the substrate with the fungus which helps it overcome plant resistance factors; (2) relatively rapid hyphal growth which allows extensive and accelerated colonization; (3) sufficient enzyme production so that there is rapid qualitative and quantitative production of nutrients; (4) production of antibiotics to reduce competition; and (5) a tolerance, or indifference, to antibiotics secreted by other microorganisms.

The behavioral and biochemical strategies developed by the ants are detailed elsewhere by Kermarrec et al. (Chapter 28) and Febvay and Kermarrec (Chapter 23). The inoculation of the substrate, which is scratched, chewed, and salivated (Weber 1956; Quinlan and Cherrett 1977, 1978b), has been determined from 61 samples to be 5.4 ± 1.6 S.D. fresh implants per mm^2. Colonization by the mycelium (Decharme 1980) occurs effectively only on the parenchyma of the leaf sections or on abraded leaf surfaces. Hyphae which are implanted superficially do not seem to be equipped to lyse the intact leaf cuticle. Treatment of the leaf fragments by the ants before they are taken into the garden only partially decontaminates their surfaces (almost none at all in pubescent areas). The continued survival of the symbiotic fungus, therefore, requires the active presence of the ants (Weber 1957b). Production of antibiotics by the fungus has been disputed (Suter 1954; Martin et al. 1969b), but we have clearly confirmed, at least qualitatively, that they are produced by the symbionts of Ac.

octospinosus, A. sexdens, and A. cephalotes (see Kermarrec et al. Chapter 28). These symbionts excrete, in vitro, substances antagonistic to bacteria and fungi. Very recently Angeli-Papa (1984) has confirmed that antibiosis occurs inside the nests. Hervey et al. (1977), Hervey and Nair (1979), and Nair and Hervey (1979) reported the presence of lepiochlorin in cultures of Lepiota cultivated by C. costatus. This new antibiotic compound was active against Staphylococcus aureus at a concentration of 60 ppm. Thus, the symbiont contributes to its own survival by supplementing the indispensable mechano-chemical work of the ants. Any shortages in antibiotic production would be compensated by extremely high implantation densities on the plant material harvested by the ants.

THE REASONS FOR THE SYMBIOSIS AND ITS PARALLELISM WITH TERMITES

Many authors have speculated about the co-evolutionary strategic and energetic reasons for the ant-fungus symbiosis; and at times, these have appeared colored by anthropomorphism. This review relies on the concept of "selective advantage" for these two phyletically distant organisms.

In this context, the 1980 meeting of the French section of the International Union for the Study of Social Insects included several key presentations. Noirot (1980) stated that the culture of a fungus maintained by macrotermitine termites confers the following selective advantages on the insects: (1) a nursery that provides a habitat and feeding area for the brood; (2) chemical control of the environment; (3) a food buffer against the risks of bad harvest; (4) optimal use of the cellulo-lignic deposit; and (5) a supply of qualitatively essential nutrients (enzymes of the symbiont and assimilable nitrogen). Noirot emphasized the great reduction in the amount of termite fecal excretions as proof of a nearly complete use of foraged material through the symbiosis. Martin and Martin (1978, 1979) showed that the cellulolytic activity produced by the symbiotic fungus is utilized by the termites Macrotermes natalensis for the digestion of ingested materials in their intestine. The origin of this could be an invasion of food reserves by a suitable fungus. Heim (1977) considered the fungus as an accidental in the homeostatic termite nest and, that from being merely a auxiliary, it became an intruder forcing the workers to continually move the primordia so that they only develop outside the nest. This mycologist also stressed that "the perfect fructifying of the Termitomyces shows the success of the fungus over the termite." In this "uncontested divorce" pattern, "the relationship between the two partners is one of equilibrium where they make use of each other's activities and where the power of development of the fungus exerts antagonism without triumphing."

Cherrett (1980) did not believe that cellulose degradation was the only purpose for the domestication of a fungus by an ant. Stating that ants receive only 5% of their energy needs from the symbiont and that plant sap remains the essential energy source, the author suggested some selective advantages which could reasonably explain or justify the symbiosis. As with termites, the symbiont brings essential biochemical elements (sterols, assimilable nitrogen) to the larvae; however, the ecological success of the attines is attributable to the high degree of polyphagy achieved by their synergistic relationship with the fungus. This fundamental "ecological liberation" is made possible by circumventing mechanisms evolved by plants to protect them against herbivorous animals. In other words, each partner brings its "savoir-faire" to the situation. For the ant, this means careful mechanical and chemical preparation and dense implantation of the fungal substrate, as well as clipping and protecting it from biohazards. For the fungus, it means by-passing vegetative restraints (indigestible polymers, repellents, anti-feedants, toxicants), providing essential nutrients for the ants by way of the gongylidia, and generating homeostatic conditions within the subterranean nest.

Brian (1978) emphasized this last point by stressing that, within their temperature, humidity, and gas-regulated nests, the processes of digestion, assimilation, growth, and reproduction proceed as well as they do in large homeothermic vertebrates. With reference to Emerson (1956) and Prigogine (in Schoffeniels 1976), we propose another causal approach for this symbiosis. An attine society obviously is an open dissipative type structure (Fig. 1) in which energy flow passing through the biological system allows for complexity and hierarchal organization. This implies a maximal accumulation of energy in the system and may find its biological expression in the complex development of attine nests where the entering vegetative biomass represents a colossal quantity of energy. The symbiosis is supposed to allow the system to support instabilities. The nests of fungus-growing ants could, therefore, be described as naturally complex systems in which the preservation of the hierarchal internal organization requires much energy.

CONTROL OF ATTINE SOCIETIES BY CONTROLLING THEIR SYMBIONT: A VAIN HOPE?

The aim of our research is the development and improvement of rational plant protection strategies through a better understanding of the colony-symbiont biological processes. The control of attine ants traditionally rests on the use of insecticides. However, knowing that an ant society cannot exist without the fungus, it seems logical to try fungicidal techniques. In particular, it needs to be determined if fungicidal control can be attained under the technical terms already defined for insecticidal baits by Cherrett et al. (1973). One may suppose that the Basidiomycete has become relatively weakened through evolution because of its symbiotic relationship and that antagonists, competitors, parasites, and mycophagous enemies might be plentiful. In the case of man's edible mushroom cultures, this is well documented and could be used as a model (Angeli-Papa and Delmas 1981).

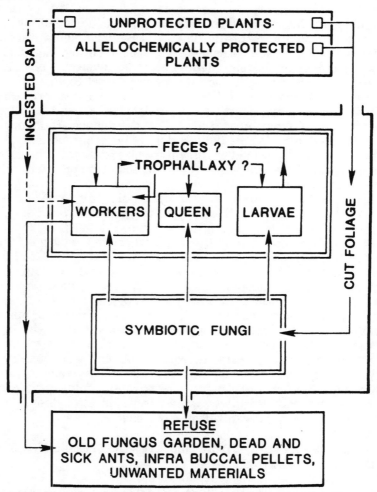

FIGURE 1. Box model of energetic material fluxes through the open ant-basidiomycetes system.

Preliminary trials with biological control have been largely negative. For example, it is very difficult to experimentally inoculate the edible mushroom, A. bisporus, with viruses; and even if it was easy, it would be necessary to find the most efficient virus compatible with the one already in place (Angeli-Papa and Eyme 1979). The most pathogenic antagonists and parasites of A. bisporus, selected for their high thermal optima and in vitro aggressiveness toward the ant's Leucocoprinus, were imported into Guadeloupe. Trials with several strains of Dactylium dendroides, Ostracoderma sp., Penicillium sp., Trichoderma viride, and Verticillium malthousei were generally positive in vitro, but all failed in functioning laboratory nests. The same results were observed with true mycophagus nematodes: Ditylenchus myceliophagus, Aphelenchus, and Aphelenchoides spp. The principal reasons for these failures are discussed by Kermarrec et al. in Chapter 28.

One possibility that remains to be explored is the use of natural fungicidal substances from plants (Lampard 1974). Atta nests have been killed by inducing ants to forage on Canavalia ensiformis (edible bean from Colombia) (Mullenax 1979). This area of research of which some aspects have been stressed by Febvay et al. (Chapter 21) can lead to two new strategies: (1) the use of these plants in baits, assuming that the ants integrate the material sufficiently into their nests in spite of normal dilution due to their diversified foraging and (2) the selection for phytochemical resistance in the most susceptible cultivated plants which should involve both partners of the symbiosis and should include repellents and feeding deterrents.

Quinlan and Cherrett (1979) stated that there was no evidence that the ants are able to detect whether or not the fungus can grow on different substrates. As worker ants obtain only a small proportion of their energy requirements from the fungus, fungal growth may not be a primary concern for the foraging workers in selecting substrate. Thus, substrate selection may not be reflected in their foraging behavior.

There remains for consideration chemical fungicides of which we have tried triforine, oxycarboxine, PCNB, benomyl, and F-Penwalt (undefined active ingredient). All of these fungicides, despite good results in vitro, are sensed by the ants with their antennae (see also Kermarrec et al., Chapter 28, and Delabie et al., Chapter 25). The fungicidal baits that are rejected by the ants are taken to their rubbish heap, and some of the ant colonies exposed to these baits moved their nests overnight. This fleeing behavior (Fowler 1981), which is a classical response in opportunistic species, cannot be ignored if one is to avoid errors in determining nest death rates in field trials.

We are convinced that present-day knowledge and the battery of fungicides available do not, as yet, permit control of attine societies based on the destruction of their fungus. Moreover, if we take into account the cost of labor associated with spot (nest-by-nest) treatments with fungicides, it becomes evident that this approach is impractical.

SOME CONCLUSIONS AND PROPOSALS FOR FURTHER RESEARCH

The taxonomic position of the symbionts could be improved if modern techniques of genetics were applied to mycology. Biochemical taxonomy (immunology, isoenzymes) can, as a first approach, be used if interpreted with caution. Insect-plant relationships are complex, and studies of their sensory and nutritional aspects might shed new light on the problem of plant protection by looking at the symbiont.

Chemical control cannot be rejected, but it must be kept in mind that nothing can be accomplished rationally without the help of the worker ants since they reduce the cost of labor by collecting and distributing the bait (see Therrien et al., Chapter 14). Once inside the nest, the workers are the only means of transport of the toxicants.

Finally, as far as biological control is concerned, the possible successful use of viruses presupposes that there are sufficient functional anastomoses to achieve spread of the particles through the fungus garden. Is it so?

GLOSSARY OF MYCOLOGICAL TERMS (Weber 1972 and Decharme 1981)

Anastomose:	functional graft between hyphae.
Bromatium:	a morel- or cheese-like mass of yeast cells cultured by Cyphomyrmex rimosus. Sometimes misused for staphylae of other ants.
Carpophore (=Sporophore):	fructifying form (or perfect, or sexual form) of the fungus.
Dolipore:	a functional hole in a mycelial septa (Basidiomycetes).
Gongylidium:	a swelling at the apex (sometimes lower) of the hypha.
Hypha:	filament of fungus.
Mycelium:	a mass of interconnected hyphae.
Staphyla:	a cluster of gongylidia.

ACKNOWLEDGMENT

This work is dedicated to the late A. Hervey.

RESUME

Cette synthese des travaux sur les champignons symbiotiques des Fourmis Attines superieures (Atta et Acromyrmex) apporte des informations sur leur origine, leur position taxonomique, la flore microbienne associee, leurs morphologie et cytologie, leurs secretions antibiotiques, la finalite de la symbiose ainsi que sur les possibilites de controler les Attines au travers de leur ecto-symbiote.

REFERENCES CITED

Angeli-Papa, J. 1984. La culture d'un champignon par les fourmis attines; mise en evidence de pheromones d'antibioses dans le nid. Cryptogam. Mycol. 5: 147-154.

Angeli-Papa, J., and J. Delmas. 1981. Technique de culture de la fourmi champignonniste: Comparaison avec celles pratiquees par l'homme, pp. 101-110. Proc. 11th Int. Sci. Cong. Cultivation of Edible Fungi. Sydney, Australia.

Angeli-Papa, J., and J. Eyme. 1979. Le champignon cultive par la "Fourmi-manioc" Acromyrmex octospinosus Reich en Guadeloupe: Resultats preliminaires sur le mycelium en culture pure et sur l'infrastructure des hyphes. C. R. Acad. Sci. Paris. 289: 21-24.

Brian, M. V. 1978. Production ecology of ants and termites. M. V. Brian (ed.). Cambridge Univ. Press. Cambridge, England. 409 pp.

Cherrett, J. M., D. J. Peregrine, P. Etheridge, A. Mudd, and F. T. Phililips. 1973. Some aspects of the development of toxic baits for the control of leaf-cutting ants, pp. 69-73. Proc. VII Congr. I.U.S.S.I. London, England.

Cherrett, J. M. 1980. Possible reasons for the mutualism between leaf-cutting ants (Hymenoptera: Formicidae) and their fungus. Biol. Ecol. Mediter. VII. 3: 113-122.

Craven, S. E., W. Dix, and G. E. Michaels. 1970. Attine fungus garden contain yeasts. Science 169: 184-186.

Decharme, M. 1978. Contribution a la connaissance de Leucocoprinus gongylophorus Moeller. Heim. champignon symbiote de la fourmi Acromyrmex octospinosus Reich. Dea de cytologie et morphogenese vegetales. Univ. Paris. 68 pp., 12 planches.

Decharme, M. 1980. Observation en microscopie electronique a balayage de la colonisation du substrat de la meule par le champignon symbiote de quelques fourmis attines. Biol. Ecol. Mediter. VII. 3: 133-136.

Decharme, M. 1981. Les champignons symbiotes de quelques fourmis attines: Differents aspects de leur cytologie et de leur evolution dans la meule. C. R. Acad. Agric. pp. 934-937.

Decharme, M., and M. Issaly. 1980. Contribution a l'etude du champignon symbiote de quelques fourmis de la tribu Attini. Cryptogam. Mycol. I: 1-18.

244

Emerson, A. E. 1956. Regenerative behaviour and social homeostasis termites. Ecology 37: 248-258.

Eyme, J., and J. Angeli-Papa. 1978. Evolution des hyphes et vacuolisation chez Agaricus bisporus et A. sylvicola. Mushrooms Science X. (Part I). Proc. 10th Int. Cong. Sci. Cultivation of Edible Fungi. France.

Febvay, G., and A. Kermarrec. 1983. Enzymes digestives de la fourmi attine Acromyrmex octospinosus Reich: Caracterisation des amylases, maltase et trehalase des glandes labiales et de l'intestin moyen. C. R. Acad. Sci. Paris. 296: 453-456.

Febvay, G., M. Decharme, and A. Kermarrec. 1984. Digestion of chitin by the labial glands of Acromyrmex octospinosus Reich (Hymenoptera: Formicidae). Can. J. Zool. 62: 229-234.

Fowler, H. G. 1981. Notes on the emigration of leaf-cutting ant colonies. Biotropica 13: 316.

Garling, L. 1979. Origin of ant-fungus mutualism: A new hypothesis. Biotropica 11: 284-291.

Garrett. S. D. 1956. Biology of root infecting fungi. Cambridge Univ. Press, Cambridge, England.

Heim, R. 1957. A propos du Rozites gongylophora Moeller. Rev. Mycol. 22: 293-299.

Heim, R. 1971. Le noyau dans la vie du champignon. Ann. Sci. Nat. Biol. Veg. 12: 465-514.

Heim, R. 1977. Termites et champignons. Boubee (ed.). Paris. 207 pp.

Hervey, A., and M. S. Nair. 1979. Antibiotic metabolite of a fungus cultivated by gardening ants. Mycologia 71: 1064-1066.

Hervey, A., C. T. Rogerson, and I. Leong. 1977. Studies on fungi cultivated by ants. Brittonia 29: 226-236.

Kreisel, H. 1972. Pilze aus Pilzgarten von Atta insularis in Kub. Allg. Mikrobiol. 12: 643-654.

Lampard, J. 1974. Demethylhomopterocarpin: An antifungal compound in Canavalia ensiformis and Vigna unguiculata following infection. Phytochemistry 13: 291-292.

Lehmann, J. 1974. Ist der Nahrungspilz der Pilzzuchtenden Blattschneiderameisen und Termiten ein Aspergillus? Waldhygiene 10: 252-255.

Lehmann, J. 1975. Ansatz zu einer allgemeinen Losung des "Ambrosiapilz" Problems. Waldhygiene 11: 41-47.

Lehmann, J. 1976. Neue Erkenstrusse uber die Nahrungspilze von Attini (Myrmicina: Hymenoptera) und Macrotermitinae (Isoptera). Waldhygiene 11: 133-152.

Lehmann, J. 1981. Die Frage nach der Eindeutigen Stellung der Ektosymbiosepilze von Macrotermitinae (Isoptera, Hentige, Termitidae) und Attini (Hymenoptera, Myrmicinae) im mykologischen System erweist sich als irrefurhrend und undeantwortbar. Alternative Mykologie 1: 1-35.

Martin, M. M., and J. S. Martin. 1978. Cellulose digestion in the midgut of the fungus-growing termite, Macrotermes natalensis: The role of acquired digestive enzymes. Science 199: 1453-1455.

Martin, M. M., and J. S. Martin. 1979. The distribution and origins of the cellulolytic enzymes of the higher termites, Macrotermes natalensis. Physiol. Zool. 52: 11-21.

Martin, M. M., R. M. Carman, and J. G. MacConnell. 1969a. Nutrients derived from the fungus cultured by the fungus-growing ant Atta colombica tonsipes. Ann. Entomol. Soc. Am. 62: 11-13.

Martin, M. M., J. G. MacConnell, and G. R. Gale. 1969b. The chemical basis for the attine ant-fungus symbiosis. Absence of antibiotics. Ann. Entomol. Soc. Am. 62: 386-388.

Michalenko, G. O., H. R. Hohl, and D. Rast. 1976. Chemistry and architecture of the mycelia wall of Agaricus bisporus. J. Gen. Microbiol. 92: 251-262.

Moeller, A. 1893. Die Pilzgarten einiger sudamericanischer Ameisen. Bot. Mitte Tropen 6: 1-127.

Mullenax, C. H. 1979. The use of Jackbean (Canavalia ensiformis) as a biological control for leaf-cutting ants (Atta spp). Biotropica 11: 313-314.

Nair, M. S., and A. Hervey. 1979. Structure of lepiochlorin, an antibiotic metabolite of a fungus cultivated by ants. Phytochemistry 18: 325-326.

Noirot, C. 1980. Donnees recentes sur la symbiose chez les termites champignonnistes (Macrotermitinae). Biol. Ecol. Mediter. 7: 123-128.

Papa, F., and J. Papa. 1982a. Etude de l'activite microbiologique dans les nids d'Acromyrmex octospinosus Reich en Guadeloupe. Bull. Soc. Path. Ex. 75: 404-414.

Papa, J., and F. Papa. 1982b. Inhibition des bacteries dans les nids d'Acromyrmex octospinosus Reich. Bull. Soc. Path. Ex. 75: 415-425.

Quinlan, R. J., and J. M. Cherrett. 1977. The role of substrate preparation in the symbiosis between the leaf-cutting ant Acromyrmex octospinosus Reich and its food fungus. Ecol. Entomol. 2: 161-170.

Quinlan, R. J., and J. M. Cherrett. 1978a. Aspects of the symbiosis of the leaf-cutting ant Acromyrmex octospinosus Reich and its food fungus. Ecol. Entomol. 3: 221-230.

Quinlan, R. J., and J. M. Cherrett. 1978b. Studies on the role of the infrabuccal pocket of the leaf-cutting ant Acromyrmex octospinosus Reich (Hymenoptera: Formicidae). Insect. Soc. 25: 237-245.

Quinlan, R. J., and J. M. Cherrett. 1979. The role of fungus in the diet of the leaf-cutting ant Atta cephalotes L. Ecol. Entomol. 4: 151-160.

Sheld, H. W., M. W. Dix, G. E. Michaels, and S. E. Craven. 1971. Bacteria distribution of attine ants and their gardens. Bact. Proc. 71: 47 (abstract).

Schoffeniels, E. 1976. Anti-chance. Pergamon Press. 123 pp.

Serzedello, A., and S. M. Tauk. 1974. Celulase de bacterias isoladas de "ninhos" de Atta laevigata Smith. Ciencia e Cultura. 10: 957-960.

Suter, V. 1954. The possibility of antibiotic control of the microflora of the bachac nest. Trans. 5th Int. Congr. Soil. Sci. 5: 113-118.

Weber, N. A. 1938. The biology of the fungus growing ants III. The sporophore of the fungus grown by Atta cephalotes and a review of reported sporophores. Rev. Entomologia 8: 265-272.

Weber, N. A. 1956. Treatment of substrate by fungus growing ants. Anat. Rev. 125: 604-605.

Weber, N. A. 1957a. Fungus growing ants and their fungi: Cyphomyrmex costatus. Ecology 38: 480-494.

Weber, N. A. 1957b. Weeding as a factor in fungus culture by ants. Anat. Rev. 28: 638.

Weber, N. A. 1958. Evolution in fungus growing ants. Proc. 10th Int. Congr. Entomol. Montreal, Canada. 1956. 2: 459-475.

Weber, N. A. 1966. The fungus growing ants. Science 153: 587-604.

Weber, N. A. 1972. Gardening ants, the attines. Mem. Am. Phil. Soc. Vol. 92, 146 pp.

Weber, N. A. 1979. Fungus culturing by ants, pp. 77-116. In L. R. Batra (ed.), Insect fungus symbiosis, mutualism and commensalism. Allanheld, Osmun and Co., Halsted Press.

Weber, N. A. 1982. Fungus ants, pp. 255-363. In H. R. Hermann (ed.), Social insects, Vol. 4. Academic Press, N.Y., NY.

Wheeler, W. M. 1907. The fungus-growing ants of North America. Bull. Am. Mus. Nat. Hist. 23: 669-807.

Wilson, E. O. 1975. Sociobiology: The new synthesis. Belknap Press of Harvard University Press. Cambridge, MA. 697 pp.

21
Prevention of Feeding
by *Acromyrmex octospinosus*
with Antifeedants from Yams

G. Febvay and A. Kermarrec

Fungus-growing ants (Attini) of the genera Atta and Acromyrmex are severe pests of natural and cultivated plants in the Neo-tropics (see Cherrett, Chapter 2 and Fowler et al., Chapter 3). These ants culture symbiotic fungi (Basidiomycetes) on fresh vegetable substrate made of flowers and leaves harvested close to their nest. According to Weber (1972) the ants feed specifically on their fungus and can be considered "monophagous." On the other hand, Littledyke and Cherrett (1976) have shown that worker ants ingest sap from the leaves they collect. Therefore, from an ecological point of view, these ants should be classified as "polyphagous" (Cherrett 1972a) since they feed on a wide range of plants (Weber 1966; Cherrett 1968). It has also been shown that the attines exhibit preferences for certain plant species (Amante 1967; Cherrett 1968; Rockwood 1976, 1977; Fowler and Robinson 1977; Pollard et al. 1983; Pintera 1983). Within a group of plants, some species appear immune to attack while other closely related species are particularly susceptible.

Complex interactions between phytophagous insects and plants are the result of a long and continuous evolutionary process which involved the development of mechanisms of defense by the plants and adaptations of the ants to these mechanisms (Beck 1965; Levin 1976). Plant defense mechanisms can involve: (a) protective anatomical structures and textures such as hairs and thorns; (b) osmotic pressure and pH of the sap that is unsuited to the insects' requirements; (c) lack of nutrients necessary for normal insect development; and (d) production of secondary metabolic components that prevent proper growth, interfere with primary metabolism, or repel the ants. The secondary chemicals that affect interspecific relationships (allelochemicals) are classified as allomones if they provide an adaptive advantage for the producer. On the other hand, the same plant substance is classified as a kairomone if the adaptive advantage is for the receiver (Pesson 1980).

Specific terminology has been defined for the allelochemicals that affect the feeding behavior of insects (Dethier et al. 1960; Munakata 1975). Two main classes of behavioral effects need to be considered: (a) recognition and location of the host plant and (b) initiation and continuation of feeding. In the first class, we need to disassociate attractants (orientation of the insect toward the food source) from repellents (orientation of the insect away from the food). These chemicals are volatile and induce responses from a distance. The second class of allelochemicals (control of feeding) functions by contact, either as feeding stimulants or deterrents.

The relationships of the attine ants with plants is complicated by the fact that they culture symbiotic fungi on the plant material they collect. According to Cherrett (1980), this ant-fungus mutualism allows polyphagy to occur on a wider scale, since each partner is able to destroy the plant defenses specifically directly against the other one. Thus, Littledyke and Cherrett (1975) showed that direct feeding on plant sap and foraging for fungus substrate are modulated by the concentrations of arrestants and/or inhibitors produced by the plants.

However, some plants are resistant to attack by attines due to physical plant defense mechanisms. For example, Waller (1982a) showed that mature leaves from Berberis trifoliata are not unpalatable to Atta texana, but they are simply too tough to cut. Intraspecific preferences for plants can also be related to differences in leaf toughness (Waller 1982b). Stradling (1978) showed that latex is a physico-chemical barrier that prevents foraging on lactiferous plant species. The amount of water or nutrients readily accessible in the leaf can be an important parameter in plant choice by the ants. Rockwood (1976) suggested that selective foraging by Atta workers in nature tends to minimize the detrimental effects of plant secondary compounds on the fungus and simultaneously maximizes the nutrient content. Bowers and Porter (1981) suggest that attines select leaves with a high water content. Finally, many non-related studies show that chemicals can affect the choice of worker ants, either by (1) enhancement as a result of attractants or stimulants in the plants harvested or (2) deterence because of the presence of repellents or deterrents (Cherrett and Seaforth 1970; Barrer and Cherrett 1972; Cherrett 1972b; Mudd et al. 1978). Littledyke and Cherrett (1978) demonstrated that under laboratory conditions old leaf extracts from several plant species inhibit foraging activity by A. cephalotes and Acromyrmex octospinosus. The inhibiting substances are concentrated in the lipid portion of the extract. Recently, the chemical nature of repellents for some attines has been described. Hubbell and Wiemer (1983) showed that Atta do not feed on the leaves of the tropical tree Hymenaea courbaril because of the presence in the leaves of a volatile repellent, caryophyllene epoxide. This compound has been shown to be a fungicide. Chen et al. (1984) found that Astronium graveolens produces trans-B-ocimene which is repellent to A. cephalotes.

In Guadeloupe, foraging by Ac. octospinosus is especially harmful to traditional polyagricuture. Yams (Dioscoreacea) are an important local crop and are very susceptible to attack. However, among the four cultivated species (Dioscorea bulbifera, D. alata, D. trifida, and D. cayenensis, subspecies cayenensis and rotundata), the yellow yam (D. c. cayenensis) is clearly resistant to defoliation.

This paper presents research results on the causes of yellow yam resistance to Ac. octospinosus.

MATERIALS AND METHODS

Biological Procedures

All experiments were carried out on colonies collected on the island of Grande-Terre (the eastern part of Guadeloupe) and maintained in the laboratory at 20 to 30°C and 70 to 90% RH. Every nest had a fertile queen and a fungus garden with an approximate volume of 0.5 to 1 liter. Outside the periods of experimentation, these nests were provided daily with a variety of fresh, palatable vegetable material.

The following five species of Dioscorea were studied in cultivated open fields free from any pesticide treatments: D. c. cayenensis (yellow yam), D. c. rotundata (grosse-caille), D. alata (white yam), D. trifida (cousse-couche), and D. bulbifera (adon). Two types of bioassays were performed: (1) the feeding preference of the ants was tested in the laboratory by exposing two samples of whole yam leaves to three ant nests. In each case, one sample of leaves was from adon (the control species) and the second sample of leaves was from one of the other species. Each exposure period lasted four hours, and three or four replicates were run with each species. At the end of any exposure period, the mass of leaves foraged was estimated by differential weighing of each sample. This measurement was corrected for evapotranspiration by monitoring the weight loss of a controlled sample. At the end of this first set of experiments, the three nests were supplied ad libitum with yellow yam leaves. (2) Leaf disks or filter paper disks treated with leaf extracts were placed on a test grid drawn on the foraging platform of the ant nests. This grid (5 x 5 latin square) allowed simultaneous placement of up to five different sets of discs. As soon as a disc was picked up and carried into the nest, it was replaced by a new disc of the same set. The number of discs of each series taken by the ants was recorded over a period of 20 minutes. If 50 discs were taken before the end of the 20-min period, the experiment was stopped, the time noted, and the results extrapolated for 20 min. A complete set of tests included two to four replicates with five different ant nests. In order to correct for variations in foraging activity (changing from one nest to another or according to the day the experiments were undertaken), the percentage of pick-up in each series was related to the total number of discs taken. After angular transformation (arcsine square root: normalization of data distribution), the results were tested with a two-way analysis of variance and a S.N.K. test.

Chemical Procedures

Two yam species (D. bulbifera and D. c. cayenensis) were used for chemical extractions. Leaves were extracted using Soxhlet extractors with solvents of increasing polarity: hexane, dichloromethane, methanol, and, finally, boiling water. The solvent from each extract was evaporated under vacuum to equivalent concentrations based on the grams of leaves. These solutions were used for bioassays with filter paper discs.

Initially, extracts of the two yam species and solvent controls were tested. Then, the extracts of D. c. cayenensis were compared to an arrestant substance (sweet orange solution Tang R at 50 g/L). In this case, two sets of control discs were used: one set was dipped in solvent alone while the other was treated with the solvent and then sprayed with the arrestant solution.

Chemical Analysis and Quantitative Determination of Sapogenins

The methanolic extracts from the two species of yams were subjected to thin layer chromatographic separation on silica gel (Merck ref. 10401) and eluted with dichloromethane and methanol (85/15). Various chemical groups were visualized using the following reagents: iodine, Dragendorf's reagent (alkaloids and other nitrogenous compounds), vanillin reagent (phenols and steroids), and antimony trichloride in concentrated hydrochloric acid (saponins and sapogenins; Brain and Hardman 1968).

A comparison of the total saponin and sapogenin content was made between the five yam species. Hydrolysis of saponins and isolation of sapogenins was performed according to Morris et al. (1958). Extracts containing sapogenins were separated by thin-layer chromatography and developed with methylene chloride (20%) in benzene. The sapogenins were detected as described by Brain and Hardman (1968). The sapogenins of the five extracts were quantified according to Sofowora and Hardman (1974) with diosgenin as a standard.

RESULTS

The relative weight of leaf fragments from five yam species collected by the ants (first bioassay) are given in Fig. 1. The ants showed similar preferences both in the laboratory and in nature in that they eagerly harvested <u>D</u>. <u>bulbifera</u> and harvested almost no <u>D</u>. <u>c</u>. <u>cayenensis</u>. In fact colonies inevitably die if they are provided only with leaves of <u>D</u>. <u>c</u>. <u>cayenensis</u>. For example, two fungus gardens of an average volume of 500 ml were eliminated within 9 to 12 days. A third nest of greater volume (between 1 to 1.5 l) took 27 days to decay. The disappearance of the fungus garden seemed to be directly related to cessation of foraging activity. The study of the behavior of the workers during this experiment showed that the leaves of <u>D</u>. <u>c</u>. <u>cayenensis</u> are free from any type of repellent; i.e., a substance that prevented the insect from harvesting the leaves. Indeed, the large workers in charge of harvesting the leaves were observed in large numbers on the yellow yam leaves. However, the few ants that started to cut the leaves usually did not finish their work. Of those pieces of leaves carried to the nest, most were dropped inside the nest and very few were used as substrate for the fungus culture. This behavior is the characteristic response of ants to a deterrent (Vigneron 1978). These substances work by contact and inhibit feeding without causing repulsion or death of the insect.

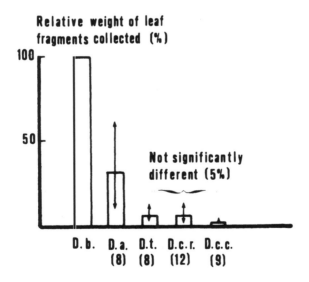

FIGURE 1. Relative weight of the leaf fragments collected by <u>Ac</u>. <u>octospinosus</u> from five yam species. Relative weight expressed in percent of <u>D</u>. <u>bulbifera</u>. Vertical bars represent the confidence interval (5%). Numbers of replicates in parentheses. Statistical analysis on angular transformed data. D.b. = <u>D</u>. <u>bulbifera</u>; D.a. = <u>D</u>. <u>alata</u>; D.t. = <u>D</u>. <u>trifida</u>; D.c.r. = <u>D</u>. <u>c</u>. <u>rotundata</u>; D.c.c. = <u>D</u>. <u>c</u>. <u>cayenensis</u>.

We considered that physical characteristics of the leaves might make cutting them into pieces too difficult for the workers. This hypothesis was tested in the leaf disc bioassay in which workers do not have to cut leaves. The ants from five nests were given a choice of leaf discs from each of the five yam species (four replicates). The results (percentage of pick-up for each series after angular transformation) showed that the nest factor was not significantly different and allowed the means calculated for each yam species to be compared (Table 1). With a 5% threshold, the S.N.K. test divided the five species into four significantly different groups as follows: D. bulbifera, D. alata, D. trifida, and a group that included the two subspecies of D. cayenensis. These results confirm that the ants' choice was not related to the physical characteristics of the yam leaves.

TABLE 1. Preference of Ac. octospinosus for five species of yams.

Yam species and percent leaf discs picked up (C.I.)[a]				
D. bulbifera	D. alata	D. trifida	D. c. rotundata	D. c. cayenensis
58.3a	26.1b	7.6c	3.6d	2.3d
(52.6–63.9)	(22.6–29.7)	(5.2–10.5)	(2.1–5.2)	(1.5–3.3)

[a]Means with the same letter are not significantly different (S.N.K. on angular transformed data, $P < 0.05$). n=20; CI=5% confidence interval.

The search for deterrent compounds in the leaves of D. cayenensis was undertaken with successive extractions with solvents of increasing polarity. This was done in parallel with the most harvested species (D. bulbifera) and the least foraged species (D. c. cayenensis). Each extract was subjected to the leaf disc behavioral test. Analysis of transformed data indicated that the nest was not a significant factor. The pick-up of extract-treated discs from the two yam species and the control was compared (Table 2). With all three organic solvents (hexane, dichloromethane, and methanol), the number of paper discs collected that had been treated with extracts of D. bulbifera was significantly greater than the number treated with extracts of D. c. cayenensis. When compared to the controls, the ants collected discs treated with dichloromethane and methanol extracts of D. c. cayenensis less frequently than discs treated with hexane extracts. Harvest of discs with water extracts of the two yam species was significantly inferior to the control. This experiment showed the existence of deterrents in the organic extracts of D. c. cayenensis or suggested the presence of feeding stimulants in those of D. bulbifera.

TABLE 2. Preference of Ac. octospinosus for extracts of two yam species obtained with four different solvents.

Solvent used to extract leaves	Pick-up (%) of filter paper discs treated with extracts of yam species[a]		
	Controls	D. c. cayenensis	D. bulbifera
Hexane	6.8a	1.0b	91.9c
Dichloromethane	4.5a	6.7a	87.2b
Methanol	19.1a	2.4a	75.1b
Water	86.4a	7.0b	4.6b

[a]n=10 to 20; means with the same letter are not significantly different (S.N.K. on angular transformed data, $P < 0.05$).

The same bioassay was used to compare organic extracts of D. c. cayenensis with and without the addition of an arrestant. The results are summarized in Table 3. The arrestant added to the methanolic extract did not result in a significant increase in the number of paper discs collected. Thus, the extract contained deterrents. With the other two solvents (hexane and dichloromethane), the extracts also contained antifeeding substances but at a lower level, since addition of the arrestant solution to the hexane extract caused a significant increase in the number of paper discs taken. This number was, however, less than that observed with arrestant-treated control discs. With the dichloromethane extract, the addition of the arrestant did not significantly increase the number of discs harvested. The pick-up was not significantly different in this case from the positive control disc. These two non-methanolic extracts will be studied with further experiments.

Thin-layer chromatographic analysis of methanol extracts from the two yam species did not show any notable difference for iodine reactive groups, alkaloids, and/or phenols. However, the saponins/sapogenins were obviously present in extracts of D. c. cayenensis and were conspicuously absent from extracts of D. bulbifera.

Acid hydrolysis of saponins to sapogenins and the extraction of the latter with petroleum ether was carried out with leaves of the five yam species. Quantitative determination of sapogenins according to Sofowora and Hardman (1974) did not indicate their presence in extracts of D. bulbifera, D. alata, and D. trifida. The concentrations of sapogenins in D. c. rotundata and D. c. cayenensis

TABLE 3. Preference of _Ac. octospinosus_ for different extracts, pure or mixed with a feeding stimulant.

Solvent used to extract leaves	Pick-up (%) of paper discs treated with material indicated[a]			
	Extracts of D. c. cayenensis	Solvent controls	Extracts of D. c. cayenensis + arrestant	Solvents + arrestant (controls)
Hexane	0.6a	1.3a	37.7b	59.2c
Dichloromethane	10.6ab	1.1a	27.5bc	53.1c
Methanol	5.1ab	1.4a	10.6b	79.5c

[a]n=10 to 15; means with the same letter are not significantly different (S.N.K. on angular transformed data, $P < 0.05$).

were 0.73 mg/g and 0.25 mg/g of dry weight, respectively, (measured in diosgenin equivalents). Thin-layer chromatography with these extracts has confirmed the results in that only D. c. cayenensis and D. c. rotundata showed the presence of saponins and sapogenins (Fig. 2).

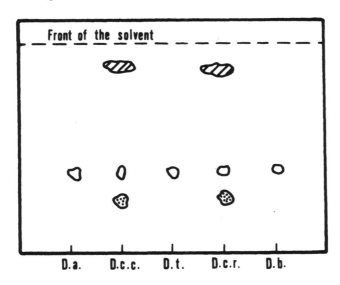

FIGURE 2. Thin layer chromatography following hydrolysis of saponins and extraction of sapogenins from five yam species leaves according to the method of Morris et al. (1958). Sapogenins are revealed by the method of Brain and Hardman (1968). D.a. = D. alata; D.c.c. = D. c. cayenensis; D.t. = D. trifida; D.c.r. = D. c. rotundata; D.b. = D. bulbifera.

DISCUSSION

Within the same plant genera Ac. octospinosus attacks certain species but not others. To our knowledge, this is the first time that it has been demonstrated that the nest of an attine can be destroyed as a result of the refusal by the ants to forage on plant material provided for them. The resistance of D. c. cayenensis to defoliation by Ac. octospinosus is of a chemical nature. However, the chemical compounds concerned do not act as repellents. Instead, they must be classified as deterrents. Saponins/sapogenins may be responsible for this resistance, and we plan to determine their chemical structure in future experiments. Saponins are glycosides with an aglucon moiety (=sapogenins) of terpenoid (penta or tetracyclic), steroid, or steroid-akaloid structure (Tschesche and Wulff 1973). These compounds, almost exclusively synthesized by higher plants,

256

are primarily toxic to fungi (Defago 1977). Numerous papers have emphasized the role played by saponins in defensive mechanisms of plants when attacked by phytopathogens (Maizel et al. 1964; Schlosser 1971; Leath et al. 1972; Defago et al. 1975). These compounds could also be responsible for the resistance (through nutritional effects) of certain plants to insect attacks (Dahlman and Hibbs 1967; Applebaum et al. 1969; Horber 1964, 1972; Horber et al. 1974; Sutherland et al. 1975).

The fungicidal properties attributed to saponins raise a basic question about attine ants: has the co-evolution of the ant-fungus ectosymbiotic complex and higher plants led to the ability of the insect to perceive and recognize compounds toxic to its Basidiomycete symbiont? This hypothesis has already been put forward by Hubbell and Wiemer (1983). These authors demonstrated that the volatile repellent for A. cephalotes found in the leaves of Hymenaea is at the same time a powerful fungicide. Investigations in this direction will add to our understanding of the unique relationship between attines and fungi.

Finally, the study of protection mechanisms in yams is of true agronomic interest. The understanding of these mechanisms can give birth to new protection methods for crops that are so often devastated by attines; e.g., genetic selection of yam species and protection of other sensitive crops by application of repellents to their leaves.

ACKNOWLEDGMENTS

The authors are grateful to Mr. G. C. Hostache for translating this paper. The studies were supported by a grant from the CORDET.

REFERENCES CITED

Amante, E. 1967. A formiga sauva Atta capiguara praga das pastagens. Biologico 33: 113-120
Applebaum, S. W., S. Marco, and Y. Burk. 1969. Saponins as possible factors of resistance of legume seeds to the attack of insects. J. Agric. Food Chem. 17: 618-622.
Barrer, P. M., and J. M. Cherrett. 1972. Some factors affecting the site and pattern of leaf-cutting activity in the ant Atta cephalotes L. J. Entomol. 47: 15-27.
Beck, S. D. 1965. Resistance of plants to insects. Ann. Rev. Entomol. 10: 207-232.
Bowers, M. A., and S. D. Porter. 1981. Effect of foraging distance on water content of substrate harvested by Atta columbica (Guerin). Ecology 62: 273-275.

Brain, K. R., and R. Hardman. 1968. An improved method of densitometric thin layer chromatography as applied to the determination of sapogenins in Dioscorea tubers. J. Chromatog. 38: 355-363.

Chen, T. K., D. F. Wiemer, and J. J. Howard. 1984. A volatile leafcutter ant repellent from Astronium graveolens. Naturwissenschaften 71: 97-98.

Cherrett, J. M. 1968. The foraging behavior of Atta cephalotes L. (Hymenoptera: Formicidae). I: Foraging pattern and plant species attacked in tropical rain forest. J. Anim. Ecol. 37: 387-403.

Cherrett, J. M. 1972a. Some factors involved in the selection of vegetable substrate by Atta cephalotes L. (Hymenoptera: Formicidae) in tropical rain forest. J. Anim. Ecol. 41: 647-660.

Cherrett, J. M. 1972b. Chemical aspects of plant attack by leaf-cutting ants, pp. 13-24. In J. B. Harborne (ed.), Phytochemical ecology. Academic Press, London, England. 272 pp.

Cherrett, J. M. 1980. Possible reasons for the mutualism between leaf-cutting ants (Hymenoptera: Formicidae) and their fungus. Biol. Ecol. Mediterr. 7: 113-122.

Cherrett, J. M., and C. E. Seaforth. 1970. Phytochemical arrestants for the leaf-cutting ants, Atta cephalotes (L.) and Acromyrmex octospinosus (Reich), with some notes on the ants' response. Bull. Entomol. Res. 59: 615-625.

Dahlman, D. L., and E. T. Hibbs. 1967. Response of Empoasca fabae to tomatine, solanine, leptine 1, tomatidine, soladine and demissidine. Ann. Entomol. Soc. Am. 60: 732-740.

Defago, G. 1977. Role des saponines dans la resistance des plantes aux maladies fongiques. Ber. Schweitz. Bot. Ges. 87: 79-132.

Defago, G., K. F. Mennen, and H. Kern. 1975. Influence du cholesterol et des saponines sur le pouvoir pathogene du Pythium paroecandrum. Phytopathol. Z. 83: 167-184.

Dethier, V. G., L. Barton-Browne, and C. N. Smith. 1960. The designation of chemicals in terms of the responses they elicit from insects. J. Econ. Entomol. 53: 134-136.

Fowler, H. G., and S. W. Robinson. 1977. Foraging and grass selection by the grass-cutting ant Acromyrmex landolti fracticornis (Forel) (Hymenoptera: Formicidae) in habitats of introduced forage grasses in Paraguay. Bull. Entomol. Res. 67: 659-666.

Horber, E. 1964. Isolation of components from the roots of alfalfa (Medicago sativa L.) toxic to the white grubs (Melolontha vulgaris F.). Proc. 12th Int. Congr. Entomol. (London), pp. 540-541.

Horber, E. 1972. Alfalfa saponins significant in resistance to insects, pp. 611-628. In J. G. Rodriguez (ed.), Insect and mite nutrition. North-Holland Publ. Co., Amsterdam, The Netherlands.

258

Horber, E., K. T. Leath, B. Berrang, V. Marcarian, and C. H. Hanson. 1974. Biological activities of saponin components from Dupuits and Lahontan alfalfa. Entomol. Exp. Appl. 17: 410-424.

Hubbell, S. P., and D. E. Wiemer. 1983. Host plant selection by an attine ant, pp. 133-154. In P. Jaisson (ed.), Social insects in the tropics. Univ. Paris-Nord, V. II. 252 pp.

Leath, K. T., K. H. Davis, M. E. Wall, and C. H. Hanson. 1972. Vegetative-growth responses of alfalfa pathogens to saponin and other extracts from alfalfa (Medicago sativa L.). Crop Sci. 12: 851-856.

Levin, D. A. 1976. The chemical defenses of plants to pathogens and herbivores. Ann. Rev. Ecol. Syst. 7: 121-159.

Littledyke, M., and J. M. Cherrett. 1975. Variability in the selection of substrate by the leaf-cutting ants Atta cephalotes (L.) and Acromyrmex octospinosus (Reich) (Formicidae: Attini). Bull. Entomol. Res. 65: 33-47.

Littledyke, M., and J. M. Cherrett. 1976. Direct ingestion of plant sap from cut leaves by the leaf-cutting ants Atta cephalotes (L.) and Acromyrmex octospinosus (Reich) (Formicidae: Attini). Bull. Entomol. Res. 66: 205-217.

Littledyke, M., and J. M. Cherrett. 1978. Defense mechanisms in young and old leaves against cutting by the leaf-cutting ants Atta cephalotes (L.) and Acromyrmex octospinosus (Reich) (Hymenoptera: Formicidae). Bull. Entomol. Res. 68: 263-271.

Maizel, J. V., H. J. Burkhardt, and H. K. Mitchell. 1964. Avenacin, an antimicrobial substance isolated from Avena sativa. I. Isolation and antimicrobial activity. Biochemistry 3: 424-426.

Morris, M. P., B. A. Roark, and B. Cancel. 1958. Simple procedure for the routine assay of Dioscorea tubers. J. Agric. Food Chem. 6: 856-858.

Mudd, A., D. J. Peregrine, and J. M. Cherrett. 1978. The chemical basis for the use of citrus pulp as a fungus garden substrate by the leaf-cutting ants Atta cephalotes (L.) and Acromyrmex octospinosus (Reich) (Hymenoptera: Formicidae). Bull. Entomol. Res. 68: 673-685.

Munakata, K. 1975. Insect antifeeding substances in plant leaves. Pure Appl. Chem. 42: 57-66.

Pesson, P. 1980. A propos de l'instinct botanique des insectes: un aspect de la co-evolution des plantes et des insectes. Ann. Soc. Entomol. Fr. 16: 435-452.

Pintera, A. 1983. Selection of plants utilized by Atta insularis in Cuba (Hymenoptera: Formicidae). Acta Ent. Bohemoslov. 80: 13-20.

Pollard, G. V., R. Riley, and E. Wattie. 1983. Preliminary investigations on the selection of citrus species by the leaf-cutting ant, Acromyrmex octospinosus (Reich) (Formicidae: Attini). Trop. Agric. (Trinidad) 60: 282-285.

Rockwood, L. L. 1976. Plant selection and foraging patterns in two species of leaf-cutting ants (Atta). Ecology 57: 48-61.

Rockwood, L. L. 1977. Foraging patterns and plant selection in Costa Rican leaf-cutting ants. J. N. Y. Entomol. Soc. 85: 222-233.

Schlosser, E. 1971. Cyclamin, an antifungal resistance factor in cyclamen species. Acta Phytopathol. Acad. Sci. Hung. 6: 85-95.

Sofowora, E. A., and R. Hardman. 1974. Chromatographic detection and spectrophotometric determination of diosgenin and other delta 5-sapogenins in crude plant extracts. Planta Med. 26: 385-390.

Stradling, D. J. 1978. The influence of size on foraging in the ant, Atta cephalotes, and the effect of some plant defense mechanisms. J. Anim. Ecol. 47: 173-188.

Sutherland, O. R. W., N. D. Hood, and J. R. Hillier. 1975. Lucerne root saponins, a feeding deterrent for the grass grub, Costelytra zealandica (Coleoptera: Scarabeidae). New Zealand J. Zool. 2: 93-100.

Tschesche, R., and G. Wulff. 1973. Chemie und biologie der saponine. Fortschr. Chem. Org. Naturst. 30: 462-606.

Vignernon, J. P. 1978. Substances antiappetantes d'origine naturelle. Ann. Zool. Ecol. Anim. 10: 663-694.

Waller, D. A. 1982a. Leaf-cutting ants and avoided plants: Defenses against Atta texana attack. Oecologia 52: 400-403.

Waller, D. A. 1982b. Leaf-cutting ants and live oak: The role of leaf toughness in seasonal and intraspecific host choice. Entomol. Exp. and Appl. 32: 146-150.

Weber, N. A. 1966. Fungus-growing ants. Science 153: 587-604.

Weber, N. A. 1972. Gardening ants: The attines. Mem. Am. Phil. Soc. Vol. 92, 146 pp.

RESUME

Les fourmis attines recoltent une grande variete de plantes pour la culture de leur champignon symbiotique. Dans leur choix, elles presentent cependant des preferences liees aux qualites physiques ou chimiques des vegetaux. Ces differences de decoupe sont observees meme pour des especes vegetales proches. En Guadeloupe, Acromyrmex octospinosus seule espece attine presente est un important ravageur des cultures vivrieres et particulierement des ignames. Cependant parmi les especes cultivees, l'une d'elle presente une nette resistance a la defoliation. Les causes en sont abordees et sont de nature chimique, les substances responsables etant extractibles par le methanol. Les analyses preliminaires permettent de les rattacher au groupe des saponines.

22
Chemical Ecology of Host Plant Selection by the Leaf-Cutting Ant, *Atta cephalotes*

J. J. Howard and D. F. Wiemer

Taken in aggregate, leaf-cutting ants are the most destructive herbivores in the New World tropics and subtropics (Weber 1972). Their overwhelming abundance and special preference for agriculturally important plants have earned them pest status from the United States to Argentina. While leaf-cutting ants severely damage a large number of cultivated plant species, in natural ecosystems they do not attack all available species with equal frequency or intensity. Under natural conditions, it is observed that (1) few plant species suffer consistent heavy defoliation; (2) many others receive only minor damage; (3) some are attacked only at certain seasons; and (4) some are avoided altogether. Our research group is investigating the host plant selection of the leaf-cutting ant Atta cephalotes in Costa Rica and searching for the reasons why some plant species escape attack. Through a unique combination of field and laboratory studies, we have begun a search for deterrent and repellent chemicals in native plants, with the belief that these natural products might be more selective and more environmentally benign than traditional synthetic insecticides. Here we will review previous studies of host plant selection and the factors which might influence ant choices, describe our ecological investigations of attine foraging, and summarize our chemical and biological characterization of the compounds we have isolated.

SELECTIVE ATTACK BY LEAF-CUTTING ANTS

Selectivity was first seriously investigated less than 20 years ago when Cherrett (1968) showed that a colony of A. cephalotes in Guyana cut some species more often and some less often than expected on the basis of natural abundance. Subsequently, A. cephalotes was found to cut 22% and A. columbica 31.4% of plant species bearing mature leaves in tropical dry forest in Costa Rica (Rockwood 1976). Different colonies of a single Atta species

260

attacked the same set of plant species at the same time, and these preferences were independent of the abundance or distance of plants from the colony (Rockwood 1976, 1978). Ants preferred different species at different times of year, and the young leaves of many species were more heavily cut than older leaves (Rockwood 1975, 1976).

The same patterns of selectivity were found in Santa Rosa National Park, Costa Rica, where we have studied the foraging of A. cephalotes since 1977 (Hubbell and Wiemer 1983). Out of approximately 200 plant species in our study area, less than a dozen are heavily cut at all times of year, and intensity of attack is unrelated to abundance (Hubbell et al., unpublished data). The lack of correlation of ant attack with abundance and distance of plants from colonies suggests that plants are selected on the basis of their vulnerability to attack and their quality as food for the ants, their fungus, or both.

PLANT CHARACTERISTICS RELATED TO SELECTIVITY

Physical Characteristics of Leaves

Mechanical defenses of leaves may prevent cutting altogether or so reduce the rate of harvest that it is unprofitable to attempt to cut leaves. Cherrett (1972) showed that A. cephalotes cuts less dense (dry weight/surface area) and less "tough" leaf material than expected from random sampling of the available plants. In similar studies with A. texana, Waller (1982a, b; also see Chapter 12) has found leaf toughness to be important in defending otherwise palatable plants. The density of leaf sections carried by ants is also correlated with the size of the ant, suggesting that larger ants might be capable of cutting denser, tougher leaves (Cherrett 1972). Since the distribution of workers on foraging trails is skewed towards smaller size classes (Stradling 1978), some plants may escape attack simply by having leaves so tough that only the largest ants can cut them. Trichomes are known to interfere with attack by herbivores (Levin 1973) but have not been investigated with specific reference to leaf-cutting ants.

Latex systems may protect plants, since latex adheres to the ants' mandibles rendering them useless. Of 117 lactiferous plants within the foraging area of one A. cephalotes colony, only 12.8% were attacked during a 10-week study period, compared with 44.5% of non-lactiferous plants (Stradling 1978).

Moisture Content of Leaves

The moisture obtained from leaves is a potentially important source of water for ants and for maintaining proper humidity levels in the fungus gardens. A. cephalotes selects leaf material of higher moisture content than that found by random sampling of leaves (Cherrett 1972). It has also been claimed that ants cut leaves of higher moisture content farther from the colony, presumably to offset loss of moisture during transport to the colony (Bowers and Porter 1981).

Nutritional Quality of Leaves

Ants imbibe sap while cutting and handling leaves (Littledyke and Cherrett 1976; Stradling 1978) and use the sugars thus obtained to satisfy their energy requirements (Quinlan and Cherrett 1979). Laboratory studies showed that leaf-cutting ants preferentially selected substrates coated with pure mono- and disaccharides (Cherrett and Seaforth 1970; Littledyke and Cherrett 1975, 1978; Mudd et al. 1978). However, A. cephalotes is not attracted to plant extracts containing natural mixtures of sugars (Cherrett and Seaforth 1970; Mudd et al. 1978). Amino acids attract the leaf-cutting ant Acromyrmex octospinosus but not A. cephalotes (Cherrett and Seaforth 1970).

Plant Secondary Chemicals

Some plants contain chemicals which are toxic to animals and microorganisms or which reduce the quality of plants as food (Feeny 1976; Rhoades and Cates 1976). Evidence suggests that leaf-cutting ants avoid those plants known to contain toxic compounds (Stradling 1978). Laboratory studies have demonstrated that tannins and the alkaloid quinine inhibit cutting and harvest of substrates by A. cephalotes (Barrer and Cherrett 1972; Littledyke and Cherrett 1975, 1976). Other studies have reported significant repellency in lipid fractions of plant extracts (Littledyke and Cherrett 1978; Mudd et al. 1978).

ECOLOGICAL INVESTIGATIONS

Previous studies offer an incomplete picture of the reasons for host plant selection by A. cephalotes. In particular, there have been few studies on the nature and role of secondary chemicals in plant defenses against attine foraging. Our group at Iowa consists of ecologists, chemists, and microbiologists working to characterize chemical defenses against leaf-cutting ants, discover the mechanisms by which they work, and understand their importance in host plant selection relative to other plant characteristics. Our program uses field and laboratory experiments to find unpalatable plants, identify active chemicals, and measure the effects of these chemicals on leaf-cutting ants and their fungi.

Field Bioassays

Rapid and repeatable field bioassays were designed to investigate the palatability of plants and compare the preferences of different colonies (Hubbell and Wiemer 1983; Hubbell et al. 1983, 1984). These bioassays eliminate the need for lengthy observation periods previously required to establish the palatability of plants and can be used to determine the palatability of plants which do not occur in the foraging area of a colony. The "pickup" bioassay is performed by cutting small disks from leaves with a paper punch and placing the disks directly on leaf-cutting ant trails. Several leaves may be tested simultaneously in a "smorgasbord" fashion by placing one disk of each leaf on the trail and replacing those taken with a fresh disk. We typically measure the number of leaf disks taken in one hour. Ants on trails with low to moderate levels of activity (10-40 loaded foragers per minute) take the most disks, while ants on highly active trails (more than 100 loaded foragers per minute) appear to ignore the smorgasbord in favor of the plant being cut (Hubbell and Wiemer 1983). When trails of very different activity levels are being compared, a standard of oat flakes coated in 10% sucrose solution is added to the smorgasbord to aid statistical analysis (Hubbell et al. 1984). Because ants do not cut these leaf disks and all ants on a trail are capable of carrying disks, this bioassay primarily measures the importance of plant chemistry (Hubbell et al. 1984).

In a second bioassay, the number of leaf sections cut from whole leaves is counted. This "cutting" bioassay uses intact leaves placed alongside trails so that the leaf tips overlap the trail edge but do not obstruct the trail itself. Leaf petioles are inserted through cotton plugs into vials of water to maintain freshness during the bioassay. Our results indicate that this bioassay may be run up to three hours before changes in palatability become apparent. This bioassay measures the effect of both chemical and physical characteristics on palatability, except for latex systems which fail once leaves are removed from plants.

Patterns of Chemical Activity

The pickup bioassay has been used to determine the relative palatability of over 100 species of woody plants in Santa Rosa National Park, Costa Rica, since 1981. The palatability of most plants, as measured in these bioassays, is moderate to low in accord with previous observations of natural foraging behavior. To begin the study of natural chemical defenses, chloroform and ethanol extracts of about 40 species of these plants were taken and tested (vide infra) on laboratory colonies. Variation in the deterrency of the lipid extracts accounts for as much as 50% of the variance in palatability in some of our pickup tests, indicating that terpenoids and other chloroform-solubles may be the primary determinants of palatability to leaf-cutting ants (Hubbell et al. 1984).

In addition to inhibitory lipid-soluble materials, leaf-cutting ants also encounter stimuli such as sugars and amino acids that may promote cutting of leaves (Cherrett and Seaforth 1970). It is clear that leaf-cutting ants are capable of perceiving these substances in the laboratory, but is there any evidence that they play a role in host plant selection in the field? We have found no correlation between total nitrogen content and palatability of 20 plant species from Santa Rosa National Park (Hubbell and Wiemer 1983). However, it is possible that attractive and inhibitory compounds interact to determine palatability. If so, then the presence of sufficient quantities of attractive compounds might offset the effect of deterrents so that the plant is palatable to some degree. To investigate this proposition, we measured the palatability and chemical characteristics of 50 species of plants in Santa Rosa National Park, Costa Rica, during 1983. Exploratory data analysis suggests that A. cephalotes may use both attractive and inhibitory chemical information to assess the quality of leaves, but a more rigorous statistical test of this possibility is underway.

Previous studies have shown that moisture content and physical characteristics of leaves may also affect host plant selection. How important are these characteristics relative to chemical repellents and deterrents, and do these factors interact to determine palatability? We are using the cutting bioassay to investigate this question in the same set of 50 species. Preliminary analysis indicates that both physical and chemical characteristics contribute to palatability; but, as described for the pickup test above, a more complete statistical analysis of data from the cutting tests is necessary to establish the relative importance of these factors.

CHEMICAL INVESTIGATIONS

Our ecological investigations have indicated the importance of plant chemistry in host plant selection by the attines. This is not surprising in view of the number of insecticides and insect antifeedants isolated from other plant sources (e.g., Kubo and Nakanishi 1977). Plant natural products, such as the pyrethroids, have been used directly as insecticides and also have served as models for the design of synthetic insecticides. Natural insect antifeedants, such as azadirachtin, isolated from the neem tree Azadirachta indica (Zanno et al. 1975) and the drimnane sesquiterpenoids (e.g., warburgenal) isolated from the African genus Warburgia (Kubo et al. 1976), demonstrate the ability of plant natural products to discourage insect herbivory.

The attine ants are a nearly ideal system for studies of plant defenses against insect herbivory; but prior to our work, there were relatively few studies of plant defenses against this specific family of herbivores (Littledyke and Cherrett 1975, 1976), and no one had characterized specific chemical repellents or deterrents. The overwhelming abundance of the leaf-cutters makes it reasonable to hypothesize that natural selection favors plants defended against their attack. The similarity in genetic makeup of ants in a colony minimizes individual variation in insect response during bioassays. Finally, the mutualistic relationship between the leaf-cutters and their associated fungus opens intriguing questions in chemical ecology. On the one hand, the fungus contributes to attine success by transforming plant material which would otherwise be indigestible, while on the other hand, their reliance on their fungus as a food source may leave the leaf-cutters vulnerable to plant defenses against fungal attack.

Our initial goals were to establish the presence of biologically active chemicals in avoided plant species, if possible; to isolate and chemically characterize any active agents; and to begin investigations on possible mechanisms of biological activity. We have now established beyond question that many native plants do contain deterrent or repellent chemicals.

Laboratory Bioassay

Our search for the chemical factors affecting leaf-cutter choices required the development of a fast and reliable laboratory bioassay which we could use to quantify the activity of plant extracts or pure chemicals (Littledyke and Cherrett 1976; Hubbell et al. 1983; Hubbell and Wiemer 1983). For this assay, a large (ca 10^5 workers) captive Atta colony housed in a series of Plexiglas boxes is connected to a foraging platform (Fig. 1). The bioassay consists of a comparison of pickup rates of control and treated rye flakes, which are distributed on the platform surface according to a computer-generated random array. The treated flakes are soaked in a solution of the plant extract, while control flakes are soaked in solvent alone. To begin the assay, the ants are allowed access to the platform surface simply by removing some of the rubber stoppers. Instead of strictly controlling the number of ants involved in each assay, we monitor the ratio of treated to control flakes removed. Typically, 60 control and 60 treated flakes are presented in a test, and the statistical significance of numbers removed is established by application of a modified binomial (Hubbell et al. 1984). At the point when half of either the treated or control flakes are removed, the assay has maximum sensitivity. When large numbers of ants are involved, the progress of the assay is recorded on videotape so that removal ratios can be accurately established. We have used this assay for several years, most often with A. cephalotes but also with

266

A. columbica and Ac. octospinosus; it has proven to be both reliable and fast.

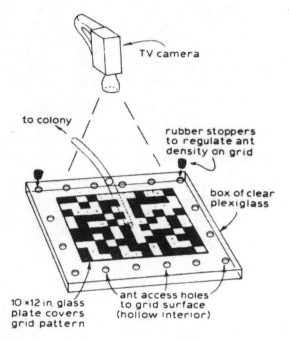

FIGURE 1. Laboratory bioassay equipment used to test plant extracts.

Isolation of Biologically Active Compounds

The laboratory bioassay is the keystone of our approach, for activity in this assay guides every step in the isolation sequence. This strategy (activity-directed isolation) involves assay of the initial plant extracts, followed by chemical fractionation of these crude mixtures if they are active in the bioassay. Each subfraction is then bioassayed, the active ones are further fractionated, and the cycle is repeated until either we have isolated pure compounds or until the activity is so dissipated (by dilution or by separation of synergistic compounds) that it is not possible to continue.

Plant samples have been collected primarily in the dry seasonal forest of Santa Rosa, Costa Rica, the site of our ecological studies, after determination of palatability in the field pick-up assay. The majority, but certainly not all, of the plants examined have yielded biologically active extracts. There are numerous explanations for the lack of activity in the minority of species.

Defense mechanisms other than secondary chemicals may be involved, some compounds may decompose in transit, or simply not be extractable. Those species which do yield active extracts can serve as sources for new biologically active chemicals.

We have isolated ant-deterrent compounds from a variety of different plant species in unrelated plant families. The structures of some of our characterized compounds, their natural source, and some bioassay data are presented in Fig. 2a, 2b. Because the isolation sequence is established for each plant by activity in the bioassays, it should not be biased for, or against, any particular class of compounds. Nonetheless, almost all of the presently character-ized deterrents are terpenoid in origin. Monoterpenoids, sesquiter-penoids, diterpenoids, and triterpenoids are already represented in this list. Beyond this generalization, it is difficult to draw conclu-sions about important molecular features until we have character-ized a larger number of active compounds.

With a better picture of the molecular features responsible for biological activity, it may be possible to prepare simpler structural analogues with enhanced activity. Our ability to synthesize large numbers of compounds expressly for testing is limited; but in some cases, we have prepared or obtained compounds structurally related to those we have isolated from plants. For example, our synthesis of lasidone (Wiemer and Ales 1981) required preparation of a number of sesquiterpenoids which can be viewed as analogues of lasidiol angelate. The starting material for this synthesis, carotol (Fig. 3), and most of the derivatives we prepared from it, show activity in the laboratory bioassay comparable or even greater than the natural product. The occurrence of cyclopropane rings in spathulenol led us to assay some related natural products (e.g., globulol, ledene, and aromadendrene), and all of these compounds also showed significant deterrent activity. We have also assayed a number of carene deriva-tives (e.g., carenone epoxide and car-2-en-4-one), and these cyclo propanoid monoterpenes also show activity in our laboratory bioassay.

CONCLUSIONS

To establish whether an isolated chemical has a defensive function in its plant vis-a-vis leaf-cutter attack is a very complex problem. It requires establishing both the natural concentration in the plant and the minimum effective concentration in the assay. In most cases, we have been so limited by the small samples of the isolated compounds that extensive studies of concentration vs activ-ity have not been feasible. However, for potential agricultural applications, it may be necessary only to establish the efficiency of a feeding deterrent. This has been done with our laboratory bioassay.

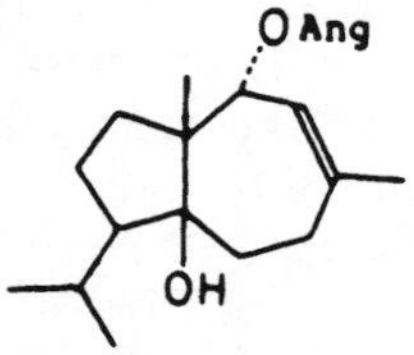

268

lasidiol angelate (35)
Lasianthaea fruticosa
(Wiemer and Ales, 1981)

trans-β-ocimene (70)
Astronium graveolens
(Chen et al., 1984)

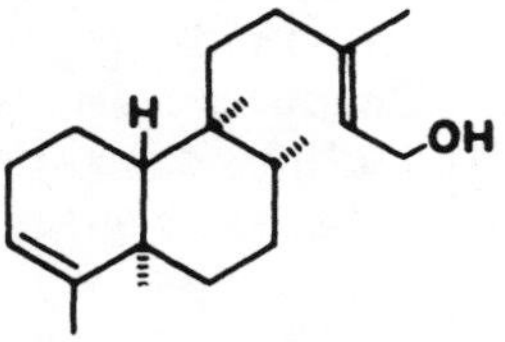

kolavenol (92)
M. divaricatum
(Hubert and Wiemer, 1985)

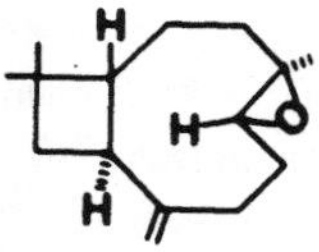

caryophyllene epoxide (50)
M. divaricatum, H. courbaril
(Hubert and Wiemer, 1985)

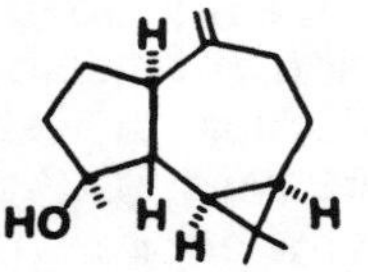

spathulenol (66)
C. alliodora, M. divaricatum
(Hubert and Wiemer, 1985)

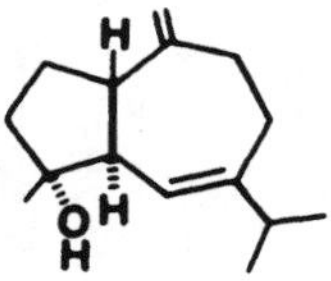

guaianol (33)
M. divaricatum
(Hubert and Wiemer, 1985)

FIGURE 2a. Natural feeding deterrents isolated from plants avoided by the leaf-cutting ant, A. cephalotes, in laboratory bioassays. Shown below each structure is the chemical name, lowest active concentration (ug/flake), plant species, and literature reference.

An intriguing question, and one of fundamental ecological importance, is whether the leaf-cutter reliance on its mutualistic fungus renders the attines vulnerable to plant defenses against fungal attack. Our field studies have shown that many plants are less palatable at the beginning of the rainy season in Costa Rica, when the risk of fungal attack might be high, than at the beginning of the dry season, when the risk might be lower (Hubbell et al. 1984). Many terpenoids have significant fungicidal activity (Stoessl 1970). We have found caryophyllene epoxide, which we have isolated from Hymenaea courbaril (Hubbell et al. 1983), Melampodium divericatum (Hubert and Wiemer 1985), and several other plants (Wiemer, unpublished), to be toxic to a variety of fungi (Cazin, unpublished). In fact, many of the compounds isolated through our approach show broad-spectrum antifungal activity. Until recently, our ability to utilize the ant fungus in direct antifungal assays has been limited by the difficulty in maintaining this species in pure culture. For the first time, we have large quantities of this fungus in culture and can begin to study the activity of these isolated compounds against the attine fungus.

To examine an alternative basis for the activity of our isolated compounds, we have begun a study of their toxicity to attine ants themselves. Using a liquid diet rich in amino acids, sugars, and vitamins, we can maintain ants for up to nine weeks in the absence of their fungus. By adding plant secondary chemicals to this diet, we can quantify the toxic effects of our isolated compounds. With the results from these two new bioassays, we hope eventually to understand the relative importance of toxicity toward ants and fungus in secondary chemicals which influence host plant selection by the attine ants.

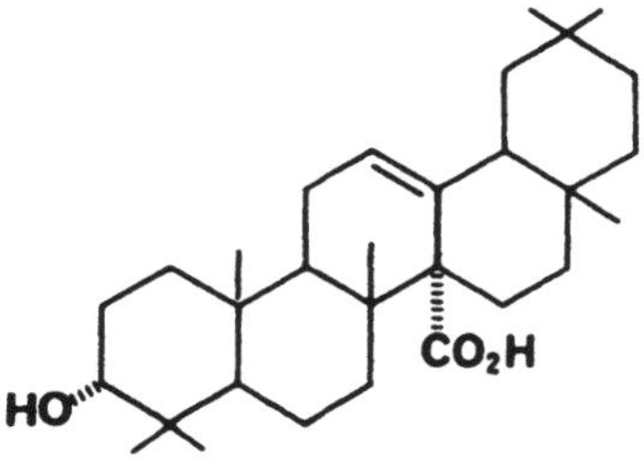

3α-hydroxyolean-12-en-27-oic acid   (108)
Cordia alliodora
(Chen et al., 1983)

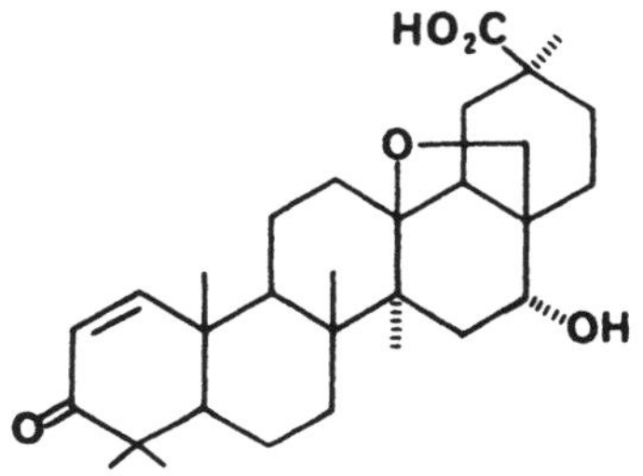

jacquinonic acid   (100)
Jacquinia pungens
(Okunade and Wiemer, 1985b)

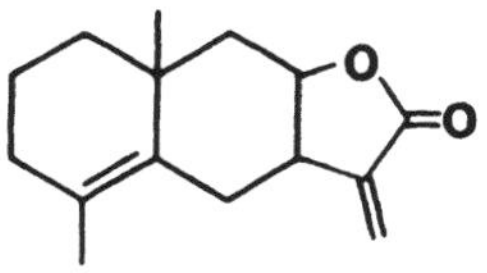

4-desoxy-8-epi-ivangulin   (17)
Eupatorium quadrangulare
(Okunade and Wiemer, 1985a)

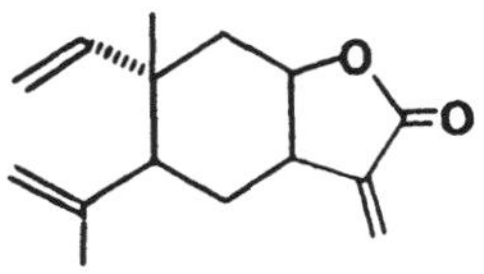

"igolan"   (8)
E. quadrangulare
(Okunade and Wiemer, 1985a)

FIGURE 2b. Natural feeding deterrents isolated from plants avoided by the leaf-cutting ant, A. cephalotes, in laboratory bioassays. Shown below each structure is the chemical name, lowest active concentration (ug/flake), plant species, and literature reference.

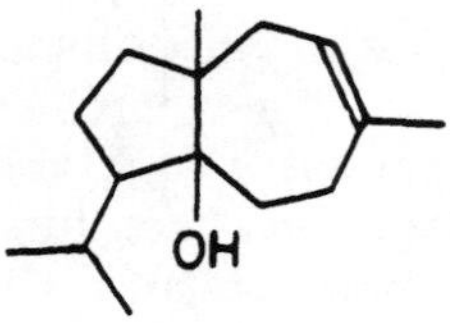

270

carotol   (17)

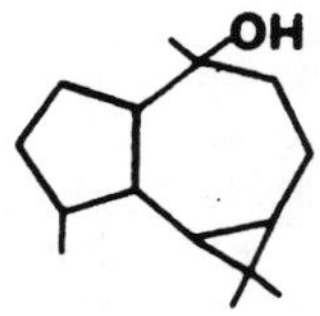

globulol   (18)

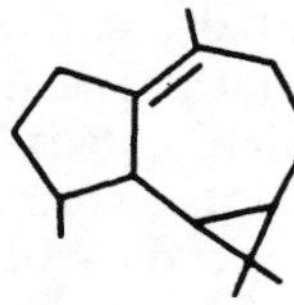

ledene   (70)

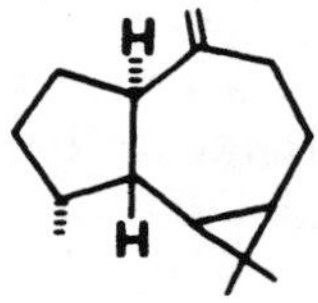

aromadendrene   (17)

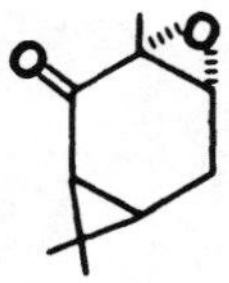

carenone epoxide   (37)
(Maas et al., 1984)

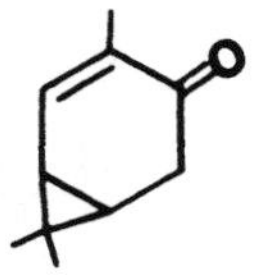

car-2-en-4-one   (25)
(Maas et al., 1984)

FIGURE 3. Synthetic and natural chemicals related to the feeding deterrents that were avoided by <u>A. cephalotes</u> in laboratory bioassays (see Fig. 2a, 2b). Shown below each structure is the chemical name, lowest active concentration (ug/flake), and literature reference.

ACKNOWLEDGMENTS

We would like to gratefully acknowledge the contributions of Professors Stephen P. Hubbell and John Cazin, Jr., the resident ecologist and microbiologist of the Iowa group. The work of a number of chemistry graduate students and postdoctorals on individual plant species, as indicated from the references, allows construction of this larger picture. The financial support of the National Science Foundation and the U.S. Department of Agriculture's Competitive Grants Program also is acknowledged with gratitude.

REFERENCES CITED

Barrer, P. M., and J. M. Cherrett. 1972. Some factors influencing the site and pattern of leaf-cutting activity in the ant Atta cephalotes (L.) J. Entomol. (A) 47: 15-27.

Bowers, M. A., and S. D. Porter. 1981. Effect of foraging distance on water content of substrates harvested by Atta columbica (Guerin). Ecology 62: 273-275.

Chen, T. K., D. C. Ales, N. C. Baenziger, and D. F. Wiemer. 1983. Ant repellent triterpenoids from Cordia alliodora. J. Org. Chem. 48: 3525-3531.

Chen, T. K., D. F. Wiemer, and J. J. Howard. 1984. A volatile leaf-cutter ant repellent from Astronium graveolens. Naturwissenschaften 71: 97-98.

Cherrett, J. M. 1968. The foraging behavior of Atta cephalotes (L.) (Hymenoptera: Formicidae). I. Foraging pattern and plant species attacked in tropical rain forest. J. Anim. Ecol. 37: 387-403.

Cherrett, J. M. 1972. Some factors involved in the selection of vegetable substrate by Atta cephalotes (L.) (Hymenoptera: Formicidae) in tropical rain forest. J. Anim. Ecol. 41: 647-660.

Cherrett, J. M., and C. E. Seaforth. 1970. Phytochemical arrestants for the leaf-cutting ants Atta cephalotes (L.) and Acromyrmex octospinosus (Reich) with some notes on the ants' response. Bull. Entomol. Res. 59: 615-625.

Feeny, P. 1976. Plant apparency and chemical defense. Recent Adv. Phytochem. 10: 1-40.

Hubbell, S. P., and D. F. Wiemer. 1983. Host plant selection by an attine ant, pp 133-154. In P. Jaisson (ed.), Social insects in the tropics, Vol. 2. University of Paris Press, Paris, France.

Hubbell, S. P., J. J. Howard, and D. F. Wiemer. 1984. Chemical leaf repellency to an attine ant: Seasonal distribution among potential host plant species. Ecology 65: 1067-1076.

Hubbell, S. P., D. F. Wiemer, and A. Adejare. 1983. An antifungal terpenoid defends a neotropical tree (Hymenaea) against attack by fungus-growing ants (Atta). Oecologia (Berlin) 60: 321-327.

Hubert, T. D., and D. F. Wiemer. 1985. Ant-repellent terpenoids from Melampodium divericatum. Phytochemistry 24: 1197-1198.

Kubo, I., and K. Nakanishi. 1977. Insect antifeedants and repellents, pp. 165-178. In P. Hedin (ed.), Host plant resistance to pests. Am. Chem. Soc. Symp. Ser. No. 62, Washington, D.C.

Kubo, I., Y. W. Lee, M. Pettei, F. Pilkiewicz, and K. Nakanishi. 1976. Potent army worm antifeedants from the East African Warburgia plants. J. Chem. Soc., Chem. Commun. 1976: 1013-1014.

Levin, D. A. 1973. The role of trichomes in plant defense. Quart. Rev. Biol. 48: 1-15.

Littledyke, M., and J. M. Cherrett. 1975. Variability in the selection of substrate by the leaf-cutting ants Atta cephalotes (L.) and Acromyrmex octospinosus (Reich) (Formicidae: Attini). Bull. Entomol. Res. 65: 33-47.

Littledyke, M., and J. M. Cherrett. 1976. Direct ingestion of plant sap from cut leaves by the leaf-cutting ants Atta cephalotes (L.) and Acromyrmex octospinosus (Reich) (Formicidae: Attini). Bull. Entomol. Res. 66: 205-217.

Littledyke, M., and J. M. Cherrett. 1978. Defense mechanisms in young and old leaves against cutting by the leaf-cutting ants Atta cephalotes (L.) and Acromyrmex octospinosus (Reich) (Hymenoptera: Formicidae). Bull. Entomol. Res. 68: 263-271.

Maas, D. D., M. Blagg, and D. F. Wiemer. 1984. Synthesis and reactions of (-)- and (+)-carenones. J. Org. Chem. 49: 853-856.

Mudd, A., D. J. Peregrine, and J. M. Cherrett. 1978. The chemical basis for the use of citrus pulp as a fungus garden substrate by the leaf-cutting ants Atta cephalotes (L.) and Acromyrmex octospinosus (Reich) (Hymenoptera: Formicidae). Bull. Entomol. Res. 68: 673-685.

Okunade, A. L., and D. F. Wiemer. 1985a. Ant repellent sesquiterpene lactones from Eupatorium quadrangulare. Phytochemistry 24: 1199-1201.

Okunade, A. L., and D. F. Wiemer. 1985b. Jacquinonic acid: An ant repellent triterpenoid from Jacquinia pungens. Phytochemistry 24: 1203-1205.

Quinlan, R. J., and J. M. Cherrett. 1979. The role of fungus in the diet of the leaf-cutting ant Atta cephalotes (L.). Econ. Entomol. 4: 151-160.

Rhoades, D. F., and R. G. Cates. 1976. Toward a general theory of plant anti-herbivore chemistry. Recent Adv. Phytochem. 10: 168-213.

Rockwood, L. L. 1975. The effects of seasonality on foraging in two species of leaf-cutting ants (Atta) in Guanacaste Province, Costa Rica. Biotropica 7: 176-193.

Rockwood, L. L. 1976. Plant selection and foraging patterns in two species of leaf-cutting ants (Atta). Ecology 57: 48-61.

Rockwood, L. L. 1978. Foraging patterns and plant selection in Costa Rican leaf-cutting ants. J. N. Y. Entomol. Soc. 85: 222-233.

Stoessl, A. 1970. Antifungal compounds produced by higher plants. Recent Adv. Phytochem. 1: 143-180.

Stradling, D. J. 1978. The influence of size on foraging in the ant Atta cephalotes, and the effect of some plant defense mechanisms. J. Anim. Ecol. 47: 173-188.

Waller, D. A. 1982a. Leaf-cutting ants and live oak: The role of leaf toughness in seasonal and intraspecific host choice. Entomol. Exp. Appl. 32: 146-150.

Waller, D. A. 1982b. Leaf-cutting ants and avoided plants: Defenses against Atta texana attack. Oecologia (Berl.) 52: 400-403.

Weber, N. A. 1972. Gardening ants, the attines. Mem. Amer. Phil. Soc. Vol. 92, 146 pp.

Wiemer, D. F., and D. C. Ales. 1981. Lasidiol angelate: An ant repellent sesquiterpenoid from Lasianthaea fruticosa. J. Org. Chem. 46: 5449-5450.

Zanno, P. R., I. Miura, K. Nakanishi, and D. L. Elder. 1975. Structure of the insect phagorepellent azadirachtin. Application of PRFT/CWD carbon-13 nuclear magnetic resonance. J. Am. Chem. Soc. 97: 1975.

23
Digestive Physiology
of Leaf-Cutting Ants

G. Febvay and A. Kermarrec

The digestion of ingested food occurs between the time an animal takes in food and the time the nutrients are absorbed and assimilated by the tissues. Digestion depends, among other factors, on the quality of the ingested food and on the digestive enzyme activities of the alimentary tract. These two aspects of digestive physiology in the attine ant, Acromyrmex octospinosus, will be discussed relative to recently published reports on this subject.

NOURISHMENT AND THE DIGESTIVE TRACT

Buckley (1860) and Lincecum (1867) thought that attine ants fed directly on harvested vegetation. Belt (1874) was the first to suspect that the leaves cut by the leaf-cutting ants did not serve as their food but as the culture substrate for their symbiotic fungus. Until 1972, this symbiont was considered the attine ants' only source of food (Muller 1874; Wheeler 1907; Weber 1966, 1972). A qualitative and quantitative study of certain nutrients derived from the fungus of Atta colombica tonsipes (Martin et al. 1969) revealed that about 56% of the dry weight of the fungus could be used as soluble nutrients; carbohydrates constituted more than 27% of this dry weight, free amino acids 4.5%, protein bound amino acids 13%, and lipids 0.2%. The carbohydrates identified were trehalose, mannitol, arabinitol, and glucose. The lipid fraction contained ergosterol as the principal sterol. Although they detected no polysaccharides, a cytological study by Decharme (1981) revealed that the mycelium cultivated by Ac. octospinosus, particularly the gongylidia, stored a large quantity of glycogen. This storage polysaccharide is frequently found in fungi. Certain constituents (triterpenoids and steroids) can come from the vegetable substrate and can vary qualitatively according to the origin of the nests (Lopes and Gilbert 1977). The fungi of the advanced attine genera (Atta and Acromyrmex) are classified as Basidiomycetes (see Kermarrec et al., Chapter 20).

According to Bartnicki–Garcia (1968), the cell walls of these fungi are made of chitin and glucans. The presence of these constituents in the symbiont of Ac. octospinosus was confirmed with histochemical methods (Decharme 1981). It can be concluded from these data that the fungus is a good quality food with a high level of easily assimilable carbohydrates along with a high level of proteins rich in essential amino acids.

Other authors have shown that attine ants consume liquids. Echols (1966) described the direct ingestion of soybean oils by A. texana from mirex bait. Weber (1972), Littledyke and Cherrett (1975), and Febvay (1981) reported that A. cephalotes and Ac. octospinosus workers fed on liquid sources; Barrer and Cherrett (1972) reported they fed on plant sap; and Littledyke and Cherrett (1976) corroborated this by means of radioactive tracers. As much as a third of the radioactivity absorbed by the leaves was ingested by the ants. Although a part of plant sap can be regurgitated on the fungus garden, these results suggest that the fungus is not the only source of food for the workers. This hypothesis was confirmed by the work of Quinlan and Cherrett (1979). These authors showed that adult ants obtain only approximately 5% of their energy requirements (based on respiration data) from the juice of staphylae (clusters of inflated hyphae). In contrast, the plant sap can satisfy the adult nutrient requirements for 24 hours. However, larvae which consume the entire staphylae can obtain sufficient nutrients for their energy and growth requirements. These conclusions are linked to the hypothesis that adults only ingest the juice of staphylae and that larvae ingest and completely digest the staphylae.

At this stage, it is worth recalling some anatomical features of the digestive tract of the ants. The adult ant is characterized by the presence of an infrabuccal pocket in the head (Janet 1905). It receives food solids (Wheeler 1910) and filters out large particles. The particle size filtered varies according to the ant species: 150 um for Camponotus pennsylvanicus (Eisner and Happ 1962), approximately 10 um for Ac. octospinosus (Quinlan and Cherrett 1978), and less than 1 um for Solenopsis invicta (Glancey et al. 1981). Therefore, only liquids reach the midgut; the excluded infrabuccal pellets can be retained by the workers for as long as 24 hours before regurgitation under normal feeding conditions (Febvay and Kermarrec 1981a). Digestion probably starts in this pocket on the infrabuccal pellet, which can contain other types of plant materials as well as cellular walls of the hyphae and staphylae of the symbiotic fungus (Febvay 1981).

The digestive system of the larvae of Ac. octospinosus is comparable to that of other ant species (Febvay 1981). The midgut is lined with a protective peritrophic membrane and receives all food.

It seems that the calculations proposed by Quinlan and

Cherrett (1979) are probably underestimated for adults, because they do not take into account the stored material in the infrabuccal pocket. However, analysis of the food ingested by the attine ants does not permit a meaningful evaluation of these different nutritional aspects. The digestive capabilities of these ants must be approached by means of a study of the enzymatic activities of the digestive tract and its associated glands.

ENZYMATIC ACTIVITY OF THE DIGESTIVE TRACT AND ASSOCIATED GLANDS

During the preparation of the vegetative substrate for inoculation with fungus, attine workers place drops of rectal fluid on it. Studies of the enzymatic activities of this liquid in different attines by Martin and Martin (1970, 1971) proved the presence of proteases, chitinase, and amylase (Martin et al. 1973). However, in A. texana, Boyd and Martin (1975a, b) showed that rectal fluid proteinases originated from the fungus they cultured, and they concluded that the adaptations of attines to their symbiont could have involved the cessation of production of their own digestive proteinases.

Recently, we started research on the enzymatic activity of the digestive tract and associated glands in Ac. octospinosus (Febvay and Kermarrec 1981b). The identification of enzymes present was made with the aid of a semi-quantitative micro-method (Api-zyn®) using artificial naphthol substrates (Monget 1978). The enzymes studied were osidases, proteinases, and lipases. For adults, the study concerned the midgut and the three pairs of salivary glands (maxillary, labial, and postpharyngeal). In the larvae, we studied the enzymes of the isolated midgut and the pair of salivary glands. Enzymatic activity of the symbiotic fungus was also analyzed since it is the only food of the larval forms and a portion of the adult food. Thus, it could also be a source of enzymes for the ants. More details on this study are given by Febvay and Kermarrec (1981b). This paper only reports some of these results.

Enzymatic Activity in Adults

Enzymatic activities detected in the different tissues of the adults are presented in Table 1. Only zymograms of worker tissues are reported; similar results were observed in the sexuals.

The postpharyngeal glands are ectodermal tubules that open into the anterior digestive tract. Our tests detected only α-glucosidase and esterase activity in this gland. The low activity of these enzymes suggests that they do not serve a digestive function. Some enzymatic studies have been reported on these glands from different species of ants: Formica polyctena (Paulsen 1971); F. integra, F. fusca, Acanthomyops claviger, Camponotus herculeanus,

C. pennsylvanicus (Ayre 1967); Messor capitatus (Delage 1968); Myrmica rubra (Abbott 1978). No osidase or endopeptidase activity was found. Three exopeptidases were found in Messor, but the author seems to attribute them to exogenous food. A lipase, usually present in the postpharyngeal glands of ants, particularly for predatory species, was not detected in Ac. octospinosus. The digestive role of this gland in ants is a matter of controversy. Forbes and McFarlane (1961) thought that its primary function was digestive, but Brugnion (1930) suggested that it was involved in the feeding of larvae by regurgitation. This view was supported by Ayre (1963, 1967). As in many species of ants (Delage 1966), the direct passage of lipids from food into these glands has also been observed in attines (Echols 1966 for A. texana; Peregrine et al. 1972 for A. cephalotes; and Peregrine et al. 1973, 1974 for Ac. octospinosus). These last authors did not observe transfer of the glands' contents to the larvae; but nevertheless, they attributed a nutritive function to the gland while emphasizing the high energy value of the triglycerides which it contained.

The maxillary glands, as the postpharyngeal glands, have not been shown to have enzymatic activity. In the ant literature (op. cit.), only two species show osidase activity (invertase, trehalase, and maltase) in this gland. No proteinases or lipases have been found. A digestive role for these glands should not be discounted; however, Abbott (1978) noted that they do not have a reservoir to store their secretions. Instead, the glands empty directly into the pharynx.

Enzyme analyses of the labial glands revealed strong α-glucosidase and N-acetyl-β-D-glucosaminidase activity. No endopeptidase was detected; however, there was weak exopeptidase activity. A trace of lipase activity was also reported. It is worth reviewing the results for other ant species obtained by the aforementioned authors. Out of nine different species, only three (M. rubra, F. integra, and F. fusca) did not show any enzymatic activity. Amylase was found in the other six species. Maltase and invertase were also present in Messor. No endopeptidase was found in the labial glands of the nine species, but Messor had three exopeptidases and one lipase.

The midgut is the site of intense digestion. As for most of the ants studied, a large range of enzymes was found in the midgut of Ac. octospinosus. As far as nitrogen metabolism is concerned, the three exopeptidases were present, and leucine arylamidase was strongly active. However, no endopeptidase was found. A lipase was found which had a greater activity than that in labial glands. Of all the osidases tested, only α-glucosidase was found, and it was strongly active.

278

TABLE 1. Zymograms of the salivary glands and the midgut of the adults of Ac. octospinosus (+ trace,   1; - no coloration; 1 to 5 semiquantitative notation: 1 = weak, 5 = strong activity) (from Febvay and Kermarrec 1981b).

Enzymatic activities	Emptied, washed midgut	Labial glands	Post-pharyngeal glands	Maxillary glands
α-D-galactosidase	-	-	-	-
β-D-galactosidase	-	-	-	-
α-D-glucosidase	5	2	1	1
β-D-glucosidase	-	-	-	-
N-acetyl-α-D-glucosaminidase	-	-	-	-
N-acetyl-β-D-glucosaminidase	-	5	-	-
α-L-fucosidase	-	-	-	-
β-L-fucosidase	-	-	-	-
β-D-fucosidase	-	-	-	-
β-D-lactosidase	-	-	-	-
α-D-mannosidase	-	-	-	-
β-D-mannosidase	-	-	-	-
α-D-xylosidase	-	-	-	-
β-D-xylosidase	-	-	-	-
trypsin	-	-	-	-
chymotrypsin	-	-	-	-
leucine arylamidase	4	1	-	-
valine arylamidase	1	+	-	-
cystine arylamidase	1	+	-	-
esterase	3	4	1	+
esterase-lipase	2	2	-	+
lipase	1	+	-	-

Enzyme Activity in Larvae

Enzyme activity found in the larvae is reported in Table 2 (see methods in Febvay and Kermarrec 1981b). The salivary glands of the larvae show β-N-acetylglucosaminidase and esterase-lipase as in the labial glands of the adult. This result was not surprising as the labial glands of the adult originate from the salivary glands of the larvae (Grasse 1951). The results obtained for the larval midgut show the presence of a large variety of enzymes. Some, such as α - and β -galactosidase, β -glucosidase, α-mannosidase and α- and β - xylosidase probably originate from food. These enzymes were also detected in the symbiotic fungus (Table 2) and in the larvae where their activity was mainly found in the ingested food enclosed in the peritrophic membrane. Besides these enzymes which have a food origin, the midgut is an important source of endogenous enzymes. As in the adult, it was observed that for the metabolism of polysaccharides there was a great deal of α-D-glucosidase activity. We also report in Table 2 the presence of β-D-fucosidase and β-D-mannosidase which were not found in the fungus. These enzymes may well play a role in the function of the digestive tract by degrading glycoproteins and mucosubstances. Other than the very strong activity of the three exopeptidases, an endopeptidase of the chymotrypsin type was also found which was not evident in our previous tests with adults. This activity is essential since larvae need large amounts of free amino acids to synthesize new tissue. Esterase and lipase were also secreted by the midgut of the larvae.

TABLE 2. Zymograms of the salivary glands and the midgut of the larvae of <u>Ac. octospinosus</u> and of the symbiotic fungus. (+ trace 1; - no coloration; 1 to 5 semiquantitative notation: 1 = weak, 5 = strong activity) (from Febvay and Kermarrec 1981b).

Enzymatic activities	Larval biological tissue			Symbiotic fungus
	Emptied midgut	Full midgut	Salivary glands	
α-D-galactosidase	2	2	-	2
β-D-galactosidase	1	1	-	2
α-D-glucosidase	5	5	-	-
β-D-glucosidase	5	3	-	3
N-acetyl-α-D-glucosaminidase	-	-	-	-
N-acetyl-β-D-glucosaminidase	-	5	5	-
α-L-fucosidase	-	-	-	-
β-L-fucosidase	-	-	-	-
β-D-fucosidase	2	4	-	-
β-D-lactosidase	-	1	-	-
α-D-mannosidase	1	1	-	-
β-D-mannosidase	+	4	-	2
α-D-xylosidase	+	4	-	1
β-D-xylosidase	+	4	-	+
trypsin	-	-	-	-
chymotrypsin	1	1	-	-
leucine arylamidase	5	4	-	2
valine arylamidase	3	1	-	1
cystine arylamidase	2	1	-	1
esterase	2	-	2	3
esterase-lipase	2	1	1	1
lipase	1	-	-	1

Enzymatic studies of ant larvae are rare, but Delage (1968), working on foraging ants (Messor), found a number of very active enzymes in the larvae. The secretion of the salivary glands contains in particular an endopeptidase of the trypsin type which was not found in the labial glands of adults.

The enzymes detected in Ac. octospinosus and their relative activities reflect the type of food normally ingested by these ants. The larvae and adults consume a varied diet and have a complete array of proteinases, lipases, and osidases, as opposed to low amounts found in insects that have a less varied diet. However, compared to other ants, Ac. octospinosus has a lesser amount of enzymatic activity. The level of lipases and proteinases in Ac. octospinosus is obviously lower than that of ants with a carnivorous diet. In the adults, no endopeptidases were found while two osidases were identified: α-glucosidase which degrades polysaccharides (glycogen, starch) and α-β-N-acetylglucosaminidase in the labial glands which functions in the digestion of chitin from the fungus walls while it is retained in the infrabuccal pocket.

GLUCOSIDASE AND CHITINASE ACTIVITY IN ADULTS

The specificity of the α-glucosidase activity was studied in the labial glands and midgut of adult ants (Febvay and Kermarrec 1983). Amylase, secreted by the labial glands, was not found in the midgut. However, 20% of the amylase activity was carried with the ingested food into the midgut. In contrast, maltase and trehalase were secreted primarily by the midgut with only 3% of the activities in the midgut being found in the labial glands. The activities of these enzymes were measured from eight extracts of stomach contents for the three worker subcastes (minor, medium, and major). The results given per animal or per milligram of fresh weight are reported in Table 3. Trehalase activity was proportional to the size of the individual ant. When expressed per milligram of fresh weight, it was not significantly different between the three groups of workers. The overall maltase activity also increased with the size of the individual but was significantly greater for the small workers when expressed relative to fresh weight.

β-N-acetylglucosaminidase activity was studied in detail; an associated chitinolytic activity in the labial glands of the adults was examined also (Febvay et al. 1984). A histoenzymological technique was used to verify that only the labial glands secreted β-N-acetyglucosaminidase (β-NAGase). Two sources of the enzyme were identified by electrophoresis (digestive glands and haemolymph). These two enzymes are clearly different regarding their biochemical characteristics (Febvay et al. 1984). The total β-NAGase activity of the labial glands for Ac. octospinosus was porportional to the weight of the ant. It varied from 1.01 ± 0.20 nmole of liberated p-nitrophenol (pNP)/min/animal for the small workers, to 17.11 ± 3.29 nmoles of pNP/min/animal for males. When activity was expressed per unit weight, it was highest for small workers and inversely proportional to the size of the animal. Fasting had no influence on β-NAGase activity.

TABLE 3. Trehalase and maltase activities of Ac. octospinosus workers (mean ± 2SE). The means with different letters (a–c; d–f; n–p) are significantly different (P=0.01; Newman and Keul's test) (from Febvay and Kermarrec 1983).

Enzyme activity in nmoles of glucose liberated per minute	Subcaste		
	Minor	Media	Major
		Trehalase	
per ant	12.8±4.2a	35.4±8.6b	55.3±7.3c
per mg of fresh weight	6.8±1.1m	6.6±1.6m	5.8±0.8m
		Maltase	
per ant	7.4±1.2d	22.1±3.0e	32.3±7.7f
per mg of fresh weight	5.84±1.13n	4.08±0.59o	2.32±0.56p

Chitinase and chitobiase activities in labial gland homogenates had pH optima of 7.0 to 7.5 and 7.0 respectively. Chitobiase activity decreased with incubation temperature, showing a marked decrease above 50°C and becoming inactive at 60°C. A gland extract maintained at 45°C for 24 h retained all of its chitobiase activity. In contrast, chitinase activity dropped to 23% of its initial activity after 1 h at 45°C. An extract of three labial glands liberated 77.4 ± 9.6 ug of N-acetylglucosamine in 24 h from approximately 200 mg (fresh weight) of symbiotic fungus. The digestion was verified by cytological observations: in vitro for hyphae incubated with gland extracts; in vivo for the walls of pieces of symbiotic fungus present in regurgitated infrabuccal pellets. The digestive chitinases are often produced by endosymbiotic microflora and are not directly secreted by the host (Jeuniaux 1950, 1954, 1956). Our observations of Ac. octospinosus showed no evidence of microflora; these chitinolytic enzymes were synthesized by the ants in their labial glands.

CONCLUSIONS

Mycophagy is not simply a modification and extension of phytophagy but a feeding strategy which demands particular digestive capacities, and Ac. octospinosus possesses the enzymes that allow it to digest its symbiont.

Degradation of proteins is a result of an endopeptidase of a chymotrypsin type in the larvae. Adults do not have this type of activity and secrete only exopeptidases. The amino acid needs of the adults are restricted to the regeneration of new tissues. These needs can be satisfied by free amino acids which come from the symbiont as well as those which result from the digestion of oligopeptides by the exopeptidases present in the midgut. It is also possible that the adults, like other Hymenoptera (Maschwitz 1965, 1966; Ishay and Ikan 1968; Wust 1973), depend on larval secretions to supplement amino acid intake.

The metabolism of polysaccharides by adults is shown schematically in Fig. 1. Martin et al. (1973) detected chitinase in the rectal fluid of several attine ants; and subsequently, Boyd and Martin (1975a, b) demonstrated that proteinases present in the rectal fluid of A. texana originally came from the symbiotic fungus. It was suggested that the ant did not synthesize any digestive enzymes but derived them from the fungus. In a review, Martin (1979) generalized this hypothesis and proposed that mycophagous insects would be dependent upon the fungus for digestion of chitin and glucans. However, in a recent study, Martin et al. (1981) conclude that in general the digestive enzymes of fungus-feeding beetles originate from a source within the beetles. Our results show that Ac. octospinosus produces its own chitinolytic enzymes in the labial glands and that the adaptation of this ant to a fungal diet did not cause the ants to lose the ability to produce their own digestive enzymes.

With the amylase activity of its labial glands, Ac. octospinosus can degrade polymers of α-1,4-glucose from plant starch ingested during the preparation of plant material for its gardens and the glycogen contained in the symbiotic fungus. The corresponding enzymes are secreted by the labial glands, and the digestion of the polymers starts when they are taken into the preoral cavity.

Trehalase and maltase are found in the midgut. These enzymes are not common in ants. Trehalase is found in the midgut of M. rubra (Abbot 1978) as well as in the labial glands of F. polyctena (Graf 1964). Trehalose is an important disaccharide in the cytoplasm of the symbiotic fungus. This disaccharide, as well as maltose, is produced from the digestion of starch and glycogen by the amylases of the labial glands and will be further degraded by midgut enzymes. The resulting glucose could then be used as a source of energy.

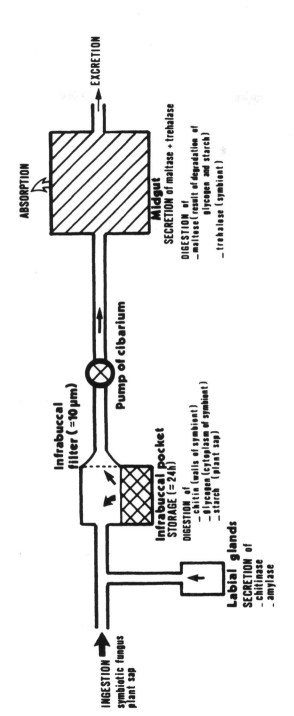

FIGURE 1. Scheme of the digestion of the polyosides by the adults of _Ac. octospinosus._

The studies carried out on the digestive physiology of Ac. octospinosus are important for research on the development of new methods of control. The description of the social organization of the attine nest has defined the ethological and ecophysiological bases of integrated control (Kermarrec et al. 1977). The destruction of foraging workers is of little importance because they represent only a tiny fraction of the total number of ants in a colony. To destroy the whole colony, it is necessary that any insecticide (chemical or biological) have a delayed effect. It is possible that this could be achieved by microencapsulation of the active agent. The poison thus being protected can be distributed to the whole ant colony by trophallaxis. Its action will begin after digestion of the wall of the capsule by digestive secretions of the ants. Our research allows us to define the chemical nature of the encapsulating material and the size of the microcapsules. Since adults do not make endopeptidases, a protein-based capsule (gelatin, casein) would not work. Diverse polysaccharides are digested (chitin, starch, and glycogen) and can be recommended for preliminary trials of the encapsulation of insecticides. The use of starch seems reasonable using the coacervation method of encapsulation (Vandegaer 1973) or an air suspension coating process (Hall and Hinkes 1973). The latter technique also allows the retention of dextrins and chitosans (deacetylated chitin). The size of microcapsules obviously remains an important parameter to define. The digestion of polysaccharides is preoral and takes place mainly in the infrabuccal pocket. The optimal diameter of the capsules must be greater than 10 um so that they will be stopped by the infrabuccal filter but no larger than 500 um, since this is the maximum size of hyphae or staphylae pellets eaten by the ants. This size range is compatible with the two encapsulation procedures mentioned above (a few microns to 1 mm).

ACKNOWLEDGMENT

The authors are grateful to Mr. Dave for translating this paper.

REFERENCES CITED

Abbott, A. 1978. Nutrient dynamics of ants, pp. 233-244. In M. V. Brian (ed.), Production ecology of ants and termites. Cambridge University Press.

Ayre, G. L. 1963. Feeding behavior and digestion in Camponotus herculeanus (L.) (Hymenoptera: Formicidae). Entomol. Exp. Appl. 6: 165-170.

Ayre, G. L. 1967. The relationships between food and digestive enzymes in five species of ants (Hymenoptera: Formicidae). Can. Entomol. 99: 408-411.

Barrer, P. M., and M. J. Cherrett. 1972. Some factors affecting the site and pattern of leaf-cutting activity in the ant Atta cephalotes (L.). J. Entomol. 47: 15-27.

Bartnicki-Garcia, S. 1968. Cell wall chemistry and taxonomy of fungi. Annu. Rev. Microbiol. 22: 87-108.

Belt, T. 1874. The naturalist in Nicaragua. John Murray. London.

Boyd, N. D., and M. M. Martin. 1975a. Faecal proteinases of the fungus-growing ant Atta texana: Properties, significance and possible origin. Insect Biochem. 5: 619-635.

Boyd, N. D., and M. M. Martin. 1975b. Faecal proteinases of the fungus-growing ant Atta texana: Their fungal origin and ecological significance. J. Insect Physiol. 21: 1815-1820.

Brugnion, E. 1930. Les pieces buccales, le sac infrabuccal et le pharynx des fourmis. Bull. Soc. R. Entomol. Egypte. pp. 85-210.

Buckley, S. B. 1860. The cutting ant of Texas (Oecodoma mexicana Sm.) Proc. Acad. Nat. Sci. Phila. 9-10: 233-236. Annu. Mag. Nat. Hist. 3, VI: 386-389.

Decharme, M. 1981. Les champignons symbiotes de quelques fourmis attines: Differents aspects de leur cytologie et de leur evolution dans la meule. C. R. Acad. Sci. Agric. France. 67: 934-937.

Delage, B. 1966. Sur une fonction particuliere des glandes pharyngiennes des fourmis. C. R. Acad. Sci. Paris 262: 1743-1744.

Delage, B. 1968. Recherches sur les fourmis moissonneuses du Bassin Aquitain: Ethologie physiologie de l'alimentation. Ann. Sci. Nat. Zool. 10: 195-265.

Echols, H. W. 1966. Assimilation and transfer of mirex in colonies of Texas leaf-cutting ants. J. Econ. Entomol. 59: 1336-1338.

Eisner, T., and G. M. Happ. 1962. The infrabuccal pocket of a formicine ant: A social filtration device. Phyche 69: 107-116.

Febvay, G. 1981. Quelques aspects (anatomie et enzymologie) des relations nutritionnelles entre la fourmi attine Acromyrmex octospinosus (Hymenoptera: Formicidae) et son champignon symbiotique. These de Docteur-Ingenieur, Univ. Lyon I, 196 pp.

Febvay, G., and A. Kermarrec. 1981a. Morphologie et fonctionnement du filtre intrabuccal chez une attine Acromyrmex octospinosus Reich (Hymenoptera: Formicidae): Role de la poche infrabuccale. Int. J. Insect Morphol. Embryol. 10: 441-449.

Febvay, G., and A. Kermarrec. 1981b. Activites enzymatiques des glandes salivaires et de l'intestin moyen d'une fourmi attine (adultes et larves): Acromyrmex octospinosus Reich (Hymenoptera: Formicidae). Arch. Biol. 92: 299-316.

286

Febvay, G., and A. Kermarrec. 1983. Enzymes digestives de la fourmi attine Acromyrmex octospinosus (Reich): Caracterisation de amylases, maltase et trehalase des glandes labiales et de l'intestin moyen. C. R. Acad. Sci. Paris. 296: 453-456.

Febvay, G., M. Decharme, and A. Kermarrec. 1984. Digestion of chitin by the labial glands of Acromyrmex octospinosus Reich (Hymenoptera: Formicidae). Can. J. Zool. 62: 229-234.

Forbes, J., and A. M. MacFarlane. 1961. The comparative anatomy of digestive glands in the female castes and the males of Camponotus pennsylvanicus (De Geer) (Formicidae: Hymenoptera). J. N. Y. Entomol. Soc. 69: 92-103.

Glancey, B. M., R. K. Vander Meer, A. Glover, C. S. Lofgren, and S. B. Vinson. 1981. Filtration of microparticles from liquids ingested by the red imported fire ant, Solenopsis invicta Buren. Insect. Soc. 28: 395-401.

Graf, I. 1964. Untersuchungen zur Physiologie von Formica polyctena Foerst. Experientia 20: 330-331.

Grasse, P. P. 1951. Traite de Zoologie. Masson (ed.). Tome 10, Fascicule II. 972 pp.

Hall, H. S., and T. M. Hinkes. 1973. Air suspension encapsulation of moisture-sensitive particles using aqueous systems, pp. 145-153. In J. E. Vandegaer (ed.), Microencapsulation. Plenum Press, New York-London.

Ishay, J., and R. Ikan. 1968. Food exchange between adults and larvae in Vespa orientalis F. Anim. Behav. 16: 298-303.

Janet, C. 1905. Anatomie de la tete du Lasius niger. Ducourtieux and Gout, Limoges, 5 pls., 2 figs.

Jeuniaux, C. 1950. Recherche de la chitinase dans les tissus glandulaires digestifs de l'escargot Helix pomatia. Arch. Int. Physiol. 3: 354-355

Jeuniaux, C. 1954. Sur la chitine et la flore bacterienne intestinale des mollusques Gasteropodes. Mem. Acad. Roy. Belg. Cl. Sci. 28: 5-45.

Jeuniaux, C. 1956. Chitinase et bacteries chitinolytiques dans le tube digestif d'un cloporte (Porcellio scaber Latr.) (Isopode, Oniscide). Arch. Int. Physiol. Biochem. 64: 583-586.

Kermarrec, A., G. Malato, and G. Febvay. 1977. Donnees nouvelles sur l'invasion de la Guadeloupe par Acromyrmex octospinosus Reich (Formicidae: Attini). II Bases ethologiques et physiologiques d'un controle integre. Nouv. Agron. Antilles Guyana 3: 486-492.

Lincecum, G. 1867. The cutting ant of Texas Oecodoma texana Buckley. Proc. Acad. Nat. Sci. Phila. 18: 24-31.

Littledyke, M., and J. M. Cherrett. 1975. Variability in the selection of substrate by the leaf-cutting ants Atta cephalotes (L.) and Acromyrmex octospinosus (Reich) (Formicidae: Attini). Bull. Entomol. Res. 65: 33-47.

Littledyke, M., and J. M. Cherrett. 1976. Direct ingestion of plant sap from cut leaves by leaf-cutting ants Atta cephalotes (L.) and Acromyrmex octospinosus (Reich) (Formicidae: Attini). Bull. Entomol. Res. 66: 205-217.

Lopes, J. C. L., and B. Gilbert. 1977. Constituintes quimicos do fungo da Atta sexdens rubropilosa, Forel, 1908 (Hymenoptera: Formicidae). Arq. Inst. Biol., Sao Paulo 44: 75-83.

Martin, M. M. 1979. Biochemical implications of insect mycophagy. Biol. Rev. 54: 1-21.

Martin, M. M., R. M. Carman, and J. G. MacConnell. 1969. Nutrients derived from the fungus cultured by the attine ant Atta colombica tonsipes. Ann. Entomol. Soc. Am. 62: 11-13.

Martin, M. M., and J. S. Martin. 1970. The presence of protease activity in the rectal fluid of attine ants. J. Insect Physiol. 16: 227-232.

Martin, M. M., and J. S. Martin. 1971. The presence of protease activity in the rectal fluid of primitive attine ants. J. Insect Physiol. 17: 1897-1906.

Martin, M. M., M. J. Gieselmann, and J. S. Martin. 1973. Rectal enzymes of attine ants: α-amylase and chitinase. J. Insect Physiol. 19: 1409-1416.

Martin, M. M., J. J. Kukor, J. S. Martin, T. E. O'Toole, and M. W. Johnson. 1981. Digestive enzymes of fungus-feeding beetles. Physiol. Zool. 54: 137-145.

Maschwitz, U. 1965. Larven als Nahrungsspeicher im Wespenvolk (Ein Beitrag zum Trophallaxieproblem). Vehr. dt. Zool. Ges. 50: 530-534.

Maschwitz, U. 1966. Das Speichelsekret der Wespenlarven und seine Biologische Bedeutung. Z. Physiol. 53: 228-252.

Monget, D. 1978. Mise au point d'une micromethode de mesure d'activites enzymatiques (Apizym). These de Docteur-Ingenieur, Univ. Lyon I, 152 pp.

Muller, F. 1874. The habit of various insects. Nat. London 10: 102-103.

Paulsen, R. 1971. Characterization of trehalase from labial glands of ants (Formica polyctena). Arch. Biochem. Biophys. 142: 170-176.

Peregrine, D. J., M. C. Perci, and J. M. Cherrett. 1972. Intake and possible transfer of lipid by the post-pharyngeal glands of Atta cephalotes (L.). Entomol. Exp. Appl. 15: 248-249.

Peregrine, D. J., A. Mudd, and J. M. Cherrett. 1973. Anatomy and preliminary chemical analysis of the post-pharyngeal glands of the leaf-cutting ant Acromyrmex octospinosus (Reich) (Hymenoptera: Formicidae). Insect. Soc. 20: 355-363.

Peregrine, D. J., and A. Mudd. 1974. The effect of diet on the composition of the post-pharyngeal glands of Acromyrmex octospinosus (Reich). Insect. Soc. 21: 417-424.

Quinlan, R. J., and J. M. Cherrett. 1978. Studies on the role of the infrabuccal pocket of the leaf-cutting ant Acromyrmex octospinosus (Reich) (Hymenoptera: Formicidae). Insect. Soc. 25: 237-245.

Quinlan, R. J., and J. M. Cherrett. 1979. The role of fungus in the diet of the leaf-cutting ant Atta cephalotes (L.). Ecol. Entomol. 4: 151-160.

Vandegaer, J. E. 1973. Encapsulation by coacervation, pp. 21-37. In J. E. Vandegaer (ed.), Microencapsulation. Plenum Press, New York-London.

Weber, N. A. 1966. Fungus growing ants. Science 153: 587-604.

Weber, N. A. 1972. Gardening ants. The attines. Mem. Am. Phil. Soc. 92: 1-146.

Wheeler, W. M. 1907. Fungus growing ants of North America. Bull. Am. Mus. Nat. Hist. 23: 669-807.

Wheeler, W. M. 1910. Ants. Columbia University Press. New York and London. 663 pp.

Wust, M. 1973. Stomodeale und Proctodeale Sekrete von Ameisen-larven und ihre biologische Bedeutung. Proc. VIIth Int. Cong. I.U.S.S.I., London. pp. 412-417.

24
The Physiology of the Imported
Fire Ants: Basic Gaps
in Our Understanding

S. B. Vinson

Our understanding of the biology and physiology of the imported fire ants, <u>Solenopsis</u> <u>invicta</u> and <u>S. richteri</u>, has expanded considerably since the first review by Lofgren et al. (1975), particularly in regard to their physiology (see reviews by Vinson 1978, 1983; Vander Meer 1983). Additional progress has been made in some aspects of fire ant physiology, primarily in the area of pheromones; but this latter subject is reviewed in this book by Fletcher (Chapter 15), Glancey (Chapter 19), and Vander Meer (Chapter 17). For these reasons I have chosen to discuss some areas of imported fire ant physiology where we either (1) lack basic information (such as mating and reproduction) or (2) have developed information on a particular subject but need additional research to fully grasp the intricacies (such as food flow). I have also selected those areas of physiology which I feel are most important in providing a background for new management techniques.

PRODUCTION OF SEXUALS

Production of sexual forms (winged or alate males and females) can occur about five months after colony founding; but after that, alates can be produced at any time during the year (Markin et al. 1973; Lofgren et al. 1975; Rhoades and Davis 1967). Alate production is greatest in the spring and to a lesser extent in early fall (Lofgren et al. 1975; Markin and Dillier 1971; Morrill 1974); the factors that influence this yearly cycle of sexual production are unknown.

Vinson and Robeau (1974) reported that the application of juvenile hormone to colonies stimulated the production of alate females. Later, Robeau and Vinson (1976) provided evidence that fertilized eggs were not predetermined to be workers or female alates since the application of juvenile hormone analogues to newly dequeened colonies produced alate females. They also found that sexuals were not produced if second or later instar larvae were the only brood present at the time of the application of the juvenile hormone analogues. While these studies suggest that juvenile hormone plays a role in alate female production during the first instar, it does not provide information about what stimulates alate production, the source of the juvenile hormone, or how juvenile hormone causes the production of alate females. For example, do workers, in response to the juvenile hormone, feed larvae a special food (perhaps analagous to the royal jelly of honey bees, Beetsma 1979), or do workers feed larvae more often? Does the juvenile hormone or analogue act directly on the larvae or are the larvae stimulated to secrete more endogenous juvenile hormone? Further, the quality and quantity of nutrients the colony receives probably plays an important but an ill-defined role in alate production.

The control of male production is also unknown. As with many other Hymenoptera, unfertilized eggs develop into males; but in social insect colonies, the production of males can be regulated in several ways. In the imported fire ant, the options may be less complicated since workers do not produce eggs and therefore cannot produce males. However, in polygynous (multiple queen) colonies (Fletcher et al. 1980; Mirenda and Vinson 1982), virgin queens could survive and produce males.

It is not known whether or not the queen can control fertilization of the eggs she releases as appears to be the situation for some Hymenoptera (Van den Assem 1971; Waage 1982). The queen may regularly produce unfertilized eggs whose survival is regulated by the workers. In the fire ant, the production of trophic eggs (Glancey et al. 1973; Voss 1981), presumably followed by larval and worker oophagy (Petralia and Vinson 1978) and/or brood cannibalism (Sorensen et al. 1983c), suggests that workers could regulate male production; but it is unclear what role workers have (cannibalism, pheromones, behavior, or nutrition) in the regulation of alate production.

Further, it is not known how long alates are tolerated in a colony if environmental conditions preclude mating flights over an extended period of time. When the colony queen is lost, some alate females will dealate (Fletcher and Blum 1981), suggesting that a substance has been removed that inhibits dealation or inhibits other colony members from producing a dealation substance. Once dealation has occurred, the substance is again produced, and the excess dealates are executed (Fletcher and Blum 1983). This substance appears to be important in the regulation of queen numbers; i.e., one queen per colony. As discussed by Fletcher and Blum (1983), the queen or her substitute appears to produce a substance that is fed to workers which causes them to execute excess queens or dealates. How the decision is made to execute one dealate rather than another is not clear. Presumably, this decision is based on the production of a queen pheromone (Glancey et al. 1981a; Fletcher and Blum 1983); and thus, the queen that produces the most pheromone is not executed. Whether the queen recognition pheromone (Lofgren et al. 1983), execution pheromone, and the dealation pheromone (Fletcher and Blum 1981) are the same or related is unknown.

Furthermore, the function of these pheromones in multiple queen colonies is far from clear. In multiple queen colonies, egg production per queen, worker size, and worker aggressiveness are all reduced (Mirenda and Vinson 1982; Greenberg et al. 1985). How the pheromones and hormones interact to produce these differences in monogynous and polygynous colonies and their evolution remains an interesting question.

MATING FLIGHTS

Very little is known about mating flights. Generally, alates accumulate in a colony until the proper conditions to initiate a mating flight occur. These conditions are generally met one or two days after a rain if the wind is not too great and the temperature is between 20° and 32°C (Rhoades and Davis 1967; Markin et al. 1971). Rhoades and Davis (1967) state that mating flights begin around 1000 h and last until 1400 h local standard time. The first sign of a mating flight is the appearance of agitated workers opening small (ca 1/8 in) holes in the mound. Generally males emerge first, climb onto the surrounding vegetation, and fly away. Males appear to be ignored by the workers as they leave the colony.

Alate females usually begin to fly one hour after males, but their flights may continue into the afternoon. Like males, the females first climb the surrounding vegetation. However, the workers crawl over the queens and tug at their legs and antennae. The significance of this behavior is not understood.

Environmental factors are at least indicative or symptomatic of the initiation of mating flights, but the actual biological/physiological causes are unknown. The frenzied activity at the time suggests that a pheromone produced by the male or female alates is involved, but there is no evidence of such a compound to date. However, it is also possible that the environmental changes that initiate flights prime the workers since mating flights occur in colonies in which only reproductive males or alate females are present (Bass and Hays 1979). It is unlikely that males and females would produce the same pheromone.

Where and how the sexes locate each other during flight is still a mystery. Markin et al. (1971), using an airplane with collecting nets, found males 150 to 300 meters above ground. Whether males attract females or vice versa, or whether both sexes are attracted to certain landmarks, is unknown. Clearly, much more data are needed regarding our knowledge of the mating activities and the physiology of reproduction in the imported fire ant.

COLONY FOUNDING

There is surprisingly little information on colony founding. Virgin queens are heaviest at the time of the mating flight (over 14 mg according to Fletcher, Chapter 15). Phillips and Vinson (1980) reported that the crop and postpharyngeal glands of queens are fully engorged with oils just before a mating flight. Thus, alates have a full complement of nutrients before flight, but the triggering mechanism for this activity is not known. Toom et al. (1976) reported that 60% of the glycogen in the thorax and abdomen is lost during the mating flight. After mating, the queen lands and casts off her wings after which her wing muscles begin to degenerate, and she begins egg production.

The factors that control wing casting, muscle degeneration, and egg production are unknown. Kearney et al. (1977) reported that juvenile hormone or CO_2 would stimulate wing casting. These compounds also stimulate muscle degeneration (Jones et al. 1978), but it is not known whether either of these compounds are actually involved. One of the first changes in the wing muscles following flight is a change in the calcium flux in the muscle tissue (Jones et al. 1981, 1982), but the significance of this change has not been determined. Jones et al. (1978), using an artificial mating technique that was described recently by Ball and Vinson (1984), found that mating initiated wing muscle histolysis in 2 hr, much earlier than juvenile hormones.

The role that flight muscle degeneration plays in the growth of the first brood of a founding queen is not clear. Proteins or amino acids are needed by larvae for growth and development. These essential nutrients may be provided by the queen's flight muscles and possibly fat body protein. Jones et al. (1978) examined the breakdown of the flight muscles which appear to go through reverse synthesis rather than the more classical lysosome and phagocytic digestion. The pathway by which the released amino acids are presumed to reach the larvae is unknown. However, there appear to be two possible sources. First, the production of trophic eggs, as noted by Voss (1981), could provide a protein source; but whether these are continually produced and fed to larvae is unknown. A second source may be either the maxillary or labial glands. Phillips and Vinson (1979) noted that both of these glands were well-developed in queens and appeared to produce a proteinaceous secretion. However, it is not known if such glands are larger or more active during initial brood production. In summary, whether the flight muscles contribute to the rearing of the first brood is not well documented for the fire ant.

The first workers produced in a new colony are referred to as minims due to their small size and have been described as a distinct worker caste (Vander Meer, see Chapter 26). Their size appears to be due to nutrition rather than the deposition of a specific type of egg by the queen. Greenberg (personal communication) has found that eggs laid by new colony founding queens will develop into minors or medias if placed in a small colony and fed by workers. These studies suggest that either the quantity or quality of nutrition, along with juvenile hormone (Vinson and Robeau 1974), may have a profound effect on the development of the various castes.

Other observations support the view that nutrition plays a major role in worker caste size. Porter and Tschinkel (1985) reported that smaller workers are more efficient in providing food to larvae than larger workers and that larger workers resulted from larvae fed by small workers. This may be particularly significant in multiple queen colonies where many queens are laying eggs. Although each queen in a multiple queen colony may lay fewer eggs than a single physogastric queen (Fletcher et al. 1980), the total number of eggs produced in a multiple queen colony exceeds that produced by a single queen colony (Greenberg et al. 1985). As Greenberg et al. (1985) have shown, the average worker size in multiple queen colonies is smaller than in single queen colonies. This finding may be explained on the basis of the worker to brood ratio although it has not yet been thoroughly investigated.

Mirenda and Vinson (1981) reported that there is an age polyethism among castes of the imported fire ant. Young workers are nurses, tending the brood and queen. As the worker ages, it spends less time as a nurse. Instead, it begins to participate in the repair of the nest and receives food from foragers, hence the name "reserve." The older ants do most of the initial foraging and, furthermore, Mirenda and Vinson (1981) found that caste size also influences worker tasks. While larger workers live longer, they spend less time as nurses and more time as foragers while the shorter-lived minor workers spend a longer time as nurses.

There is increasing evidence that age polyethism also influences the workers' response to pheromones. The queen pheromone elicits worker attraction and dragging of the queen or other pheromone-treated objects into the mound (Glancey et al. 1984). However, nurses will respond to the queen pheromone in this way but not foragers (Glancey et al. 1983). Sorensen et al. (1985d) demonstrated that the older workers execute supernumerary queens in single queen colonies. It is likely that age not only influences the ants' responses to various pheromones but may also influence what pheromones the ants produce. Careful attention should be given to the ant's age and size in relation to polyethism.

FOOD FLOW AND PROCESSING

We have considerable information on the type of food materials that act as phagostimulants, how these foods are handled by the ants, their flow into the colony, and the castes that receive the food (Vinson 1978, 1983). However, there are several areas where we lack information, such as nutrition. Ants in the laboratory are not always as vigorous as those in the field. Ants in colonies reared on some diets may be lighter in color, their size may be more uniform, and they may be less aggressive. Whether these differences are due to the lack of specific nutrients or are influenced by the quantity or overall quality of the food source is unknown.

The developing larvae require amino acids for growth, and the queen needs amino acids for egg production. However, proteins in the form of pieces of insect tissue or particles of food above 1 micron in size cannot be ingested by workers due to an oesophageal screen that filters out larger particles (Glancey et al. 1981b). Thus, proteinaceous food above 1 micron in size is either carried into the colony in toto or formed into a buccal pellet by the workers (Petralia and Vinson 1978). The adult worker does not appear to produce much proteinase (Ricks and Vinson 1972a; Sorensen et al. 1983c) and, therefore, is less able to digest proteins. Larvae, on the other hand, have large and active labial glands containing protein-ases (Petralia et al. 1980). Petralia and Vinson (1978) found that workers placed buccal pellets or pieces of food in depressions on the venter of larvae near the head (the food basket or praesaepium) upon which the larvae secrete digestive enzymes. Through extra-oral digestion the buccal pellets and food pieces are dissolved, and the liquid is consumed by the larvae or adult ants. As demonstrated by Sorensen et al. (1983c), the proteinase activity of workers increases in the presence of larvae. W. R. Tschinkel (personal communication) found that the presence of larvae also has an influence on egg production. Queens need a source of amino acids in order to produce large numbers of eggs. This may be due to the increased availability of amino acids to the colony in the presence of larvae.

In social insects the habit of coprophagy is not uncommon and is a means of recycling nutrients, pheromones, and microorganisms important to the biology and physiology of the colony. Nothing is known about adult ant excretions, but larvae have received some study. O'Neal and Markin (1973) reported that larvae secreted two materials: a clear fluid which the adults pick up and discard outside the nest; and an opaque white fluid on which adults appear to feed. Petralia et al. (1982) determined that the white, opaque fluid consisted primarily of uric acid while the clear liquid was water and salt. However, Petralia et al. (1982) reported that neither material was consumed by adults unless under water stress. His results suggest that larval excrement played little, if any, role in colony integration. The role of the excretory products of the worker or queen is yet to be determined. As larvae pupate, a meconial pellet is excreted (Petralia and Vinson 1979); but whether the meconium provides information or nutrition to the colony is unknown.

Food flow in an imported fire ant colony is complex, but progress has been made in unravelling some of it. It is not only important for our understanding of colony nutrition, but it is also essential if we are to understand and predict which control agents and foods are compatible. The types of food, colony makeup, and physiological status of the ants can affect the distribution and ultimate fate of control agents.

The types of food eaten by the ants are distributed differently within the colony. Carbohydrates are utilized primarily by workers while proteins (and amino acids) are preferentially utilized by larvae and the queen. Utilization of oils (lipids) is intermediate compared to carbohydrates and proteins (Howard and Tschinkel 1981a; Sorensen and Vinson 1981). Food form may also be important. For example, proteins in water are distributed, almost equally, between larvae and workers in 48 h while over 80% of the solid protein is distributed to larvae over the same period (Sorensen et al. 1980). This change in distribution is explained by the inability of workers to ingest solid food and the ability of fourth-instar larvae to eat solid food as well as liquids (Petralia and Vinson 1978; Glancey et al. 1981b).

Food form and type are not the only factors affecting food flow. Worker age, colony size, and nutritional history play a role. Sorensen et al. (1981) found that the amount of food handled and consumed and its internal distribution varied between the temporal castes (Sorensen et al. 1983b). Using honey, Sorensen et al. (1985b) found that older workers tend to donate food more readily than other castes, and nurses tend to be receivers. Foragers tend to engage more often in trophallaxis than reserves or nurses. When oil was used instead of honey, Sorensen and Vinson (1985) found that oil was passively donated to other workers and, thus, the movement of oil is controlled by receivers rather than donors.

Since age affects a worker's food handling, the flexibility of workers in their tasks was examined. Sorensen et al. (1985a) found that foragers could readily perform nurse duties, but nurses were less flexible requiring several days before adjusting to foraging duties. While workers appear "flexible," they readily revert to their proper age-related role if given the choice (Sorensen et al. 1985c).

The ability of the imported fire ant to store food is unclear. Wilson (1978) reported that smaller workers were more specialized for sugar (carbohydrate) storage than large workers while Glancey et al. (1973) reported that larger workers tend to be a replete caste in regards to oil storage. Howard and Tschinkel (1981b) examined the internal distribution of liquid food in different sizes of workers. They reported that small ants are less suited for the storage of sugar or amino acids than large workers; but even in large workers, sugar and amino acids are readily passed from the crop to the mid-gut. Oil is retained the longest in the crops of major workers. The role of food form and worker age on food storage still needs much study.

Colony size and starvation can also affect food flow, and starvation can result in cannibalism of the brood (Sorensen et al. 1983a). Howard and Tschinkel (1980) reported that workers took up greater amounts of sugar and shared it with more nestmates in starved colonies. They also found that the distribution of food differed in large and small colonies, primarily in the size of the ant receiving food, and that starvation stimulated food dispersal but at different intensities for different foods. Sorensen et al. (1985a) examined the role of starvation on the division of labor and found that starved foragers increased their carbohydrate intake while starved nurses increased their foraging for oil and protein. The results suggest that foragers respond to the nutritional needs of the colony and are regulated by the nurses through the reserves.

Food preferences in the field have been examined by several researchers. Ricks and Vinson (1972b) found differences in the nutrient content of workers during the year indicating that seasonal variation in their food preference, or at least intake, may occur. However, efforts to examine preference have been disappointing. Glunn et al. (1981) found the greatest recruitment in the field was to oil followed by carbohydrate and protein, but laboratory preferences were carbohydrates first, followed by oil. Both S. A. Phillips (personal communication) and D. Bogar (unpublished data) found that food preferences in the field changed with the time of year. Like Glunn et al. (1981), they found a great degree of variation. However, these are important questions since the use of baits for control depends on stimulating the ants to forage on the bait in competition with the naturally available resources. Further, the integration of the colony via the release of pheromones may be via trophallaxis (Sorensen et al. 1985d).

Issues of particular interest now are the factors that influence the development of reproductives, changes in food flow when alates are present, changes in food flow prior to the mating flight, and the effect of polygyny on food flow. Also, the competition for food resources and movement of food between colonies of the fire ant in the field needs to be studied. Wilson et al. (1971) reported that the fire ant has distinct foraging territories while Summerlin et al. (1975) found food flow between field colonies in both Florida and Texas. Whether this is due to polygyny or monogyny is under study.

SUMMARY

While considerable progress has been made in regard to the physiology of the imported fire ant (Vinson 1978, 1983; Vander Meer 1983), there is much yet to be done. I have concentrated on problems concerning the production of sexuals, mating activities, colony founding, and food flow because of their importance to our understanding of the imported fire ant and ultimate management of this insect. With regard to these problems, we either know very little (e.g., how the sexes locate each other) or we have begun to make some inroads into understanding them; but we lack many pieces of the puzzle (e.g., food flow).

I also have not discussed the physiology of pheromones and caste determination in any detail due to previous discussions (Fletcher, Chapter 15; Vander Meer, Chapter 17; Glancey, Chapter 19); important aspects of the physiology and behavior of imported fire ants, such as neurophysiology and endocrinology, have not been addressed largely because of an almost complete lack of research. It is hoped that this review will provide a basis for direction of research by others.

ACKNOWLEDGMENTS

Approved TA 20792 by the Director of the Texas Agricultural Experiment Station. The author wishes to express his appreciation to the Texas Department of Agriculture. This work was supported in part by the Texas Department of Agriculture Cooperative Agreement IAC 84-85 (1018). Appreciation is extended to Les Greenberg for his extensive editorial help.

REFERENCES CITED

Ball, D. E., and S. B. Vinson. 1984. Anatomy and histology of the male reproductive system of the imported fire ant, Solenopsis invicta. Int. J. Insect Morph. Embryol. 13: 283-294.
Bass, J. A., and S. B. Hays. 1979. Nuptial flights of the imported fire ant in South Carolina. J. Ga. Entomol. Soc. 14: 158-161.
Beetsma, J. 1979. The process of queen-worker differentiation in the honeybee. Bee World 60: 24-29.
Fletcher, D. J. C., and M. S. Blum. 1981. Pheromonal control of dealation and oogenesis in virgin queen fire ants. Science 212: 73-75.
Fletcher, D. J. C., and M. S. Blum. 1983. Regulation of queen number by workers in colonies of social insects. Science 219: 312-314.
Fletcher, D. J. C., M. S. Blum, T. V. Whitt, and N. Temple. 1980. Monogyny and polygyny in the fire ant, Solenopsis invicta. Ann. Entomol. Soc. Am. 73: 658-661.
Glancey, B. M., A. Glover, and C. S. Lofgren. 1981a. Pheromone production by virgin queens of Solenopsis invicta Buren. Sociobiology 6: 119-127.
Glancey, B. M., C. S. Lofgren, J. R. Rocca, and J. H. Tumlinson. 1983. Behavior of disrupted colonies of Solenopsis invicta towards queens and pheromone-treated surrogate queens placed outside the nest. Sociobiology 7: 283-288.

Glancey, B. M., J. Rocca, C. S. Lofgren, and J. Tumlinson. 1984. Field tests with synthetic components of the queen recognition pheromone of the red imported fire ant, Solenopsis invicta. Sociobiology 9: 19-30.

Glancey, B. M., C. E. Stringer, Jr., C. H. Craig, P. M. Bishop, and B. B. Martin. 1973. Evidence of a replete caste in the fire ant, Solenopsis invicta. Ann. Entomol. Soc. Am. 66: 233-234.

Glancey, B. M., R. K. Vander Meer, A. Glover, C. S. Lofgren, and S. B. Vinson. 1981b. Filtration of microparticles from liquids ingested by the red imported fire ant, Solenopsis invicta Buren. Insect. Soc. 28: 395-401.

Glunn, F. J., D. F. Howard, and W. R. Tschinkel. 1981. Food preferences in colonies of the fire ant, Solenopsis invicta. Insect. Soc. 28: 217-222.

Greenberg, L., D. J. C. Fletcher, and S. B. Vinson. 1985. Monogyny and polygyny in the imported fire ant: Correlation with worker size and mound location. J. Kans. Entomol. Soc. 58: 9-18.

Howard, D. F., and W. R. Tschinkel. 1980. The effect of colony size and starvation on food flow in the fire ant, Solenopsis invicta. Behav. Ecol. Sociobiol. 7: 293-306.

Howard, D. F., and W. R. Tschinkel. 1981a. The flow of food in colonies of the fire ant, Solenopsis invicta: A multifactorial study. Physiol. Entomol. 6: 279-306.

Howard, D. F., and W. R. Tschinkel. 1981b. Internal distribution of liquid food in isolated workers of the fire ant, Solenopsis invicta. J. Insect. Physiol. 27: 67-74.

Jones, R. G., W. L. Davis, H. K. Hagler, and S. B. Vinson. 1981. Calcium and muscle degeneration in Solenopsis: Histochemistry and electron microprobe analysis. J. Cell Biol. 95: 385a.

Jones, R. G., W. L. Davis, A. C. F. Hung, and S. B. Vinson. 1978. Insemination induced histolysis of the flight musculature in fire ants (Solenopsis spp.). Am. J. Anat. 151: 603-610.

Jones, R. G., W. L. Davis, and S. B. Vinson. 1982. An electron microscopic histochemical and analytical x-ray microprobe study of calcium changes in insect flight muscle degeneration in Solenopsis, the queen fire ant. J. Histol. Cytochem. 30: 273-304.

Kearney, G. P., P. M. Toom, and G. L. Blomquist. 1977. Induction of dealation in virgin female Solenopsis invicta with juvenile hormones. Ann. Entomol. Soc. Am. 70: 699-701.

Lofgren, C. S., W. A. Banks, and B. M. Glancey. 1975. Biology and control of imported fire ants. Annu. Rev. Entomol. 20: 1-30.

Lofgren, C. S., B. M. Glancey, A. Glover, J. Rocca, and J. Tumlinson. 1983. Behavior of workers of Solenopsis invicta to the queen-recognition pheromone: Laboratory studies with an olfactometer and surrogate queens. Ann. Entomol. Soc. Am. 76: 44-50.

Markin, G. P., and J. H. Dillier. 1971. The seasonal life cycle of the imported fire ant, Solenopsis saevissima richteri, on the gulf coast of Mississippi. Ann. Entomol. Soc. Am. 64: 562-565.

Markin, G. P., J. H. Dillier, and H. L. Collins. 1973. Growth and development of colonies of the red imported fire ant, Solenopsis invicta. Ann. Entomol. Soc. Am. 66: 803-809.

Markin, G. P., J. H. Dillier, S. O. Hill, M. S. Blum, and H. R. Hermann. 1971. Nuptial flight and flight ranges of the red imported fire ant, Solenopsis saevissima richteri. J. Ga. Entomol. Soc. 6: 145-156.

Mirenda, J. T., and S. B. Vinson. 1981. Division of labor and specification of castes in the red imported fire ant, Solenopsis invicta Buren. Anim. Behav. 29: 410-420.

Mirenda, J. T., and S. B. Vinson. 1982. Single and multiple queen colonies of imported fire ants in Texas. Southwest. Entomol. 7: 135-141.

Morrill, W. L. 1974. Production and flight of alate red imported fire ants. Environ. Entomol. 3: 265-271.

O'Neal, J., and G. P. Markin. 1973. Brood nutrition and parental relationships of the imported fire ant, Solenopsis invicta. J. Ga. Entomol. Soc. 8: 294-303.

Petralia, R. S., and S. B. Vinson. 1978. Feeding in the larvae of the imported fire ant, Solenopsis invicta: Behavior and morphological adaptations. Ann. Entomol. Soc. Am. 71: 643-648.

Petralia, R. S., and S. B. Vinson. 1979. Developmental morphology of larvae and eggs of the imported fire ant, Solenopsis invicta. Ann. Entomol. Soc. Am. 72: 472-484.

Petralia, R. S., A. A. Sorensen, and S. B. Vinson. 1980. The labial gland system of larvae of the imported fire ant, Solenopsis invicta: Ultrastructure and enzyme analysis. Cell Tissue Res. 206: 145-156.

Petralia, R. S., H. J. Williams, and S. B. Vinson. 1982. The hindgut ultrastructure and excretory products of larvae of the imported fire ant, Solenopsis invicta Buren. Insect. Soc. 29: 332-345.

Phillips, S. A., Jr., and S. B. Vinson. 1979. Comparative morphology of glands associated with the head among castes of the red imported fire ant, Solenopsis invicta Buren. J. Ga. Entomol. Soc. 15: 215-226.

Phillips, S. A., Jr., and S. B. Vinson. 1980. Sources of the post-pharyngeal gland contents in the red imported fire ant, Solenopsis invicta. Ann. Entomol. Soc. Am. 75: 257-261.

Porter, S. D., and W. R. Tschinkel. 1985. Fire ant polymorphism: Factors affecting worker size. Ann. Entomol. Soc. Am. 78: 381-386

Rhoades, W. C., and D. R. Davis. 1967. Effects of meteorological factors on the biology and control of the imported fire ant. J. Econ. Entomol. 60: 554-558.

Ricks, B. L., and S. B. Vinson. 1972a. Digestive enzymes of the imported fire ant, Solenopsis richteri. Entomol. Exp. Appl. 15: 329-334.

Ricks, B. L., and S. B. Vinson. 1972b. Changes in nutrient content during one year in workers of the imported fire ant. J. Econ. Entomol. 65: 135-138.

Robeau, R. M., and S. B. Vinson. 1976. Effects of juvenile hormone analogues on caste differentiation in the imported fire ant, Solenopsis invicta. J. Ga. Entomol. Soc. 11: 198-202.

Sorensen, A. A., and S. B. Vinson. 1981. Quantitative food distribution studies within laboratory colonies of the imported fire ant, Solenopsis invicta Buren. Insect. Soc. 28: 129-160.

Sorensen, A. A., and S. B. Vinson. 1985. Behavior of temporal subcastes in the fire ant, Solenopsis invicta, in response to oil. J. Kans. Entomol. Soc. 58: 586-596.

Sorensen, A. A., T. M. Busch, and S. B. Vinson. 1983a. Factors affecting brood cannibalism in laboratory colonies of the imported fire ant, Solenopsis invicta Buren. J. Kans. Entomol. Soc. 56: 140-150.

Sorensen, A. A., T. M. Busch, and S. B. Vinson. 1983b. Behavior of worker subcastes in the imported fire ant, Solenopsis invicta, in response to proteinaceous food. Physiol. Entomol. 8: 83-92.

Sorensen, A. A., T. M. Busch, and S. B. Vinson. 1985a. Control of food influx by temporal subcastes in the fire ant, Solenopsis invicta. Behav. Ecol. Sociobiol. 17: 191-198.

Sorensen, A. A., T. M. Busch, and S. B. Vinson. 1985b. Trophallaxis by temporal subcaste in the fire ant, Solenopsis invicta, in response to honey. Physiol. Entomol. 10: 105-111.

Sorensen, A. A., T. M. Busch, and S. B. Vinson. 1985c. Behavioral flexibility of temporal subcastes in the fire ant, Solenopsis invicta, in response to food. Pysche 91: 319-331.

Sorensen, A. A., D. J. C. Fletcher, and S. B. Vinson. 1985d. Distribution of inhibitory queen pheromone among virgin queens of an ant, Solenopsis invicta. Pysche 92: 57-69.

Sorensen, A. A., R. S. Kamas, and S. B. Vinson. 1980. The biological half-life and distribution of 125-iodide and radio-iodinated protein in the imported fire ant, Solenopsis invicta. Entomol. Exp. Appl. 28: 247-258.

Sorensen, A. A., R. S. Kamas, and S. B. Vinson. 1983c. The influence of oral secretions from larvae on levels of proteinases in colony members of Solenopsis invicta Buren. J. Insect Physiol. 29: 163-168.

Sorensen, A. A., J. T. Mirenda, and S. B. Vinson. 1981. Food exchange and distribution by functional subcastes of the imported fire ant, Solenopsis invicta Buren. Insect. Soc. 28: 383-394.

Summerlin, J. W., W. A. Banks, and K. H. Schroeder. 1975. Food exchange between mounds of the red imported fire ant. Ann. Entomol. Soc. Am. 68: 863-866.

Toom, P. M., E. W. Cupp, and C. P. Johnson. 1976. Amino acid changes in newly inseminated queens of Solenopsis invicta. Insect Biochem. 6: 327-331.

Van den Assem, J. 1971. Some experiments on sex ratio and sex regulation in the pteromalid, Lariophagms distinguendus. Neth. J. Zool. 21: 373-401.

Vander Meer, R. K. 1983. Semiochemicals and the red imported fire ant, Solenopsis invicta Buren. Fla. Entomol. 66: 139-161.

Vinson, S. B. 1978. The physiology of the imported fire ant. Proc. Tall Timbers Conf. Ecol. Anim. Control Habitat Manage. 7: 67-85.

Vinson, S. B. 1983. The physiology of the imported fire ant revisited. Fla. Entomol. 66: 126-139.

Vinson, S. B., and R. Robeau. 1974. Insect growth regulator: Effects on colonies of the imported fire ant. J. Econ. Entomol. 67: 584-587.

Voss, S. H. 1981. Trophic egg production in virgin fire ant queens. J. Ga. Entomol. Soc. 16: 437-440.

Waage, J. K. 1982. Sibmating and sex ratio strategies in scelionid wasps. Ecol. Entomol. 7: 103-112.

Wilson, E. O. 1978. Division of labor in fire ants based on physical castes (Solenopsis). J. Kans. Entomol. Soc. 51: 615-636.

Wilson, N. L., J. H. Dillier, and G. P. Markin. 1971. Foraging territories of imported fire ants. Ann. Entomol. Soc. Am. 64: 660-665.

25
Neurobiological Basis
of Chemical Communication
in the Fungus-Growing Ant,
Acromyrmex octospinosus

J. Delabie, C. Masson,
and G. Febvay

The antennae of an ant are essential for its social behavior, since they are the location of receptors that detect multimodal chemical (olfactory or gustatory), tactile, vibratory, proprioceptive, and/or thermal information. Once information is detected by these specialized antennal organs, it is coded by the neurons of the peripheral nervous system, transmitted by the antennal nerve to the brain (the deutocerebrum in particular) where it is redistributed, and combined with other information before being directed to the protocerebrum. It is at this point that possible behavioral responses are elaborated. This sensory mechanism has been named the "antennal system" (Masson 1973). The present study is concerned with the role of the antennal system in chemical communication.

As with most social insects (Wilson 1963a; Parry and Morgan 1979; Bergstrom 1981; Blum 1982; Fonta 1984), chemical communication plays an essential part in the survival of fungus-growing ants. The chemical signals on which communication is based are, for the most part, complex molecular mixtures that are generally referred to as allelochemicals or pheromones according to whether they elicit interspecific or intraspecific communication. Allelochemicals are subdivided into allomones that are emitted to the advantage of the individual emitting them and kairomones that are emitted to the advantage of the individual that perceives them (Karlson and Luscher 1959). They govern behaviors of the "plant-insect" or "insect-insect" type (for example, the odor of the fungus grown by the ant or the odor of harvested plants).

Pheromones are emitted by individual ants to elicit behavioral responses from other individuals of the same species, either by inducing a rapid modification in the behavior of the individual receiver (releaser pheromone) or by inducing a modification of its physiology (primer pheromone). Pheromones of the attine ants have been the object of numerous studies. The best known are the alarm (Crewe and Blum 1972; Parry and Morgan 1979) and trail pheromones

(Moser and Blum 1963; Tumlinson et al. 1971).

Individual ants are capable of reacting simultaneously to diverse inter- and intraspecific communication signals. In order to filter appropriate information from this complex mass of chemical signals, the ant must possess an elaborate sensory system capable of detecting signals that trigger each specific behavior.

We concentrated our research with Acromyrmex octospinosus on an analysis of this system from an anatomical-functional perspective by studying the flow of afferent information into the two successive stages of the nervous system: (1) the peripheral stage, the level of information receptors or sensilla on the antenna and (2) the central stage, particularily at the deutocerebral level. The analysis of the antennal system entailed two parameters—polymorphism and development. The polymorphic approach allowed us to study both anatomical and functional peculiarities of each group of adult ants; i.e., major and minor workers, queens, and males. The developmental study of the sensory system (last larval stage to adult) enabled us to understand the principles by which the antennal system was formed.

MATERIALS AND METHODS

The ants were collected from specific nests of Ac. octospinosus in the field in Guadeloupe (French West Indies) and then maintained in a rearing room at 25° to 30°C. The experiments were conducted at the Station de Zoologie in the Centre de Recherche Agronomique des Antilles et de la Guyane (I.N.R.A.) at Petit-Bourg, Guadeloupe and in the Laboratoire de Neurobiologie Sensorielle de l'Insecte I.N.R.A./C.N.R.S. at Bures sur Yvette (France).

Two anatomical techniques were used: (1) scanning electron microscopy (SEM) for studying the antennal sensilla of the imagoes and (2) classical histological and antennal afferent marking procedures with cobalt chloride (Pitman et al. 1973). The latter method allowed us, after specific treatment, to determine specific pathways of the central nervous system. With this technique, the cobalt ions migrate only into those severed nerve axons which it contacts, and they are able to pass through the synapses. This technique was used on ants of all ages, dated from the end of larval life to the adult stage. Some complementary tests were conducted with antennectomized ants.

For the functional studies of the antennal system, we used the technique of electroantennography. This technique permits the evaluation of variations in the global electric activity of the antennal nerve when the antenna is subjected to a given stimulus (Schneider 1962). An electroantennogram (EAG) appears in the form of a slow-wave potential whose amplitude can vary with stimulation, in that it is the result of the summation of the receptor potentials

304

from the stimulated neuroreceptors. These studies were conducted on live imagoes held in place by a retention system. The recording electrode, made of glass, was placed at the far end of the antenna while the indifferent electrode was placed on the insect's body. The chemical compound to be tested was placed in a glass tube through which nitrogen gas was passed for two-second intervals. During non-stimulation intervals the nitrogen flow was diverted to a neutral tube. In both cases, the flow was mixed with a constant quantity of nitrogen such that the antenna always experienced a constant gas flow. A variety of substances were tested; i.e., pheromone components (3-octanol, 3-octanone), various alarm-releasing compounds (1-octanol, carbon dioxide, formic acid), plant constituents (lemongrass essence, isoamyl acetate), and odors of the fungus symbiont.

RESULTS

Polymorphism-Peripheral Level

Morphological Study. The odor detection mechanism of the adult ant is located exclusively on the antenna and is made up of chemoreceptor sensilla unequally distributed on the antennal surface. The chemoreceptor sensilla of Ac. octospinosus belong to the following types: sensilla trichoidea, sensilla trichoidea curvata, sensilla basiconica, sensilla coeloconica, sensilla ampullacea, and sensilla chaetica. They have been identified on the basis of similar studies published on other species of ants using transmission electron microscopy (TEM) (Masson 1973; Walther 1981).

Our work has demonstrated a great similarity between different species of ants, in particular the virtual absence of the sensilla placodea which are found in the majority of other Hymenopteran families (Esslen and Kaissling 1976; Walther 1979; Fonta and Masson 1982). The similarities also were evident in the close proximity of the olfactory sensilla basiconica to the gustatory sensilla chaetica in Ac. octospinosus females. We were able to notice similar associations in the figures of previous publications on the antenna of various other ant families (Callahan 1975; Fresneau 1976; Walther 1981), but none of these authors had made specific reference to this apparent generality. It must also be pointed out that in certain female worker classes the sensilla chaetica occur alone, especially in minor workers characterized by a group of about 50 of these sensilla at the apex of the antenna. The males are devoid of sensilla basiconica and have only isolated sensilla chaetica. The number and distribution of these two types of sensilla are characteristic for each type of ant.

We also analyzed in detail the distribution of the olfactory sensilla trichoidea curvata. These sensilla, taken segment by segment, are arranged following an exponential function with the maximum number at the extremity of the antenna. Our observations agree with Masson's hypothesis (1973) that there is a gradient of antennal olfactory sensitivity with a maximum at the extremity of the antenna. This is not true for mechanical sensitivity. The distribution of the sensilla is also characteristic for each group of ants.

Certain data regarding the distribution of the sensilla and size of the antenna suggest that the sensory equipment varies not only in terms of sex and caste but also in terms of the size of the individual ant (Delabie 1984). This suggests that it is directly linked to the social role of the individual (polyethism) (e.g., studies on Atta by Wilson, 1963b).

Variation in the sensory equipment of the antenna among the castes has also been observed in other species of Hymenoptera (Esslen and Kaissling 1976 and Agren 1978 for the Apoidea; Jaisson 1970 and Walther 1981 for the Formicoidea), but there are some exceptions as shown by the sensilla placodea of Bombus sp. (Fonta and Masson 1982).

Based on the sensilla distribution of Ac. octospinosus, a hypothesis regarding their function may be advanced: (1) the gustatory sensilla chaetica may intervene in the recognition of the food substrate (fungus) since they are to be found mostly in the minor workers that cultivate the fungus and in the males which do not leave the nest before the mating flight; (2) the high number of sensilla trichoidea curvata in the males suggest that they have a role in the olfactory perception of females during the mating flight; and (3) the sensilla basiconica–sensilla chaetica association, which is more frequent in females of a large size (queens or major workers), suggests that it may play a part in the perception of odors outside the nest (trails, plant odors).

Functional Study. The hypotheses listed above were confirmed in some cases by electrophysiological recordings obtained for different odorous compounds. The EAG recorded for males showed that stimulation with a plant odor yielded a relatively low response with respect to that of other groups of ants suggesting that males, who are not adapted for cutting plant material, also do not take part in the search for, or preparation of, plant material. The fungus odor induces responses in males at a level comparable to that in the females.

It is particularly notable that female sensitivity to plant odors is more pronounced in the major workers that supply the vegetative materials to the colony than in the other females.

Polymorphism–Central Level

The deutocerebrum of Ac. octospinosus (second cerebroid ganglion belonging to the antennal metamere) has the characteristic structure of insect ganglions with a fibrous neuropile at the center and clusters of neurons at the periphery. It is formed by two lobes (Fig. 1): (1) the antennal or sensory lobe receives only sensory afferents from the antenna which, according to Suzuki's nomenclature (1975), are distributed in the deutocerebrum in two large characteristic bundles called tractus (T1 and T3) and in two minor bundles (T2 and T4); and (2) the dorsal lobe or antennal motor center from which the antennal musculature starts and where a bundle of sensory afferents called T5 end. It is, moreover, the site for a group of sensory fibers, T6, that pass to other regions of the protocerebrum and the suboesophageal ganglion.

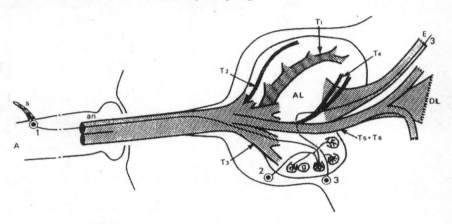

FIGURE 1. Organization of the antennal sensory projections in the deutocerebrum of Acromyrmex octospinosus (sagittal view). A = antenna; AL = antennal lobe; an = antennal nerve; DL = dorsal lobe; E = efferences; g = glomerulus; s = sensilla; T_1 to T_6 = sensory tracts T_1 to T_6. 1 = sensory neuron; 2 = interneuron; 3 = efferent deutoneuron.

The distribution of the sensory fibers emitting from the antenna appears similar to that described for the honey bee by Pareto (1972), Suzuki (1975), Mobbs (1982), and Arnold et al. (in press) and for Bombus sp. by Fonta (1984). This observation thus suggests a considerable homogeneity in the Hymenoptera.

The connections between afferent neurons and deutoneurons are in the glomeruli. These egg-shaped structures contain an association of several hundred afferent elements with only a few efferent deutoneurons and local interneurons. The main region of the association, where the synaptic connections are established, is at the periphery of the glomerulus and is called the "glomerular cortical layer" in Ac. octospinosus, as in other hymenopterans studied from this point of view (see Masson 1984; Delabie 1984; Fonta 1984;

Arnold et al., in press). Presumably, information coming from the antenna is grouped, integrated, and amplified at the level of the glomeruli before being transmitted to the association centers of the protocerebrum by the efferent deutoneurons. Cobalt chloride marking reveals the pathways through which information travels before or after the deutocerebral relay.

In the females, a more precise study of the number and position of the glomeruli in the deutocerebrum appeared to show caste differences. It is possible at this stage to note differences related to the size of the neuropile and the estimated number of antennal afferences. The former is linked to the number of neuroreceptors on the antennae of the different types of ants.

On the other hand, in the ventral region of the neuropile of the males, we were able to observe the existence of a large glomerular complex. This type of complex is known in the antennal lobe of the deutocerebrum of males of various species of insects, particularly cockroaches, moths, and certain Hymenoptera including the honey bee (Jawlowski 1948, 1957; Arnold et al. 1983, 1984, in press) and the bumble bee (Fonta 1984; Fonta and Masson, in press) and has been termed "macroglomerulus" or "macroglomerular complex."

It has been demonstrated that the macroglomerular complex of certain moths is involved in the recognition of pheromone signals (Boeckh and Boeckh 1979; Matsumoto and Hildebrand 1981). This is thought to be the case with Ac. octospinosus. This glomerular complex could function as part of a specialized system that involves the recognition of species-specific messages (pheromones) while the other glomeruli receive generalized messages involved in the recognition of odors such as food (Masson and Brossut 1981).

DISCUSSION

According to Wilson (1953), polymorphism is an adaptation to the division of labor, a factor that improves the efficiency of a colony. If this adaptation is clear for certain behavioral tasks, then our analysis also shows that its effects can be found in the intricate organization of the nervous system where it is linked to sensory functions at the level of the receptor system and the first relay of the antennal afferent sensory system.

Developmental Processes

The study of developmental processes in the antennal system between the end of larval life and the adult stage was undertaken on major workers using anatomical and physiological approaches. These two aspects of development are of a highly complex nature and are situated at the interface of two very different modes of life—larval life which is totally assisted by the adult members of the colony and is almost completely passive, and imaginal life which is autonomous.

The Antennal System of Larvae. The mode of odor perception in ant larvae is not well known; however, an organ given the name "larval antenna" was described for Ac. octospinosus and appears to have three chemoreceptor sensilla (Wheeler and Wheeler 1976). This organ, whose function is poorly defined, corresponds to a cupule visible in the frontal region of the cephalic capsule of the larvae. The imaginal disk of the antenna is attached under this cupule. Inside the disk numerous cells, different from either the epidermal cells or the innervation elements connected with the brain, can be observed. The latter is barely developed, and the antennal lobe of the deutocerebrum is nonexistent.

Morphogenesis of the Antennal System During Metamorphosis. The start of pupal morphogenesis (Fig. 2) was correlated with ejection of the meconium at the end of the larval stage. This event occurs simultaneously with the eversion of the pupal antenna. Between the ejection of the meconium and ecdysis to the pupal stage, the insect is called a "prepupa." After about five days the insect molts and enters the pupal stage. In major workers the stage lasts about 25 days at 26°C.

Simultaneously with the progression of morphogenesis of the nervous system in the antennal region is the development of structures of the peripheral nervous system. These make excellent reference points for histological studies.

The first two days of the prepupal period are best characterized by intense mitotic activity both at the level of the nerve cells of the deutocerebrum and the protocerebral association centers as well as the neuroglia and epidermal cells. Following this period, cellular multiplication decreases rapidly, and in three-day-old prepupae there is little mitotic activity except in the "globuli-cells" of the mushroom bodies. In this tissue, mitotic activity continues beyond pupal ecdysis as described by Masson (1970) for the ponerine ant, Mesoponera caffraria.

The axonal endings of the afferent neurons begin to colonize the neuropile of the antennal lobe just after eversion of the antennal imaginal disk. The organization of the antennal nerve into two main bundles is evident from the outset of growth. The neuropile of the antennal lobe, which then begins to grow, appears at this time to be a homogeneous structure without any differentiation.

During pupal ecdysis, the larval cuticle is shed and the antenna assumes its permanent position. Up to the age of four days, continuous growth of the antennal nerve occurs. The axonal endings of the sensory neurons gradually reach the deutocerebrum as does the growth of the neuropile of the antennal lobe. The two main fiber bundles, T1 and T3, are soon identifiable in the neuropile of this lobe. The texture of the neuropile remains homogeneous without any differentiation.

309

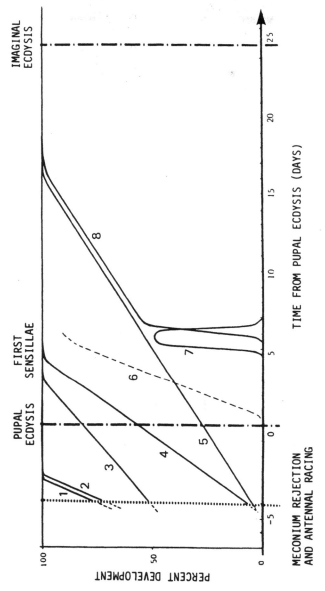

FIGURE 2. Ontogenesis of some structures belonging to the nervous system, during pupation: Mitotic activity (1) in the deutoneurons, (2) in the glial cells, (3) in the globuli cells in the mushroom bodies; (4) growth of the antennal nerve; (5) growth of the neuropile of the antennal lobe; (6) pigmentation of the compound eye; (7) densifications in the neuropile; (8) "pre-glomeruli" differentiation and glomeruli formation.

Antennal nerve growth is completed by the fourth day. This was shown in pupae of different ages by measuring sectioned portions of the nerve at the point where it penetrates the brain. The most important morphogenic phenomena of differentiation of the antennal system appear from the fifth day onwards as, for example, the elaboration of the antennal sensilla from the trichogenic cells. This phenomenon is always subsequent to the axonal elongation of the sensory neuron (Wigglesworth 1953).

In five-day-old pupae, the peripheral regions of the neuropile of the antennal lobe seem to condense and form clusters, while the neuropile center maintains a homogeneous structure. Very small, sharply outlined glomeruli, called "pre-glomeruli" in the honey bee (Arnold and Masson 1983), gradually individualize in these clusters. At seven days of age, this process is extended to the entire neuropile. At eight days, the general arrangement of the antennal lobe is the same as that in the adult. However, a volumetric growth study shows that it continues to grow until the pupa is about 16 days old. Differentiation of the glomerular cortical layer seems to take place during this period.

Consequences of Antennal Ablations. Several antennectomy experiments were conducted at different developmental stages in order to evaluate the respective roles of neurons in the morphogenesis of glomeruli. In every case, regardless of the time the operation was performed, ablation of the afferences resulted in a cessation of neuropile growth and blockage of the differentiation of the glomeruli. This indicates that the sensory afferences play an essential part in the morphogenesis of the deutocerebrum. If antennectomy occurs after glomeruli differentiation, no notable change in the organization of this region of the brain occurs.

Functional Aspects

Electrophysiological recordings of very young imagoes of known ages (hours after ecdysis) were made to study antennal variations in olfactory sensitivity (Delabie 1984; Delabie et al., unpublished). The electroantennograms indicated that when the imago emerges, weak electrical activity is detectable in the antennal nerve, suggesting that the antennal nerve originates prior to ecdysis. This observation is similar to that of Masson and Arnold (1984) for the honey bee and effectively corresponds with observations for the end of the pupal stage in certain moths (Payne et al. 1970; Schweitzer et al. 1976).

Electrical activity comparable to that of old major workers was obtained with three chemicals tested (3-octanone, isoamyl acetate, lemon-grass essence) after the second or third day of imago life (Delabie 1984; Delabie et al., unpublished).

Discussion

The small amount of data on the development of the antennal system of Ac. octospinosus can be summarized in the following way. Positioning of the imago structures (appendices, sensory neurons, sensilla) precedes the setting up of the synaptic connections with the central nervous system during the process of differentiation. Our research shows a period of olfactory maturation which, if similar to that reported for honey bees (Masson and Arnold 1984), may represent a sensitive or "critical period." Modification of the chemical environment during this "critical period" could permanently affect the ability of the adult insects to react to their allelochemicals by interfering with the stabilization of the afferent synaptic connections.

CONCLUSIONS

A thorough knowledge of the neurobiology of attine ants, both from polymorphic and developmental points of view, is needed to define new approaches to the rational control of these pests. This knowledge could provide information that would lead to:

1. Utilization of odorous compounds to modify or induce specific behaviors in order to disturb their social life;
2. A means to discover attractants for baits (Cherrett et al. 1973; Robinson and Cherrett 1974; Robinson et al. 1982);
3. Adaptations of control methods to certain growth phases of specific societal groups; and
4. Selection, within the framework of a plant genetic program, plant varieties with repellent or antifeedant compounds.

ACKNOWLEDGMENTS

The authors would like to thank Alain Kermarrec and Jean-Paul Torre-Grossa for discussions, Lydia Rousseau for her technical assistance, and Teresa Keane for the English translation.

REFERENCES CITED

Agren, L. 1978. Flagellar system of two species of Andrena (Hymenoptera: Andrenidae). Int. J. Insect Morphol. Embryol. 7: 73-79.

Arnold, G., and C. Masson. 1983. Mise en place des connexions synaptiques de la voie afferente antennaire au cours du developpement nymphal de l'ouvriere d'abeille Apis mellifica L. C. R. Acad. Sci. Paris 296: 131-136.

312

Arnold, G., C. Masson, and S. Budharugsa. 1983. Organisation spatiale du systeme nerveaux antennaire de l'abeille etudiee au moyen d'une technique de marquage aux ions cobalt. Apidologie 14: 127-135.

Arnold, G., C. Masson, and S. Budharugsa. 1984. Demonstration of a sexual dimorphism in the olfactory pathways of the drones Apis mellifica L. (Hymenoptera: Apidae). Experientia 40: 723-725.

Arnold, G., C. Masson, and S. Budharugsa. Comparative study of the antennal lobes and their afferent pathway in the worker bee and the drone (Apis mellifera L.). Cell Tissue Res. (In press).

Bergstrom, G. 1981. Chemical aspects of insect exocrine signals as a means for systematic and phylogenetic discussions in aculeate Hymenoptera. Entomol. Scand. Suppl. 15: 173-184.

Blum, M. S. 1982. Pheromonal bases of insect sociality: Communications, conundrums and caveats. Les Mediateurs Chimiques Agissant sur le Comportement des Insectes Versailles, 16-20, Nov. 1981. INRA publ., Les Colloques de l'INRA. 7: 1-7.

Boeckh, J., and V. Boeckh. 1979. Threshold and odor specificity of pheromone-sensitive neurons in the deutocerebrum of Antheraea pernyi and A. polyphemus (Saturnidae). J. Comp. Physiol. 132: 235-242.

Callahan, P. S. 1975. Insect antennae with special reference to the mechanism of scent detection and the evolution of the sensilla. Int. J. Insect Morphol. Embryol. 4: 381-430.

Cherrett, J. M., D. J. Peregrine, P. Etheridge, A. Mudd, and F. T. Phillips. 1973. Some aspects of the development of toxic baits for the control of leaf-cutting ants. Proc. VII Congr. IUSSI. London. pp. 69-75.

Crewe, R. M., and M. S. Blum. 1972. Alarm pheromone of the Attini, their phylogenetic significance. J. Insect Physiol. 18: 31-42.

Delabie, J. 1984. La communication chimique chez la fourmi champignonniste Acromyrmex octospinosus: Polymorphisme et developpement du systeme antennaire. These de Doctorat Troisieme Cycle. Univ. Paris VI, 251 pp.

Esslen, J., and K. E. Kaissling. 1976. Zahl und Verteilung antennaler Sensillen bei der Honigbiene (Apis mellifera L.). Zoomorphologie 83: 227-251.

Fonta, C. 1984. La communication chimique chez les bourdons (Bombus sp.): Une approche neurobiologique pluridisciplinaire. These de Doctorat de Troisieme Cycle, Univ. Paris VI. 228 pp.

Fonta, C., and C. Masson. 1982. Analyse de l'equipement sensoriel antennaire du bourdon Bombus hypnorum L. Apidologie 13: 247-264.

Fonta, C., and C. Masson. Organisation neuroanatomique de la voie afferente antennaire chez les bourdons males et femelles (Bombus sp.). C. R. Acad. Sci. Paris. (In press).

Fresneau, D. 1976. Recherche du role sensoriel de l'antenne dans l'ethogenese des soins aux cocons chez Formica polyctena par la technique des antennectomies. Doctorat de Troisieme Cycle, Psychosociologie animale, Paris. 99 pp.

Jaisson, P. 1970. Note preliminaire sur le polymorphisme sensoriel et l'existence d'un nouveau type de sensillum chez la fourmi champignonniste Atta laevigata Fred. Smith. C. R. Acad. Sci. Paris. 271: 1192-1194.

Jawlowski, H. 1948. Studies on the insect brain. Ann. Univ. M. Curie Slodowska C. 3: 1-30.

Jawlowski, H. 1957. Nerve tracts in the bee (Apis mellifica) running from the sight and antennal organs to the brain. Ann. Univ. M. Curie Slodowska D. 22: 307-323.

Karlson, P., and M. Luscher. 1959. "Pheromones," a new term for a class of biologically active substances. Nature 183: 55-56.

Masson, C. 1970. Mise en evidence, au cours de l'ontogenese d'une fourmi primitive (Mesoponera caffraria F. Smith), d'une proliferation tardive au niveau des cellules globuleuses ("Globuli-cells") des corps pedoncules. Z. Zellforsch. 106: 220-231.

Masson, C. 1973. Contribution a l'etude du systeme antennaire chez les fourmis. Approche morphologique, ultrastructurale et electrophysiologique du systeme sensoriel. These de Doctorat d'Etat, Universite de Provence, Marseille. 332 pp.

Masson, C. 1984. Neural basis of olfaction in insects, pp. 245-255. In L. Bolis, R. D. Keynes, S. H. P. Maddress (eds.), Comparative physiology of sensory systems. Cambridge Univ. Press.

Masson, C., and G. Arnold. 1984. Ontogeny, maturation and plasticity of the olfactory system in the worker bee. J. Insect Physiol. 30: 7-14.

Masson, C., and R. Brossut. 1981. La communication chimique chez les insectes. La Recherche. 12: 7-14.

Matsumoto, S. G., and J. G. Hildebrand. 1981. Olfactory mechanisms in the moth Manduca sexta: Response characteristics and morphology of central neurons in the antennal lobes. Proc. Roy. Soc. London. B 213: 249-277.

Mobbs, P. G. 1982. The brain of the honey bee Apis mellifera. I - The connections and spatial organization of the mushroom bodies. Phil. Trans. R. Soc. Lond. B 298: 309-354.

Moser, J. C., and M. S. Blum. 1963. Trail marking substance of the Texas leaf-cutting ant: Source and potency. Science 3572: 1228.

Pareto, A. 1972. Die zentrale Verteilung der Fuhlerafferenz bei Arbeiterinnen der Honigbiene Apis mellifera L. Z. Zellforsch. 131: 109-140.

Parry, K., and E. D. Morgan. 1979. Pheromones of ants: A review. Physiol. Entomol. 4: 161-189.

314

Payne, T. L., H. H. Shorey, and L. K. Gaston. 1970. Sex pheromones of noctuid moths: Factors influencing antennal responsiveness in males of Trichoplusia ni. J. Insect. Physiol. 16: 1043-1055.

Pitman, R. M., C. D. Tweedle, and M. J. Cohen. 1973. The form of nerve cells: Determination by Cobalt impregnation, pp. 83-97. In S. B. Kater and C. Nicholson (eds.), Intracellular staining in neurobiology. Springer Verlag.

Robinson, S. W., and J. M. Cherrett. 1974. Laboratory investigations to evaluate the possible use of brood pheromones of the leaf-cutting ant Atta cephalotes L. (Formicidae: Attini) as a component in an attractive bait. Bull. Entomol. Res. 63: 519-529.

Robinson, S. W., A. R. Jutsum, J. M. Cherrett, and R. J. Quinlan. 1982. Field evaluation of methyl 4-methyl pyrrole-2 carboxylate, an ant trail pheromone, as a component of baits for leaf-cutting ant (Hymenoptera: Formicidae) control. Bull. Entomol. Res. 72: 345-356.

Schneider, D. 1962. Electrophysiological investigation of insect olfaction, pp. 85-103. In Y. Zotterman (ed.), Olfaction and taste 1. Pergamon Press, Oxford, England.

Schweitzer, E. S., J. R. Sanes, and J. G. Hildebrand. 1976. Ontogeny of electroantennogram responses in the moth Manduca sexta. J. Insect. Physiol. 22: 955-960.

Suzuki, H. 1975. Antennal movements induced by odour and central projection of the antennal neurones in the honey bee. J. Insect Physiol. 21: 831-847.

Tumlinson, J. H., R. M. Silverstein, J. C. Moser, R. G. Brownlee, and J. M. Ruth. 1971. Identification of the trail pheromone of a leaf-cutting ant, Atta texana (Hymenoptera: Formicidae). Nature 234: 348-349.

Walther, J. R. 1979. Vergleichende morphologische Betrachtung der antennalen Sensillenfelder einiger ausgewahlter Aculeata (Insecta: Hymenoptera). Z. Zool. Syst. Evol.-forsch. 17: 30-56.

Walther, J. R. 1981. Die Morphologie und Feinstruktur der Sinnesorgane auf den Antennengeisseln der Mannchen, Weibchen und Arbeiterinnen der roten Waldameise Formica rufa (L.) mit einen Vergleich der antennalen Sensillenmuster weiterer Formicoidea (Hymenoptera). Inaugural Dissertation zur Erlangtung der Doktorwurde am Fachbereich Biologie der freien Universitat Berlin, 309 pp.

Wheeler, G. C., and J. Wheeler. 1976. Ant larvae: Review and synthesis. Mem. Entomol. Soc. Wash. 7, 108 pp.

Wigglesworth, V. B. 1953. The origin of sensory neurones in an insect, Rhodnius prolixus (Hemiptera). Quart. J. Microbiol. Sci. 94: 93-112.

Wilson, E. O. 1953. The origin and evolution of polymorphism in ants. Quart. Rev. Biol. 28: 136-156.

Wilson, E. O. 1963a. The social biology of ants. Ann. Rev. Entomol. 8: 345-368.

Wilson, E. O. 1963b. Caste and division of labor in leaf-cutter ants (Hymenoptera: Formicidae: Atta). I. The overall pattern in A. sexdens. II. The ergonomic optimization of leaf-cutting. Behav. Ecol. Sociobiol. 7: 143-165.

26
Chemical Taxonomy as a Tool
for Separating *Solenopsis* spp.

R. K. Vander Meer

Chemotaxonomy had its beginnings in 1886 when Helen Abbott published a paper relating the chemical constituents of plants to their morphology and evolution (citation from Smith 1976). Chemotaxonomy has since grown in importance; it is a particularly useful tool for plant systematics (Smith 1976), but has also been used in studies of animals (Wright 1974) and microorganisms (Ragan and Chapman 1978). The purpose of bringing the two seemingly divergent disciplines of chemistry and taxonomy together is to find objective characters which can aid in determining the genetic distances between organisms. Smith (1976), Wright (1974), and Ragan and Chapman (1978) have reviewed the extensive literature in this field. Howse and Bradshaw (1980) recently discussed several aspects of social insect chemical systematics.

We have found chemotaxonomy useful also for differentiating fire ant species and castes. In this review, I will discuss the venom alkaloids and cuticular hydrocarbons we have found useful in distinguishing fire ant species and then deal in more detail with how these characters have been used to address a variety of fire ant research problems. Finally, I will show how the use of the alkaloids and hydrocarbons, in conjunction with Dufour's gland components, have been used to demonstrate hybridization between Solenopsis invicta and S. richteri.

VENOM ALKALOIDS

Chemistry

The poison gland contents of Solenopsis species are primarily alkaloids rather than the proteins which are typical of most stinging Hymenoptera. Less than 1% of the venom is proteinaceous (Baer et al. 1979), with the remainder consisting of alkaloids in aqueous suspension. The venom alkaloids are perhaps the most well-known and studied class of chemicals from fire ants. Their basic structure was first identified correctly by MacConnell et al. (1970) as 2-methyl-6-alkylpiperidine (Fig. 1A, B). Depending upon the species, the methyl and alkyl groups can exist in cis or trans conformations relative to the piperidine ring; also the 6-alkyl group can have several chain lengths and may be saturated or contain a single double bond. For ease of reference, I will use a shorthand notation that specifies the conformation and chain length of the compound and whether it is saturated or unsaturated. For example, cis-2-methyl-6-n-undecylpiperidine is equivalent to cis-$C_{11:0}$ and trans-2-methyl-6-(cis-6-n-pentadecenyl)-piperidine simplifies to trans-$C_{15:1}$.

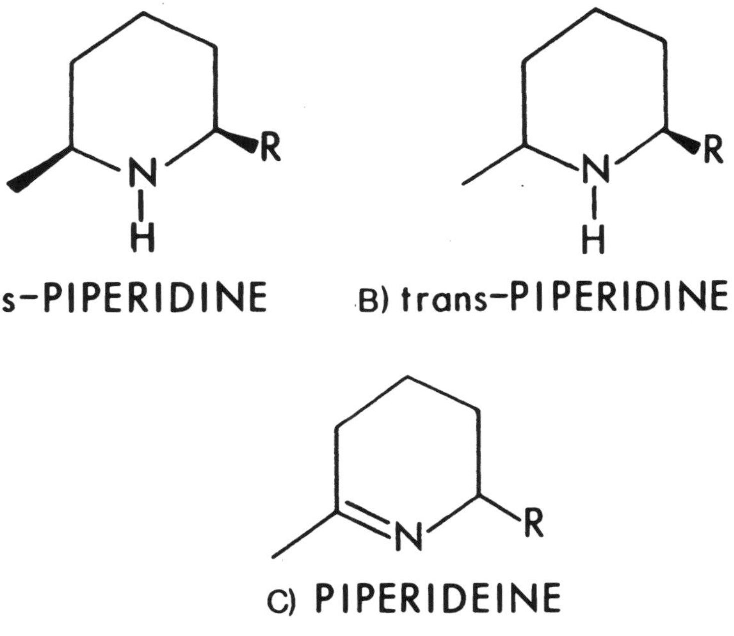

A) cis-PIPERIDINE   B) trans-PIPERIDINE

C) PIPERIDEINE

FIGURE 1. General chemical structures of the alkaloids produced by native and imported Solenopsis species in the United States.

Taxonomic Value

Comparative analyses of venom alkaloids from two native Solenopsis species (S. geminata and S. xyloni) and the two imported species (S. invicta and S. richteri) showed alkaloid patterns (Table 1) that clearly differentiated all four species (Brand et al. 1972; MacConnell et al. 1971). It is interesting to note that S. richteri and S. invicta are in the same S. saevissima complex (Buren 1972) and have qualitatively similar alkaloid patterns although they can be readily distinguished by the absence of C_{15} alkaloids in S. richteri venom. S. geminata and S. xyloni are in different species complexes (Creighton 1930); and although the cis $C_{11:0}$ alkaloid dominates both chromatograms, S. xyloni is distinctive due to the structurally unique $C_{11:0}$ piperideine (Fig. 1C). In contrast to the species-specific worker venom patterns, alate queens of all four species have similar alkaloid compositions consisting primarily of cis $C_{11:0}$ and trans $C_{11:0}$ alkaloids (Brand et al. 1973).

TABLE 1. Approximate piperidine alkaloid composition for four species of fire ants.[a]

Species	Percent piperidine alkaloid in venom					
	Cis $C_{11:0}$	Trans $C_{11:0}$	Trans $C_{13:1}$	Trans $C_{13:0}$	Trans $C_{15:1}$	Trans $C_{15:0}$
S. invicta		minor	20	15	40	20
S. richteri		20	70	10		
S. geminata	60	40				
S. xyloni[b]	70	30				

[a] Adapted from Brand et al. 1972 and MacConnell et al. 1971.
[b] S. xyloni has a distinctive minor component, $C_{11:0}$ piperideine (Fig. 1).

    S. geminata is the only Solenopsis species in the U.S. that has two non-sexual morphological castes, generally referred to as workers and soldiers (soldiers may be a misnomer since their modified large head and mandibles are used primarily for milling seeds). Brand et al. (1973) found that soldiers had a consistently higher ratio of cis to trans $C_{11:0}$ isomers than workers; however, there was considerable variation in this ratio within both castes. A positive correlation between soldier size and the cis to trans ratio led to overlap in the alkaloid isomer ratios of alates and soldiers.
    The variation in worker alkaloid ratios in the four major Solenopsis species in the U.S. does not detract from the chemotaxonomic value of venom alkaloids because the differentiating elements are distinguished by their presence or absence. It is not a question of how much.
    MacConnell et al. (1976) directly addressed the chemotaxonomic possibilities of venom alkaloids by analyzing the venom composition of 29 populations of 13 New World fire ant species. They concluded that the alkaloids do have taxonomic value but must be used in conjunction with all other taxonomic characters. Analysis of multiple venom samples from the four Solenopsis species in the

U.S. showed almost complete quantitative and qualitative homogeneity. Venom alkaloid samples from the saevissima complex in South America in most cases showed a strong correlation with morphological differences; however, there were a few instances where there were differences in morphology but identical alkaloid patterns. Like any taxonomic character, the venom alkaloids cannot be used alone.

Brand (1978) and Brand et al. (1973) used the differences among venoms of the four fire ant species in the southern U.S. to suggest a scheme for the evolution of venom components that placed the cis $C_{11:0}$ alkaloid closest to the ancestral form. They postulated that this was followed in evolutionary time by a biosynthetic switch to the trans isomer and then to an increase in the alkyl chain length. This model is based on only a few species, but it certainly suggests future research directions.

Minims as a Distinct Caste

The first workers produced by colony-founding queens (called either minims or nanitic workers) have been characterized as more timid (Oster and Wilson 1978) and much smaller (Dumpert and Johnson 1981) than their mature colony counterparts. These general observations led to an investigation of potential biochemical differences between nanitic workers and mature colony workers. In S. geminata, alates and the two worker castes have the same venom alkaloids but in different ratios. The most dramatic caste differences within a species are found in S. invicta female sexuals and workers. The venom of female alates is mainly cis-$C_{11:0}$ and trans-$C_{11:0}$ alkaloids, whereas the four trans C_{13} and C_{15} isomers dominate worker venom. Thus, the alkaloid patterns are characteristic of each caste. Surprisingly, minims from newly established colonies had their own distinct pattern consisting of about 94% of the $C_{13:1}$ alkaloid. This is in direct contrast to patterns of both the workers and female alates.

Since minims are the smallest workers produced in a colony, an argument could be made that worker alkaloid patterns vary with worker size and that minims are simply at the extreme end of this variation. Analysis of alkaloids in a range of mature colony worker sizes showed a strong correlation (r = 0.94) between worker head-width and the percent $C_{13:1}$ alkaloid. Extrapolation to the minim head-width clearly showed that the new colony minim alkaloid pattern did not fit the mature colony correlation (Vander Meer, unpublished results). These data clearly illustrate that the alkaloid pattern in minims is distinctly different from that of female alates and workers and that minims are not simply small versions of mature colony workers. The behavioral (timid) and morphological (size) factors, coupled with our biochemical studies, led to the conclusion that minims represent a distinct, although transient, caste.

HYDROCARBONS

Taxonomic Value

Cuticular hydocarbons have been used as a taxonomic tool in Diptera (Carlson and Service 1980; Carlson and Walsh 1981) and Hymenoptera (Carlson and Bolton 1984). In addition, hydrocarbons have been demonstrated to have a variety of physiological and behavioral roles in insect life (Howard and Blomquist 1982). Nelson et al. (1980) were the first to identify fire ant hydrocarbons, although they did not have chemotaxonomy in mind. They identified the cuticular hydrocarbons of S. invicta and S. richteri as saturated normal, mono, and dimethyl branched hydrocarbons. Although each had the same chromatographic peaks, the relative amounts were so distinctive that they could be diagnostic of the species.

Thompson et al. (1981) identified the five major hydrocarbons in the postpharyngeal gland of S. invicta as the same major components found by Nelson et al. (1980) in cuticular rinses. Further, Vander Meer et al. (1982) found that the postpharyngeal gland of S. invicta female alates and newly-mated queens contain large (15-50 ug) amounts of the species-specific hydrocarbons and that there is a significant quantitative increase shortly after mating. This increase is accompanied by a distinct change in the qualitative pattern. The function of these unusual amounts of hydrocarbons in the post-pharyngeal gland is not fully understood.

The characteristic S. invicta pattern of five major components appears to be ubiquitous in S. invicta colonies (Vander Meer et al. 1982) having been found in fat body, larvae, eggs, crop, and even in nest soil. In addition, there appears to be little phenotypic variability since colonies across the southern U.S. maintain the same distinctive qualitative pattern (Vander Meer, unpublished).

The cuticular hydrocarbons of S. geminata and S. xyloni are normal and monounsaturated compounds (Vander Meer, unpublished) with their own distinctive patterns. Therefore, hydrocarbon analysis can be used to differentiate the four major Solenopsis species in the U.S.

Myrmecophile Integration Mechanism

Several myrmecophilous beetles are associated with fire ant colonies. These beetles live in harmony with the ants despite the predaceous habits of the beetles and ants. One of these, Myrmecaphodius excavaticollis, has been found in nests of all four fire ant species. Analysis of cuticular hydrocarbons from beetles collected from a S. richteri colony showed a pattern that was qualitatively identical to that of its host's hydrocarbons (Vander Meer and Wojcik 1982). In addition, the beetles contained high molecular weight hydrocarbons not associated with their host. Analysis of beetles that had been removed from the ant host for 14 to 21 days showed that the high molecular weight compounds were innate to the beetle and that the host pattern was probably acquired by the beetles. To test this hypothesis, beetles were removed from S. richteri colonies and after two weeks introduced into S. invicta colonies. After a few days, the surviving beetles were analyzed for hydrocarbons; and, as predicted, they showed a combination of their own innate hydrocarbons and the species-specific hydrocarbon pattern of their new host, S. invicta. Based on these and other experiments, we postulated that the overall mechanism used for integration of M. excavaticollis into its host colonies involved an initial passive defensive behavior (armored exterior and death-feigning) that enabled it to survive long enough to acquire the species odor of its host, as well as the environmental part of the host's colony odor.

Hydrocarbons and Nestmate Recognition

The species-specific hydrocarbons, which account for over 70% of cuticular lipids (Nelson et al. 1980), can be diagnostic of colony odor transfer and may directly contribute to colony odor (Vander Meer and Wojcik 1982). There is not space to discuss nest-mate recognition in S. invicta at this time (see Jaffe, Chapter 18, for information on this subject) except to say that multivariate analysis of hydrocarbon patterns indicates that significant differences exist between different colonies and that S. invicta samples can be assigned to their respective colonies based on their cuticular hydrocarbon chromatographic patterns (Vander Meer, unpublished).

CHEMOTAXONOMIC EVIDENCE FOR HYBRIDIZATION

S. invicta and S. richteri are both indigenous to South America. S. richteri is known to occur in central Argentina, Uruguay, and extreme southern Brazil, whereas, S. invicta occurs in southwestern Brazil, Paraguay, and extreme northeastern Argentina (Lofgren et al. 1975). Their distribution in South America and their potential areas of overlap have not been fully defined. S. richteri was the first to arrive in the U.S. in the early 1900s; it was quickly displaced by the more aggressive S. invicta, which arrived in the late 1930s. Presently, S. richteri only occupies enclaves in northern Alabama and Mississippi (see Lofgren, Chapter 4). Although interaction between the two species in South America is not documented, it is certain that under the conditions of their accidental introduction into the U.S., they were forced into direct competition and contact with each other.

The taxonomy of the S. saevissima complex has had an interesting past and, as we shall see, will have an equally interesting future. Creighton (1930) described 16 forms in the saevissima complex and lumped these into 8 subspecies. Wilson (1952) further consolidated the complex into two species (S. interrupta and S. saevissima) and one subspecies (S. saevissima richteri). Wilson's view (1953, 1958) of the imported fire ant problem was that in South America the red form (S. invicta) and black form (S. richteri) represented two geographically separated races of the same species. When introduced into the U.S., the two forms interbred, producing a population explosion with the red form dominating and replacing the black form except in northeastern Mississippi.

Buren (1972) expanded the saevissima complex to six species. Most importantly, he elevated the red form to species status and named it S. invicta. The main criteria for the classification of S. richteri and S. invicta as separate species were: (1) constant phenetic characters of the two color forms and (2) no evidence for hybridization. However, biochemical evidence based on comparisons of venom alkaloids, Dufour's gland components, and cuticular hydrocarbon patterns demonstrate that hybridization does indeed occur (Vander Meer et al. 1985).

S. richteri is easily differentiated from S. invicta by the absence of C_{15} alkaloids (Brand et al. 1972). The venom alkaloid pattern of the hybrid contains the C_{15} alkaloids characteristic of S. invicta, yet its venom pattern is clearly not that of S. invicta either due to the large amount of trans $C_{11:0}$ alkaloid characteristic of S. richteri. Thus, the hybrid exhibits a blend of S. richteri and S. invicta alkaloid patterns.

As mentioned previously, the cuticular hydrocarbon patterns of S. invicta and S. richteri are also diagnostic of the species. Our data show that hybridization is reflected in the hydrocarbon patterns, since the hybrid exhibits a combination of what has been defined as the two pure species patterns.

Comparisons of the Dufour's gland trail pheromone components provides the most dramatic chemical evidence for hybridization. Gas chromatograph traces of extracts of S. invicta and S. richteri Dufour's glands are very different (Barlin et al. 1976). S. richteri has only one major peak (4 ng/WE) designated C-1/C-2 (see Vander Meer, Chapter 17). These components are found in minute quantities in S. invicta (75 pg/WE). We have not been able to detect S. invicta's major Dufour's gland component, Z,E-α-farnesene, in S. richteri Dufour's glands. In the Dufour's gland profile of the hybrid, S. invicta components dominate; however, the comparatively large amount of C-1/C-2 clearly differentiates the hybrid from pure S. invicta.

Behaviorally, a recruitment or point source bioassay demonstrates two-way species specificity in S. invicta and S. richteri. However, bioassays of the Dufour's gland extracts of hybrid workers and the two "pure" imported Solenopsis species (Table 2) clearly show that the hybrid responds to the trail recruitment pheromones of both species, and both species respond to the hybrid's trail recruitment pheromone.

TABLE 2.   Percent response to Dufour's gland extract in the trail recruitment bioassay.

| Test species | Dufour's gland source and percent response | | |
	S. invicta	S. richteri	Hybrid
S. invicta	100.0	14.4	93.5
S. richteri	13.3	100.0	107.5
Hybrid	91.1	113.6	100.0

We have monitored the distribution of S. invicta, S. richteri, and the hybrid in Mississippi and Alabama and found that S. invicta dominates in all areas except northeastern Mississippi (north of Meridian) and northwestern Alabama. The hybrid dominates north of Meridian to the area around Starkville. Further north, S. richteri becomes the major form.

We have developed techniques to analyze venom alkaloids from alcohol-preserved specimens. This has allowed us to analyze alkaloid patterns from ants collected in 1964 by Professor E. O. Wilson. These collections spanned the state of Mississippi from north to south and included areas where the black and red form interfaced at and around Meridian, Mississippi. Our analyses of these ant samples clearly demonstrated that hybridization occurred whenever and wherever S. invicta and S. richteri populations overlapped.

Our chemical analyses of the three biochemical characters unambiguously illustrate hybridization between S. richteri and S. invicta. Behavioral evidence to substantiate the biochemical data was obvious in a bioassay that showed that the hybrid responded to Dufour's gland extracts of both species. In addition, the gross morphological characteristics of the hybrid are those of S. richteri, and the hybrid generally has the color and gaster spot characteristic of S. richteri.

One important question raised by this research is whether or not S. richteri and S. invicta are indeed separate species, since the two major criteria used to assign species status were the lack of both hybridization and phenotypic variability among the two populations. These criteria are negated by our findings, since hybridization is expressed biochemically and not obviously through morphological characters.   The paradox of the situation is that if the ants are indeed separate species, then hybridization is occurring; however, if they are simply variations of a single species, then hybridization is not occurring. A definitive answer to this question may be found in a search for pre-mating isolation mechanisms in South America. This discovery provides a unique opportunity to study the ecology, taxonomy, behavior, and biochemistry of naturally occurring hybridization (at least non-laboratory induced).

CONCLUSIONS

Chemotaxonomy is a useful tool for addressing a wide range of questions in addition to those associated with the general taxonomy of Formicidae. We have identified three taxonomically useful biochemical characters and have demonstrated how these characters can be used to solve a variety of problems ranging from a myrmecophile host integration mechanism to the discovery of a new caste. These and other species-specific biochemical characters will aid in the solution of other diverse problems in the future.

REFERENCES CITED

Barlin, M. R., M. S. Blum, and J. M. Brand. 1976. Fire ant trail pheromones: Analysis of species specificity after gas chromatographic fractionation. J. Insect Physiol. 22: 839-844.

Baer, H., T. Y. Liu, M. C. Anderson, M. Blum, W. H. Schmid, and F. J. James. 1979. Protein components of fire ant venom (Solenopsis invicta). Toxicon 17: 397-405.

Brand, J. M. 1978. Fire ant venom alkaloids: Their contribution to chemosystematics and biochemical evolution. Biochem. Syst. Ecol. 6: 337-340.

Brand, J. M., M. S. Blum, and M. R. Barlin. 1973. Fire ant venoms: Intraspecific and interspecific variation among castes and individuals. Toxicon 11: 325-331.

Brand, J. M., M. S. Blum, H. M. Fales, and J. G. MacConnell. 1972. Fire ant venoms: Comparative analyses of alkaloidal components. Toxicon 10: 259-271.

Brand, J. M., M. S. Blum, and H. H. Ross. 1973. Biochemical evolution in fire ant venoms. Insect Biochem. 3: 45-51.

Buren, W. F. 1972. Revisionary studies on the taxonomy of the imported fire ants. J. Ga. Entomol. Soc. 7: 1-26.

Carlson, D. A., and A. B. Bolton. 1984. Identification of Africanized and European honey bees using extracted hydrocarbons. Bull. Entomol. Soc. Am. 30: 32-35.

Carlson, D. A., and M. W. Service. 1980. Identification of mosquitoes of Anopheles gambiae species complex A and B by analysis of cuticular components. Science 207: 1089-1091.

Carlson, D. A., and J. F. Walsh. 1981. Identification of two West African black flies (Diptera: Simuliidae) of the Simulium damnosum species complex by analysis of cuticular paraffins. Acta Tropica 38: 235-239.

Creighton, W. S. 1930. The new world species of the genus Solenopsis (Hymenoptera: Formicidae). Proc. Am. Acad. Arts Sci. 66: 39-151.

Dumpert, K., and C. Johnson. 1981. The social biology of ants. Pitman Advanced Publishing Program, Boston, MA. 298 pp.

Howard, R. W., and G. L. Blomquist. 1982. Chemical ecology and biochemistry of insect hydrocarbons. Annu. Rev. Entomol. 27: 149-172.

Howse, P. E., and J. W. S. Bradshaw. 1980. Chemical systematics of social insects with particular reference to ants and termites, pp. 71-90. In F. A. Bisby, J. G. Vaughan, and C. A. Wright (eds.), Chemosystematics: Principles and practice. Syst. Assoc. Spec. Vol. No. 16, Academic Press, N.Y., NY.

Lofgren, C. S., W. A. Banks, and B. M. Glancey. 1975. Biology and control of imported fire ants. Annu. Rev. Entomol. 20: 1-30.

MacConnell, J. G., M. S. Blum, W. F. Buren, R. N. Williams, and H. M. Fales. 1976. Fire ant venom: Chemotaxonomic correlations with alkaloidal composition. Toxicon 14: 69-78.

MacConnell, J. G., M. S. Blum, and H. M. Fales. 1970. Alkaloid from fire ant venom: Identification and synthesis. Science 168: 840-841.

MacConnell, J. G., M. S. Blum, and H. M. Fales. 1971. The chemistry of fire ant venom. Tetrahedron 26: 1129-1139.

Nelson, D. R., C. L. Fatland, R. W. Howard, C. A. McDaniel, and G. J. Blomquist. 1980. Re-analysis of the cuticular methyl-alkanes of Solenopsis invicta and S. richteri. Insect Biochem. 10: 409-418.

Oster, G. F., and E. O. Wilson. 1978. Caste and ecology in the social insects. Princeton University Press, Princeton, NJ. 352 pp.

Ragan, M. A., and D. J. Chapman. 1978. A biochemical phylogeny of the protists. Academic Press, N.Y., NY. 317 pp.

Smith, P. M. 1976. The chemotaxonomy of plants. Edward Arnold, London, England. 313 pp.

Thompson, M. J., B. M. Glancey, W. E. Robbins, C. S. Lofgren, S. R. Dutky, J. Kochansky, R. K. Vander Meer, and A. R. Glover. 1981. Major hydrocarbons of the post-pharyngeal glands of mated queens of the red imported fire ant Solenopsis invicta. Lipids 16: 485-495.

Vander Meer, R. K., and D. P. Wojcik. 1982. Chemical mimicry in the myrmecophilous beetle Myrmecaphodius excavaticollis. Science 218: 806-808.

326

Vander Meer, R. K., B. M. Glancey, and C. S. Lofgren. 1982.
Biochemical changes in the crop, oesophagus and post pharyn-
geal gland of colony-founding red imported fire ant queens
(Solenopsis invicta). Insect Biochem. 12: 123-127.

Vander Meer, R. K., C. S. Lofgren, and F. M. Alvarez. 1985.
Biochemical evidence for hybridization in fire ants. Fla.
Entomol. 68: 501-506.*

Wilson, E. O. 1952. The Solenopsis saevissima complex in South
America (Hymenoptera: Formicidae). Mem. Inst. Oswaldo Cruz
50: 49-68.

Wilson, E. O. 1953. Origin of the variation in the imported fire ant.
Evolution 7: 262-263.

Wilson, E. O. 1958. Recent changes in the introduced population of
the fire ant Solenopsis saevissima (Fr. Smith). Evolution 12:
211-218.

Wright, C. A. 1974. Biochemical and immunological taxonomy of
animals. Academic Press, N.Y., NY. 490 pp.

27
Diseases of Fire Ants:
Problems and Opportunities

D. P. Jouvenaz

During a 1973 taxonomic study of fire ants from South America, William F. Buren noticed subspherical, cyst-like bodies in the gasters of alcohol-preserved workers of Solenopsis invicta (Fig. 1). These "cysts" proved to be membrane-bound masses of spores of a microsporidium—the first specific pathogen known from fire ants[1] (Allen and Buren 1974). Buren's observation revitalized interest in biological control research on fire ants. Earlier, Broome (1974) and Federici (personal communication, B. F. Federici, Div. Biol. Control, Dept. Entomol., Univ. Cal., Riverside) had detected only non-specific, facultative pathogens in imported fire ants (IFA) in the United States. The virtual absence of specific pathogens of IFA in this country was later confirmed in an extensive survey by Jouvenaz et al. (1977).

The literature on the pathobiology and microbial control of fire ants consists of just 22 titles, of which 3 are non-primary articles and 4 concern non-specific microbial insecticides. Six of the remaining papers are notes or short papers that merely document observations of pathogens or report negative data. Thus, the study of the diseases of fire ants is largely in an exploratory stage.

SUMMARY OF FIRE ANT PATHOBIOLOGY

Pathogens and other natural enemies of IFA and the tropical fire ant, S. geminata (a nearctic species we have studied as a model) have been reviewed by Jouvenaz (1983). For the convenience of the

---

[1]Holldobler (1929) observed the haplosporidium Myrmecinosporidium durum infecting Solenopsis (Diplorhoptrum) fugax. This host is a thief ant; fire ants sens. str. are members of the subgenus Solenopsis.

reader, a summary of fire ant pathobiology is given here. The known pathogens are listed by type in Table 1.

FIGURE 1. Alcohol-preserved worker of S. invicta infected with Thelohania solenopsae. Note the white, spherical cyst-like mass in the partially cleared gaster.

TABLE 1. Specific pathogens of fire ants, Solenopsis spp.

Type of Pathogen	S. invicta[a]	S. geminata	Total
Virus	1	1	1 or 2
Bacterium	1?	–	1?
Fungus	1	–	1
Protozoa			
Microsporidia	2	4	6
Neogregarinida	2	1	2
Nematodes	1	1	2

[a]S. invicta and other members of the S. saevissima complex in South America.

Microsporidia

Only two of the six microsporidia, Thelohania solenopsae (Knell et al. 1977) and Burenella dimorpha (Jouvenaz and Hazard 1978), have been described. The description of a third species is in manuscript.

Thelohania solenopsae. Our knowledge of this species is essentially limited to the description by Knell et al. (1977). Other papers (Allen and Buren 1974; Allen and Silviera-Guido 1974; Jouvenaz et al. 1980) merely document its occurrence or incidence in host populations.

T. solenopsae infects the fat body of workers and sexuals, and the ovaries of queens. Infected cells hypertrophy, forming the cysts observed by Buren in alcohol-preserved specimens. These spore masses typically number 4 to 6 per gaster, although Knell et al. (1977) observed as many as 22 in single workers.

T. solenopsae is dimorphic, producing two morphologically distinct types of spores which develop simultaneously in the same tissues. The numerically predominant spore type is uninucleate and occurs in octets bound by a membrane (a sporophorous vesicle, previously called a "pansporoblast"). These "octospores" arise from plasmodia by endogenous budding. Spores of the secondary type are binucleate, are not bound by a membrane ("free spores"), and arise from disporous sporonts.

Ants infected with T. solenopsae exhibit no gross pathological signs or changes in behavior. Knell et al. (1977) state that "infected colonies may be as large as healthy ones in the field, but cannot be maintained in the laboratory as long as non-infected colonies. Thus, the effect of this microsporidium appears to be one of debilitation caused by destruction of the adult fat body."

It is uncertain whether T. solenopsae is one species or a complex of sibling species of microsporidia. It has been detected in a dozen or more described and undescribed Solenopsis spp. in Brazil, Argentina, Uruguay, and Paraguay. Of the 865 colonies (primarily S. invicta) that I have examined from Mato Grosso, Brazil, 67 (7.8%) were infected with this parasite. Attempts to transmit the infection perorally and by placing brood from heavily infected colonies in healthy colonies have not been successful (conspecific brood is adopted).

Burenella dimorpha. By far the best known pathogen of fire ants, B. dimorpha, has served as a model for my basic studies of pathobiology and for the development of protocols. This host-specific pathogen of S. geminata is locally available near Gainesville, Florida, can be transmitted perorally, and infected pupae exhibit pathognomonic signs.

As the specific name indicates, B. dimorpha also produces two morphologically distinct spores: binucleate free spores develop from disporous sporonts in the hypodermal tissues; uninucleate octospores develop in sporophorous vesicles from plasmodia in the fat body. The free spores develop before the octospores and (in contrast to T. solenopsae) predominate numerically. The development of octospores is temperature-dependent; at optimum temperature (28°C) they constitute ca 35% of the total spore population (Jouvenaz and Lofgren 1984).

Pupae infected with B. dimorpha develop clear, blister-like areas in the vertex of the head and in the petiole (Fig. 2). In sexual pupae, clearing may also develop in the dorsal thorax. The eyes appear sunken and irregular in outline, with deranged facets. These signs are pathognomonic for B. dimorpha infections and are the direct result of destruction of the developing cuticle. The blisters result from tissue fluids seeping between denuded hypodermal tissues and the pupal sheath. The brain and fat body atrophy and recede from the hydraulically distended pupal sheath. In areas where blisters do not occur, cuticle development is slowed or arrested, but the cuticle is not destroyed. The malformation of the eyes is due to destruction of the cuticular lenses, which leaves the ommatidia unanchored distally to become tangled, amorphous masses. The lamina ganglionaris is also destroyed, as is much other neural tissue (Jouvenaz et al. 1984). Infected pupae have never been observed to mature.

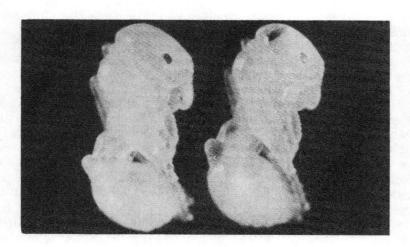

FIGURE 2. Pathognomonic signs of B. dimorpha infection in a pupa of S. geminata (right). Note clear areas in the vertex of the head and in the petiole, and compare the eye to the normal pupa at left (X25).

The intracolonial cycle of transmission of B. dimorpha is from pupae to fourth instar larvae. As the disease progresses, the blisters become more extensive, the cuticle becomes very fragile, and eventually it ruptures. The adult nurses cannibalize the ruptured pupae but do not ingest the spores into their crops. Instead, the spores and all other particulate matter are filtered, diverted to the infrabuccal cavity, and molded into a pellet. The infrabuccal pellet is expelled and placed on the praecipium of a fourth-instar larva, which consumes it ad lib. This larval instar is the only stage in the life cycle of S. geminata that is vulnerable to infection by B. dimorpha. The intracolonial transmission of infection is thus vectored mechanically by adult workers and is facilitated by the destruction of the cuticle (Jouvenaz et al. 1981a).

Only the free spores of B. dimorpha are infectious per os; the octospores are expelled unextruded in the meconium upon pupation. The function of the octospore is unknown, as is the mode of intercolonial transmission of infection. Intracolonial infection rates in field colonies are usually less than 5%, but may approach 100%.

Other microsporidia. One other microsporidium, a dimorphic species, is known to infect S. invicta in Brazil (Jouvenaz and Ellis, in press). Three undescribed species of microsporidia infect S. geminata in Florida, but they have not been studied.

Neogregarines

Mattesia geminata (Apicomplexa: Neogregarinida) infects S. geminata in Florida and Solenopsis spp. in Brazil (Jouvenaz and Anthony 1979). This genus is characterized by cycles of micronuclear and macronuclear merogony, and by gametogeny resulting in the development of two octonucleate spores within a membrane (a gametocyst). In M. geminata, the gametocyst membrane is transient, and the lemon-shaped spores are confined to the hypodermal tissues.

The signs of M. geminata infection occur in pupae and are pathognomonic. The developing eyes become blurred and irregular (much like those of pupae infected with the microsporidium B. dimorpha). The cuticle then melanizes abnormally, beginning in the legs and posterior margins of the sclerites of the gaster. The pupa progresses from a "sooty" appearance to almost solid black (Fig. 3). As in the case of B. dimorpha, infected pupae have never been observed to mature.

The mode of transmission of M. geminata is unknown; our attempts to transmit infection perorally and by placing diseased pupae in healthy colonies have failed. The intracolonial infection rates are usually less than 5% but may exceed 90%.

A second (undescribed) neogregarine infects S. invicta in Brazil. The spores of this species are morphologically distinct from

332

those of M. geminata and develop in fat body rather than hypodermal tissues. There are neither physical nor behavioral signs of infection, and infected ants survive into adulthood (Jouvenaz, unpublished).

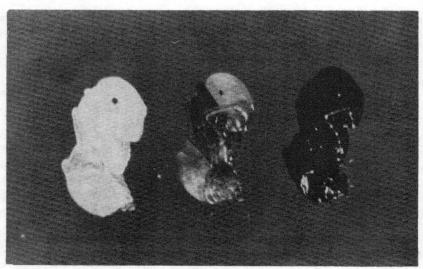

FIGURE 3. Pathognomonic signs of Mattesia geminata infection in pupae of Solenopsis geminata. The pupae are (left to right) normal, diseased, and moribund or dead (X18).

Fungus

The only microorganism that appears to be specifically associated with IFA in the United States is an unidentified, unicellular fungus that occurs in the haemolymph of S. invicta. The cells are club-shaped and multiply by budding. A mycelial form develops in vitro and in the haemolymph of lepidopterous larvae (which are susceptible to infection by injection only). The fungus can be transmitted perorally to healthy S. invicta colonies. Although the cells may become very numerous in the haemolymph, there are usually no physical or behavioral signs of infection. On occasion, however, high mortality with death preceded by tremors has occurred in infected laboratory colonies. It was not unequivocally established that the mortality was due to the fungus. Field populations of S. invicta are not reduced obviously even in areas where the infection rates approach 50%. Jouvenaz et al. (1977) found this organism in 93 (9.23%) of 1,007 colonies of S. invicta from six states.

Bacterium

A possible bacterial infection was observed in one colony of S. invicta in Brazil by Jouvenaz et al. (1980). The bacterium was motile, with a subterminal spore and a parasporal body that remained attached to the spore after disintegration of the sporangium. Unfortunately, it was not possible under field conditions at that time to isolate and culture this organism.

Virus

Virus-like particles (VLP) were found by Avery et al. (1977) in an unidentified species of fire ant of the S. saevissima complex from Brazil. The host colony was infected with an undescribed microsporidium and some individuals had dual infections. The colony was, in fact, collected because of the protozoan infection, and the VLP were discovered in the course of electron microscope studies of the protozoan. Similar VLP were subsequently detected in S. geminata. The VLP were rod-shaped and slightly tapered at both ends, and enclosed by double membranes. There was no evidence of occlusion of the VLP in a protein matrix.

The VLP occurred in both nuclei and cytoplasm of cells in the adipose tissue. The nuclei of infected cells greatly hypertrophied, and contained considerably less condensed nucleoprotein than normal cells. In a specimen in which ca 75% of the fat body cells were infected, both cytoplasm and nuclei were completely disrupted. The mortality rate in this colony was elevated; however, because of the dual infection, the cytopathology and mortality could not be attributed exclusively to either infection. There were no signs of infection in living ants, and the infection rate could not be estimated.

Nematodes

An unidentified nematode was found in the gasters of several alcohol-preserved workers of S. geminata by Mitchell and Jouvenaz (1985). The hosts had been collected months earlier as part of an ecological study and attempts to collect fresh material in the same area were unsuccessful. Jouvenaz and Wojcik (unpublished) subsequently found nematodes, apparently of the family Tetradonematidae (Mermithoidea), in Solenopsis spp. in Brazil. Studies on the latter nematode have just been initiated.

PROBLEMS AND OPPORTUNITIES

In their native land in South America, IFA are beset by a complex of natural enemies, including specific diseases, parasites, social parasites, live-in predators, and competitors. I have listed all of the known diseases in the first part of this paper, and undoubtedly, many more remain to be discovered. Information on the other types of organisms can be found in papers by Jouvenaz et al. (1981b), Jouvenaz (1983), Williams (1980), and Wojcik (Chapter 8). Almost all of these enemies were left behind when IFA were introduced into the United States. The obvious possibility of biological control presented by this situation demands investigation. The primary question raised by the current status of our knowledge is whether or not there is a future for the use of diseases of fire ants for biological control. While some of these may be very pathogenic, their incidence in natural populations is generally low. Essentially, nothing is known about intercolonial transmission of ant diseases. The potential for augmentative releases does not appear bright since none of the known organisms appear easily adaptable to mass propagation techniques. However, the existence of as many agents of disease as have been reported indicates opportunities do exist and their utilization for control of IFA populations in the United States depends upon our ability to unlock complex biological, pathological, and epizootiological relationships between the pathogens and their host. In the remainder of my paper, I will outline the biological and technical problems that lie ahead; however, the adage that problems are only opportunities awaiting discovery is appropos also for biocontrol of IFA.

BIOLOGICAL PROBLEMS

In Brazil, as in the United States, fire ants are most abundant in environments which have been disturbed by man (see Wojcik, Chapter 8). The fact that fire ants are common in many disturbed areas in Brazil may lead one to conclude that prospects for biological control in the United States are poor. This is not necessarily true. Environmental factors that limit natural enemies may exist in Brazil, but not in the United States. For example, if the extended dry season in the State of Mato Grosso (the homeland of S. invicta) is indeed a period of stress, the mortality differential between healthy and diseased colonies at that time may be accentuated, reducing the intercolonial incidence of disease. The dry season may also suppress putative vectors of disease. The wet season, with its lush vegetation, abundant food and water, and mating flights, would logically be a period in which fire ant populations increase. It would also—after a lag period—be a time in which density-dependent natural enemies increase. The return of the dry season, however, would reverse these population trends. The antithesis of this hypothesis is also possible; i.e., stress of dry season makes IFA more susceptible to disease. These hypothetical oscillations of populations of fire ants and diseases may actually prevent disease from reaching epizootic levels in South America. In the infested area of the United States, however, extended dry seasons do not occur; instead, there are relatively mild and short winters. Could this difference in the environment allow natural enemies to become more abundant in the United States than in Mato Grosso? Such seasonal effects, it must be emphasized, are purely hypothetical at present.

Also of concern for biological control is the tremendous reproductive potential of IFA which tends to mitigate the effects of biological control agents. In the absence of other enemies, pathogens may only serve to increase the rate of colony turnover without significantly affecting population density. Perhaps the primary value of pathogens in IFA management may be as stressors; colonies debilitated by disease may be less able to compete with native ants or deliberately introduced competitive exotic species (Jouvenaz et al. 1981b). Our goal should be to establish and encourage a complex of natural enemies of all types, and to develop pest management techniques that are compatible with these enemies.

The development of a microbial insecticide[2] is not a primary goal of our research. To effectively control an ant colony, a formicide must kill the queen(s) or stop reproduction. This requirement eliminates currently known microorganisms from candidacy as microbial formicides since they appear to affect only larvae or adult workers. Adult insects are typically refractory to infection by protozoa, viruses, and bacteria (bacteremia may result from injury). Certain fungi are able to infect adult insects; however, fire ants apparently fumigate their nests with venom (Obin and Vander Meer 1985), which has antimicrobial—including antifungal—properties (Jouvenaz et al. 1972; Cole 1975). Fire ant queens are also groomed meticulously and fed only regurgitated, filtered liquids. Thus, they appear to be invulnerable to most pathogens. The non-specific nematode, Steinernema feltiae (=Neoaplectana carpocapsae), and perhaps the recently discovered and as yet unidentified nematode from fire ants in Brazil may prove useful for single-mound, biological insecticide treatments. It must be noted, however, that effective and inexpensive chemical toxicants are readily available for this purpose.

Two of the more intriguing problems in the area of basic pathobiology may be related. These are (1) the mode of intercolonial transmission of infection and (2) the function of the octo-

---

[2]Microbial insecticides are applied in lieu of, or in conjunction with, toxicants to achieve rapid pest control in the treated area. The pathogen is not expected to become permanently established in the target pest population. Thus, microbial insecticides are pathogens that are employed in the manner of chemical toxicants.

spores of dimorphic microsporidia. Summerlin et al. (1975) demonstrated extensive food and possibly worker exchange between colonies of S. invicta. Such exchanges alone may affect transmission of those infections that occur in adult workers. Ants infected with B. dimorpha and M. geminata, however, do not mature. These infections are confined to the brood chambers; and, in the case of B. dimorpha, the ruptured pupae are cannibalized by nurses who feed the solids, including the spores, to fourth-instar larvae. In view of this behavior, one wonders how spores are transferred to brood in conspecific colonies frequently enough to sustain the pathogen.

Only the free spores of B. dimorpha are infective perorally for fire ant larvae. What then is the function of the octospore? Surely these morphologically very distinct spores have a function. And the raison d'etre of spores is to infect something. Since B. dimorpha octospores do not infect fire ants, they evidently infect something else—an intercolonial vector, perhaps. Or perhaps the vector is only mechanical, but "primes" the spore for extrusion by passage through the gut. In any event, this putative vector is proving elusive.

TECHNICAL PROBLEMS

The first technical problem encountered in microbial control of IFA research was the efficient screening of colonies for pathogens. This problem is still not completely solved. The phase-microscopic examination of aqueous mass extracts of samples of ants described by Jouvenaz et al. (1977) is quite sensitive for the detection of fungi, spore-forming protozoa, and nematodes. Pathogenic bacteria are difficult, though not impossible, to detect by this procedure. Viruses (with the possible exception of occluded viruses) are undetectable.

The first virus known from fire ants was seen during the electron microscopic study of protozoan-infected specimens from Brazil, as mentioned earlier (Avery et al. 1977). Avery et al. subsequently detected virus-like particles in phosphotungstic acid negative stains of ultracentrifuge pellets of extracts of S. geminata. Since electron microscope facilities are not available to us in Brazil, we must develop protocols for preserving partially processed extracts to send to the United States. Care must be taken to eliminate bacteria from these samples, or confusion of bacteriophage with ant viruses could result. Even if satisfactory procedures for preparing such samples are developed, the time lapse between collection of the colonies in Brazil and their examination in the United States presents obvious difficulties. Tissue culture techniques have not yet been developed for IFA.

Other technical problems that will be encountered before diseases may be established in the United States include: (1) the development of more rigorous laboratory and field evaluations of the

potential of candidates for introduction; (2) the safety of the
proposed introductions must be ascertained; (3) methods of mass
production and dissemination (including inoculum size, formulation,
mechanics of dispersal, schedule of dissemination) will have to be
developed; and (4) procedures for monitoring the establishment,
spread, and impact of introduced pathogens must be developed.

I have attempted to introduce the known diseases of fire ants
and to point out some of the problems and opportunities they present
for basic research in ecology and pathobiology, as well as the possibility of using them in biological control. It is evident that this
research is in its infancy. We are not yet acquainted with all of the
diseases of fire ants; we understand little or nothing of the role they
play in IFA population dynamics. Much work remains before we may
even estimate the probability of using pathogens to help manage IFA
populations.

REFERENCES CITED

Allen, G. E., and W. F. Buren. 1974. Microsporidian and fungal
diseases of Solenopsis invicta Buren in Brazil. J. N. Y.
Entomol. Soc. 82: 125-130.
Allen, G. E., and A. Silviera-Guido. 1974. Occurrence of micro-
sporidia in Solenopsis richteri and Solenopsis spp. in Uruguay
and Argentina. Fla. Entomol. 57: 327-329.
Avery, S. W., D. P. Jouvenaz, W. A. Banks, and D. W. Anthony. 1977.
Virus-like particles in a fire ant, Solenopsis sp. (Hymenoptera:
Formicidae) from Brazil. Fla. Entomol. 60: 17-20.
Broome, J. R. 1974. Microbial control of the imported fire ant,
Solenopsis richteri Forel (Hymenoptera: Formicidae). Ph.D.
dissertation, Mississippi State University, Mississippi State,
MS. 66 pp.
Cole, L. K. 1975. Antifungal, insecticidal, and potential chemo-
therapeutic properties of ant venom alkaloids and ant alarm
pheromones. Ph.D. dissertation, University of Georgia, Athens,
GA. 172 pp.
Holldobler, K. 1929. Uber eine merkwurdige Parasitenerkrankung
von Solenopsis fugax. Z. Parasitenkunde 2: 67-72.
Jouvenaz, D. P. 1983. Natural enemies of fire ants. Fla. Entomol.
66: 111-121.
Jouvenaz, D. P., and D. W. Anthony. 1979. Mattesia geminata sp. n.,
(Neogregarinida: Ophrocystidae) a parasite of the tropical fire
ant, Solenopsis geminata (Fabricius). J. Protozool. 26: 354-356.
Jouvenaz, D. P., and E. A. Ellis. Vairimorpha invictae, n. sp.
(Microspora: Microsporida), a parasite of the red imported fire
ant, Solenopsis invicta Buren (Hymenoptera: Formicidae). J.
Protozool. (In press).

338

Jouvenaz, D. P., and E. I. Hazard. 1978. New family, genus, and species of Microsporida (Protozoa: Microsporida) from the tropical fire ant, Solenopsis geminata (Fabricius) (Insecta: Formicidae). J. Protozool. 25: 24-29.

Jouvenaz, D. P., and C. S. Lofgren. 1984. Temperature-dependent spore dimorphism in Burenella dimorpha (Microspora: Microsporida). J. Protozool. 31: 175-177.

Jouvenaz, D. P., G. E. Allen, W. A. Banks, and D. P. Wojcik. 1977. A survey for pathogens of fire ants, Solenopsis spp., in the southeastern United States. Fla. Entomol. 60: 275-279.

Jouvenaz, D. P., W. A. Banks, and J. D. Atwood. 1980. Incidence of pathogens in fire ants, Solenopsis spp., in Brazil. Fla. Entomol. 63: 345-346.

Jouvenaz, D. P., M. S. Blum, and J. G. MacConnell. 1972. Antibacterial activity of venom alkaloids from the imported fire ant, Solenopsis invicta Buren. Antimicrob. Agents Chemother. 2: 291-293.

Jouvenaz, D. P., E. A. Ellis, and C. S. Lofgren. 1984. Histopathology of the tropical fire ant, Solenopsis geminata, infected with Burenella dimorpha (Microspora: Microsporida). J. Invertebr. Pathol. 43: 324-332.

Jouvenaz, D. P., C. S. Lofgren, and G. E. Allen. 1981a. Transmission and infectivity of spores of Burenella dimorpha (Microsporida: Burenellidae). J. Invertebr. Pathol. 37: 265-268.

Jouvenaz, D. P., C. S. Lofgren, and W. A. Banks. 1981b. Biological control of imported fire ants: a review of current knowledge. Bull. Entomol. Soc. Am. 27: 203-208.

Knell, J. D., G. E. Allen, and E. I. Hazard. 1977. Light and electron microscope study of Thelohania solenopsae n. sp. (Microsporida: Protozoa) in the red imported fire ant, Solenopsis invicta. J. Invertebr. Pathol. 29: 192-200.

Mitchell, G. B., and D. P. Jouvenaz. 1985. Parasitic nematode observed in the tropical fire ant, Solenopsis geminata (F.) (Hymenoptera: Formicidae). Fla. Entomol. 68: 492-493.

Obin, M. S., and R. K. Vander Meer. 1985. Gaster flagging by fire ants (Solenopsis spp.): functional significance of venom dispersal behavior. J. Chem. Ecol. 11: 1757-1768.

Summerlin, J. W., W. A. Banks, and K. H. Schroeder. 1975. Food exchange between mounds of the red imported fire ant. Ann. Entomol. Soc. Am. 68: 863-866.

Williams, R. N. 1980. Insect natural enemies of fire ants in South America with several new records. Proc. Tall Timbers Conf. Ecol. Anim. Control Habitat Manage. 7: 123-134.

28
Protection of Leaf-Cutting Ants from Biohazards: Is There a Future for Microbiological Control?

A. Kermarrec, G. Febvay, and M. Decharme

"Everything that debilitates our enemy must be looked into, may it even be a chariot wheel " Batra and Batra, 1978.

There is an urgent need for a complete re-thinking of the control methods for leaf-cutting ants (see Cherrett, Chapter 29), not only because of the cost of damage by ants to man's cultivated crops but also with respect to the often overlooked and difficult to evaluate costs of unrational control strategies. Also, because leaf-cutting ants (LCA) remain specifically a problem of tropical and developing countries, new plant protection strategies are urgently needed. In this context, biological control measures need to be examined closely to evaluate their biological fitness, phytosanitary power, and social costs. Biological control certainly is a very promising research area; but in actual practice, it is clear that more basic biological knowledge is needed to allow it to be applied successfully on LCA.

This work summarizes almost ten years of observations and unpublished experimental results on the parasites and pathogens of the only Guadeloupean LCA, Acromyrmex octospinosus, and its symbiotic fungus. It emphasizes the difficulties encountered in the experimental processes and tends to correlate them to particular biological adaptations and special morphological features of LCA.

First, we will discuss LCA-nematode relationships, which have already been described by Kermarrec (1975) and Laumond et al. (1979) who tried to control Ac. octospinosus with Neoaplectana carpocapsae, and its symbiotic fungus with mycophagous nematodes (see Kermarrec et al., Chapter 20). Secondly, we will analyze some ecopathological aspects of several strains of entomopathogenous fungi (Deuteromycotines, Hyphomycetes) used against LCA and as potential parasites of the symbiotic Basidiomycete. Only the very first stage of these studies concerning an entomopathogenous (but dangerous for man) Entomophthorale (Entomophthora coronata) has been published (Kermarrec and Mauleon 1975).

MATERIALS AND METHODS

The experimental nests of Ac. octospinosus were maintained in the laboratory under controlled environmental conditions on various plants. Mycophagous (Aphelenchoides composticola, Ditylenchus myceliophagus), saprophagous (Rhabditis spp.) and entomoparasitic (Neoaplectana/=Steinernema carpocapsae, strains of "agriotos") nematodes were obtained from the live collections at INRA-Antibes and were reared according to standard methods (Kermarrec 1973, 1975; Laumond et al. 1979). The symbiotic fungus was cultivated according to the technique described by Decharme and Issaly (1980). Three entomophagous Hyphomycetes from the INRA-la Miniere collection (Metarrhizium anisopliae, Beauveria bassiana, Paecilomyces fumoso-roseus), as well as the fungus, Trichoderma viride, that is parasitic to the symbiont, were cultivated on simple nutritive agar medium favorable to good sporulation (Barnes 1975). The three strains of Trichoderma (INRA-Bordeaux collection) were pre-selected for their high thermal optima and activity against higher Basidiomycetes, such as the Psalliota that is cultivated by man and is phyletically very close to the symbiont. The sensitivity of the ants to pathogens and parasites was tested in vitro according to a method similar to that of Laumond et al. (1979) and Riba et al. (1982). They used Petri dishes containing absorbent paper to which the inoculum was applied in 1 ml of water. Ten healthy worker ants were placed in the dish; a small container of glucose-water (2%) was added as an energy source for the ants. The histological fixatives and the procedures for the examination of the fungus-infected ants were those used by Amargier and Vago (1966). The paraffin sections were cut (7 um thick) with a Wild microtome.

Extracts from the salivary glands of adult worker ants were prepared and their chitinolytic activity measured according to the procedure of Febvay et al. (1984a). The antibiotic secretions of the symbiotic fungi were examined after ethyl-acetate extraction from a liquid culture (Decharme 1978). Details are given in other parts of this text.

The behavior of workers encountering diverse biohazardous situations was examined as they walked along laboratory trails and in foraging areas that were easy to observe and control. The rate of food pick-up was based on the confetti method used by Febvay and Kermarrec (1984b). Necrophorism (with fungus-infected corpses) was tested in the same way by making observations on general hygienic behavioral patterns. The details of each test are described in the text. Electroantennograms (EAG) were made with an apparatus similar to that described by Delabie et al. (Chapter 25).

EXPERIMENTAL PARASITOLOGY WITH NEMATODES

The mycophagous or saprophagous nematodes that either sting mycelia of the fungi or introduce undesirable bacteria were not effective in controlling the fungi or LCA. Populations of nematodes in the fungus–gardens decreased with time in a quasi–exponential manner.

The infesting stage of the entomoparasitic nematode N. carpocapsae (designated as "dauer larven" or "winkenden larven" (Bovien 1937)) are reported to penetrate the insect through the anus, mouth, trachea, or intersegmental membranes. Within the insect they release bacteria that cause a septicemia that results in the insect's death (Boemare et al. 1982, 1983). Approximately 10^5 "dauer" larvae of N. carpocapsae were placed in contact with 20 ant larvae (3rd and 4th instars), 20 young pupae (recently molted), and 20 old pupae (close to the final molt). A similar batch, killed by heat, was exposed to the same parasitic conditions. Dissections performed five days later clearly showed the ability of the nematode to penetrate ant brood as 100% of the 3rd and 4th instar larvae (dead or alive) were infected. The penetration rate decreased from 80% in the young pupae to 5% in the old pupae. Host suitability (= the development of the parasite up to the fertile adult stages) also decreased sharply with the age of the brood (not more than 10% of the parasitized pupae contained adults of the parasite against 95% to 100% of the 3rd and 4th instar larvae). Up to ten large first generation females of the nematode have been counted in parasitized ant larvae.

One hundred workers were exposed to either a single initial inoculum or exposed to a fresh inoculum every seven days. The mortality due to parasitism was slow to develop but was nevertheless real and higher under sustained inoculation. However, dissections clearly revealed that no nematodes developed in the corpses and, that in most cases (70-80%), the infesting stages were found concentrated in the buccal sphere; i.e., the infra-buccal pocket and other mouth parts. One parasitic larvae was introduced by microsurgery into the thoracic cardiac sinus of each of several adult workers. The cuticle was closed with synthetic glue. The ants continued normal activity for a while; all the parasites developed to the adult stage.

Other ants of Guadeloupe (Camponotus of the senex and picipes-fumidus groups; Solenopsis geminata) placed in the same experimental conditions showed a lesser resistance to penetration. The infesting larvae developed satisfactorily at least up to the fourth or pre-adult stage. The thorax of Camponotus was particularly affected.

About 3×10^6 dauer larvae of N. carpocapsae were injected at days 0 and 5 into four fungus gardens of about 2 liters in volume.

342

The size of the comb was reduced by one-third, but no dead brood were removed. Intense social grooming combined with cleaning and building activities (closing of openings) followed the inoculations. After ten days no living nematodes were found in the fungus gardens after extraction by the Baermann (1917) method.

EXPERIMENTAL PATHOLOGY WITH FUNGI

Kermarrec et al. (Chapter 20) present a number of ecological aspects of the Basidiomycete symbionts of LCA. Ponchet (1982) portrayed T. viride as having the weapons of the perfect antagonist for the Basidiomycete: antibiotic, competitive, lysogenic, and parasitic. The three strains selected (Nos. 19 and 285 and "oisan") were thermophilic and monosporous. Their antagonistic action had been proven in vitro on the symbiont which was rapidly invaded and killed by true parasitism. The antagonistic substances excreted by the Trichoderma are thermostable at 110°C for 20 min. In our tests, however, none of the three strains had a destructive effect when injected directly into the fungus gardens of two nests (10 ml of suspension containing 10^{12} spores/ml dispersed by Tween 80).

A preliminary survey list of important collections of viruses, bacteria, fungi, and protozoa was made, but it is too long to list here. However, certain Deuteromycotina (Hyphomycetes) were selected as potential candidates for biological control of the attines. These were B. bassiana (Bb) (Fig. 1), M. anisopliae (Ma), and P. fumoso-roseus (Pf). These species sporulate abundantly and a strain of each (Bb 32; Ma 115; Pf 3) was selected from 8-year-old cultures for pathogenicity tests. They were tested at concentrations of 10^9 spores in 1 ml of water. The thermal optima of these strains is between 25° and 28°C. The maximum diameter of growth was reduced by one-half for Pf 3 when the culture temperature increased from 27° to 29°C; whereas, for Bb 32 and Ma 115, a temperature of 32°C was necessary to achieve the same effect. The dose-response of LCA to the strains has been shown in vitro, and the lethal time diminishes regularly with an increased concentration of propagules. In general, the tests showed that the conidiospores (surface cultures) possess a greater virulence than the blastospores of the same strain (shaken liquid medium). In addition, after ten contacts with host ants, the pathogenic efficiency was enhanced considerably compared to the cultures kept in the mycotheque. The penetration of these mycelia by perforation of the cuticle and the classic formation of an appressorium have been followed by suitable histological techniques. A humid atmosphere is necessary for high spore virulence to develop on the cuticle. After six days, the mortality rate is reduced two-fold if the environment is kept dry.

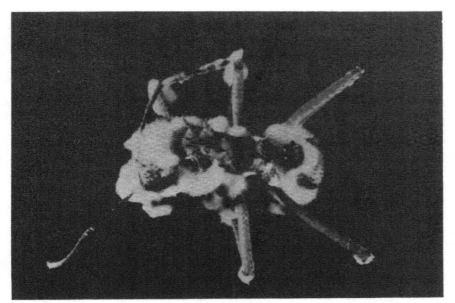

FIGURE 1. Worker of <u>Ac</u>. <u>octospinosus</u> killed by <u>Beauveria</u> <u>bassiana</u>. Mycelia sporulate outside the body through inter-segmental membranes.

Inoculations of laboratory and active field nests with these pathogenic strains never led to tangible results in spite of using maximum infesting spore concentrations in direct and sudden applications onto the fungus gardens.

MORPHOLOGICAL, SENSORY, AND BIOCHEMICAL PROTECTIVE FACTORS

The extraordinary resistance of the attine-Basidiomycete symbiont to epizootic and epiphytic diseases is due to many factors linked to various actions of the two partners in maintaining the internal hygiene of the nest. This homeostasis is studied here through some of its most obvious aspects which, in no way, precludes the existence or the importance of other as yet unknown factors.

Protective Morphology

Laumond et al. (1979) have already reported the resistance of adult <u>Acromyrmex</u> to penetration by entomoparasitic rhabditids. The experiments described above confirm this initial observation and bring the causes into focus. Scanning electron microscope and histological sections showed that there are few possibilities for penetration of the ants (Fig. 2). The dimensions of a <u>Neoaplectana</u> larvae are 618 ± 27 um in length by 32 ± 8 um in diameter. The potential points of entry on ant larvae are: the cuticle, the mouth (it does not have a filtering mechanism as in the adults), the orifice of the labial gland, the anus, and the spiracles. The cuticle represents an important protective barrier for the insect. Its overall rigidity and hardness give the cuticular surface structures an essential protective role (Locke 1974), certainly impenetrable by the nematode. All measurements were recorded from major worker larvae of the last instar to maximize the dimensions: buccal orifice (44 um), pharynx (18 um), sericteries' opening (4 um), anus (10 um) and rectum (4 um), and peritremes of the spiracles (9.2 ± 2.0 um, n = 102). The atria, or spiracular chambers, possess spinules (Fig. 2B). The size of these openings shows that a nematode of 30 um diameter would have difficulty entering the larvae, and only then by the mouth or membranous cuticular areas.

In adult ants, certain morphological features provide protection from invasion. For example, the infra-buccal filter (Quinlan and Cherrett 1978; Febvay and Kermarrec 1981) does not permit the nematodes to enter the pre-pharynx but retains them in the infra-buccal pocket. Moreover, the mouth can be completely closed if there is an overly strong exposure to nematodes (Fig. 2A). The anus may be protected by its pilosity. The abdominal spiracles are only open on maximum dilations of the abdomen. The only two possible points of nematode entry are the buccal area, by forcing through the pharyngeal glands, and the articulating cuticular membranes.

Sensory Aspects of Protective Behavior

Kermarrec (1975) states there is antennal sensing (probably by mechano-reception) of the nematodes by the ants, followed by alarm and active cleaning of the contaminated parts, as well as inter-individual licking. The tarsal comb is often used against the dauer larvae which behave phoretically and climb onto the antennae.

The largest dimensions of the conidiospores of pathogenic Hyphomycetes range from 2 to 7 um. The sensitivity of the ants' antennae is illustrated by the fact that attractive baits containing spores are not picked up, and cultures sporulating on agar are immediately cut up and taken outside the nest area. It seems that agar processed by worker ants no longer permits the germination of spores. Moreover, "dangerous" items are clearly distinguished from "non-dangerous" ones since workers will cover pathogenic Hyphomycetes in a Petri dish with various materials (clean agar, paper, vegetable debris). This activity is accelerated by the proximity of the pathogen to the symbiotic fungus. In a box containing two species of fungi (for example, <u>Metarrhizium</u> and a wholly saprophytic fungus), the bio-hazardous one will be covered by cuttings from the other. Colored confetti were used to check this behavior. In this experiment, spore-contaminated confetti was covered with incredible care with clean material. This action was accompanied by very precise movements of the fore-legs to flatten down the neutral material onto the "dangerous" object. Only clean confetti was integrated into the nesting space, while the contaminated confetti was taken to the refuse area along a new separate trail. Licking and use of the tibio-tarsal comb was intense.

The hygiene of an animal society depends to a large extent on an efficient management of unhealthy individuals and infected corpses. Necrophoric behavior is prominent among Acromyrmex, and a non-infected corpse (a worker killed by freezing) may even be exchanged with a fungus-infected one (Fig. 1) on the way to the refuse area. A sporulating corpse may be cut up on the spot and the infectious spores agglutinated in small pellets in the infra-buccal pocket and removed one by one from the nesting zone. Dry spore powder placed in minute quantities on a trail causes arrestant behavior and brings about antennal exploration and licking.

An electrophysiological study of the ability of LCA to perceive fungal spores has been attempted (Table 1). In another trial, 2-heptanone caused twice as much depolarization as that of the specific alarm pheromone (3-octanone). Confetti impregnated with an n-pentane wash of Metarrhizium spores was 2 to 3 times more readily collected than control confetti (pure n-pentane).

Biochemical Aspects

Kermarrec (1975) observed that rectal fluids and fungus garden extracts of Ac. octospinosus had nematocidal effects.

Worker ants isolated with sporulating fungal cultures (entomopathogenic or phytoparasitic) develop a white wax-like cuticular deposit three to four times more frequently than control workers. Whitened ants soaked in a concentrated Metarrhizium spore suspension have only a 10% mortality rate compared to 50 to 100% for the control. It must also be noted, without explanation, that the cuticle of small workers is shiny compared to the dull cuticle of large workers.

Spore pellets formed by the infra-buccal pocket are coated with salivary secretions containing thermoresistant chitinolytic enzymes (Febvay et al. 1984a) from the labial glands. Quantitative analysis showed that small workers secrete three times as much N-acetyl- β -D-glucosaminidase as larger workers (1.68 vs 0.64 nmoles of pNP/min/mg). Febvay et al. (1984a) demonstrated digestion of the chitin of the fungal walls by the contents of the salivary glands. An "anti-germinating power" was revealed when sterile inoculations of Metarrhizium spores were made in 1 ml of glucose-water (2%)

together with 20, 50, and 100 crushed and filtered labial glands of large workers. The frequency of germination was decreased to zero after a 12-hour period of agitated incubation with 50 glands. An effect also appeared on the first hyphae produced which were three times thinner and 20% shorter than control hyphae.

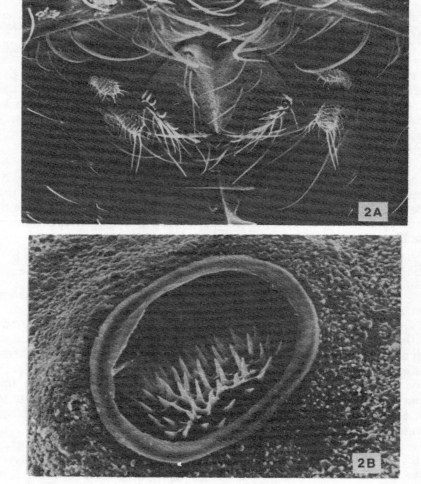

FIGURE 2. Some of the morphological features contributing to the protection of Ac. octospinosus from parasitic nematode penetration. A) Tightly closed mouth opening and sensorial mechanoreceptors. B) Close view of an abdominal spiracle with spinules in the atrium.

TABLE 1. Electroantennographs (in mV, absolute maximum observed values) obtained with various products on antennae of medium workers. (Stimulation: 1 sec; circuit impedance: 7 mega-OHM; oscilloscope sensitivity: 0.15 mV).

Product	Antennal nerve depolarization	Behavioral response
3-Octanone	1.8	alarm pheromone
3-Octanol	0.5	alarm pheromone (secondary product)
Crushed heads	1.0	alarm inducing
2-Heptanone	1.5	alarm inducing
Fungus garden	1.8	strong attraction
Beauveria spores	1.0	arrestant
Metarrhizium	0.6	arrestant
Paecilomyces spores	1.5	arrestant
Nitrogen gas	0.0	neutral carrier gas (4 l/hr)

The antibiotic secretions of the symbiont (see Chapter 20) have been evaluated by a purely qualitative, but classical and sensitive, confetti test. Ethyl acetate extracts (Decharme 1978) of washed mycelia or liquid culture filtrates of Leucocoprinus contain antibiotic materials that produce growth-inhibiting halos of more than 20 mm in cultures of Metarrhizium and Paecilomyces and almost 10 mm in Beauveria cultures. The Leucocoprinus, to a certain extent, may play a role in the control of exogenous entomopathogenic fungi. This action seems, however, to be rather fleeting on fungi but persistent (9 days) on several bacteria (Bacillus megatherium, B. subtilis, and Sarcinea lutea.

DISCUSSION

The observations and experiments presented here show that the attine, Ac. octospinosus, together with its ecto-symbiont, has several strategies available to control pathological and parasitic biohazards. These strategies are morphological, mechanical, and biochemical in nature.

Parasitological Aspects

Acromyrmex larvae are more sensitive in vitro to Neoaplectana than adults. However, because they are buried in the fungus garden and receive the attention of the nurse ants, their actual contact with exogenous biohazards is very indirect or even

348

non-existent. Nest soil is biologically complex, and the rhabditid nematodes have developed parasitic forms through a long co-evolution with terrestrial arthropods. Generally, geobiological homeostatic systems cause lasting spatial interactions favoring these trophic shifts. Some tens of millions of years of sympatric development in the hot and humid climate of the Amazonian plateau would have induced, through co-evolution, adaptations for survival. The morpho-physiological and eco-ethological aspects underlined in this work illustrate the probable ant-nematode co-evolution process that may have begun in the Cretaceous-Oligocene period.

Few cases of nematode parasitism of ants are reported: Janet (1893), Vandel (1930), Wahab (1962), and Lofgren et al. (1975). These latter authors point out that Neoaplectana dutkyi (DD136) parasitizes Solenopsis invicta but without encouraging results in the field. Kermarrec (1975) and Laumond et al. (1979) suggest a possible use of the "agriotos" strain of N. carpocapsae in controlling Ac. octospinosus; however, more recent work clearly showed that, in practice, it does not work.

Pathological Aspects

Similar to nematodes, certain soil-borne saprophytic fungi have adapted, during a long period of sympatric co-development with subterranean insects, to a purely parasitic life style. The slow co-evolution of sporulating Hyphomycetes and attines seems evident here and might explain the negative ecopathological results. Thus, the ant-Basidiomycete association, under adaptive pressures from the geobiological system, is able to control their respective pathogens.

Ants are subject to many fungal diseases (Evans 1974; Lofgren et al. 1975; Evans and Samson 1984; also see Jouvenaz, Chapter 27). The attines, however, have received little attention in this respect (Gosswald 1938; Steinhaus and Marshall 1967; Weber 1972; Kermarrec and Mauleon 1975). The pathogens most frequently referred to are Hyphomycetes (Beauveria, Metarrhizium, and Aspergillus). The pathogenicity of this latter genus has not been proven.

Sensory and Behavioral Aspects

The electroantennogram conducted in parallel with qualitative behavioral analysis shows that Ac. octospinosus is very sensitive to 2-heptanone. Behaviorally the compound causes an alarm reaction similar to that of the ants' specific alarm pheromone. Some authors (Gehrig and Knight 1961; Collins and Kalnins 1966; Franzke and Thurm 1970) demonstrated the existence of, and studied the production of, 2-heptanone by the spores of filamentous fungi (Aspergillus,

Penicillium, Paecilomyces, Scaputariopsis) from capric acid. Blum (1981) points out that the Hymenoptera are amazing "ketone virtuosi" (Table 2), but "the precise selective value of these important secretions will continue to remain terra incognita."

TABLE 2. Defensive allomones from the mandibular glands of some leaf-cutting ants (from Blum 1981).

Chemical	Ant	Reference
2-Heptanone	A. bisphaerica A. capiguara, A. laevigata A. robusta, A. sexdens A. texana	Moser et al. 1968
	A. columbica	Blum et al. 1968
	A. cephalotes	Riley et al. 1974
3-Octanone	Ac. octospinosus	Crewe and Blum 1972
	A. cephalotes, A. texana	Riley et al. 1974

In the same way, Blum et al. (1968) and Moser et al. (1968) believe that the production of 2-heptanone (alarm pheromone of the phyletically very distant Dolichodorines) by Atta cannot be explained. Crewe and Blum (1972) state that Ac. octospinosus does not secrete this molecule. Parry and Morgan (1979) hypothesized that the symbiont secretes 2-heptanone. Other authors (Bevan et al. 1961; Shearer and Boch 1965) demonstrated the presence and repellency of 2-heptanone for black cocktail ants and domestic bees, respectively. Moser et al. (1968) cannot fully explain the role of certain pheromones and suspects the existence of multiple functions for the molecules. Behavioral tests on Ac. octospinosus showed 2-heptanone to be an arrestant, attractant, and alarm-inducer and caused aggression or repulsion according to the concentration. Cole et al. (1975), Gochnauer et al. (1979), and Gupta et al. (1979) show that 2-heptanone possesses strong anti-germination effects on spores of Hyphomycetes among which is Paecilomyces marquandii. We believe that this molecule possesses a fundamental selective value in relation to constant pressure from pathogens in geobiological systems where attines are evolving. It indicates the presence of dangerous propagules, sets off a sequence of hygienic or alarm-defense behaviors, and can even be anti-germinating according to dosage. It is clear that adaptive advantages, as well as an economy of semiochemicals, are real for the attine-Basidiomycete duo.

Antibiotic Aspects

As early as 1956, Weber postulated the existence of micro-biostatic substances in the anal and salivary excretions of attines. Martin et al. (1969) did not confirm this hypothesis with regard to A. cephalotes. In 1970, Maschwitz et al. showed that the metathoracic glands of A. sexdens secrete an antiseptic (phenylacetic acid). Schildknecht and Koob (1971) discovered in A. sexdens the secretion of myrmicacin (1- β-hydroxydecanoic acid) which, depending on dosage, stimulated the symbiont's growth or inhibited that of exogenous germs (Schildknecht et al. 1973). Iizuka et al. (1979) showed that myrmicacin inhibits spore germination of Alternaria and Alcaligenes. The thoracic glands of Ac. octospinosus secrete myrmicacin but not phenylacetic acid.

The large number (Torregrossa et al. 1981) and the ethological importance (Wilson 1980a, 1980b) of the smallest worker ants must be emphasized. Their salivary lysosomal hydrolases, such as β-N-acetylglucosaminidases, inhibit fungal and bacterial growth (Muzzarelli 1977). The analysis of polyethism among the higher attines and the recent definition of alloethism (Wilson 1980b) provide acceptable explanations for our data on the production and anti-germinating function of the minor worker salivary secretions. The clear antibiohazard function, particularly well-developed in this group of workers in relation to their dominant task, is linked to the functional morphology of the infra-buccal pocket which has been described by Febvay and Kermarrec (1981). The hygienic behavior of social Hymenoptera seems fairly uniform in its expression. Necrophoric (Ataya and Lenoir 1984) and grooming behaviors (Gilliam et al. 1983) entail the use of the infra-buccal pocket. The function of this organ (Quinlan and Cherrett 1978; Glancey et al. 1981; Jouvenaz et al. 1984; the latter two for Solenopsis) is not only to filter solid particles but also to act as a mechano-chemical mixer. Its task is repeated millions of times a day within the nest, thus contributing to the homeostasis of this fungus-growing society. The sizes of Beauveria or Paecilomyces conidiospores are well below the filter threshold (about 10 um according to Quinlan and Cherrett 1978), but these propagules are immediately imbedded in a sticky compressed pellet impregnated with fungus-inhibiting salivations.

The reality of the antibiotic secretions of the symbiotic fungus was confirmed by Decharme (1978). Despite the negative tests of Hervey et al. (1977), Hervey and Nair (1979) and Nair and Hervey (1979) showed that the symbiont of Cyphomyrmex costatus excretes lepiochlorine, a lactol structure, active on Staphylococcus aureus. This work confirms the possibility of an active participation of the Acromyrmex symbiont, Leucocoprinus, in the control of exogenous micro-flora.

CONCLUDING REMARKS

Wilson (1980a), emphasizing the importance of social acts such as grooming, defines an amplifier effect linked to polyethism. The result of the idiosyncratic action of each individual, at the level of the group, is considerably enhanced by stigmergic amplification (Grasse 1959). Thus, all the workers of a sub-caste would seem to be interchangeable as long as they are in phase and stimulated by a tangible result of their activity. This extraordinary homeostasis, a concept analyzed by Emerson (1956) and Lindauer (1977), appears in attine nests as a consequence of a polymorphism that is disconcerting in complexity. Wallis (1965) claims, moreover, that there are, within ant societies, groups of individuals forming "permanent strategic reserves." These inactive workers are available for any eventuality. Therefore, social vigilance and altruism are the basis of hygiene in the ant nests and hence cause the failure of microbiological control. Antennal detection of microorganisms is efficient at the molecular level, and many other devises have been developed by attine ants to control biohazards. Could this, in fact, be an indication of the reduction of the chance element by the progressive building of anticipatory strategies?

There is no single way of protecting crops from phytophagous arthropod attacks. Taking into account environmental factors makes it essential to develop integrated pest controls wherein biological action remains a potentially powerful weapon. Highly developed social insects such as attines pose a technical challenge of which only a few aspects have been emphasized in this review.

ACKNOWLEDGMENT

This work is dedicated to our late Master and Friend, Emile Biliotti.

REFERENCES CITED

Amargier, A., and G. Vago. 1966. Coloration histologique pour etude de mycoses d'invertebres. Mikroskopie 21: 271-275.

Ataya, H., and A. Lenoir. 1984. Le comportement necrophorique chez la fourmi Lasius niger L. Insect. Soc. 31: 20-33.

Baermann, G. 1917. Eine einfache Methode zur Auffindung von Ankylostomum (Nematoden) larven in Erdproben. Meded. Geneesk Lab. Weltevr. pp. 41-47.

Barnes, G. L. 1975. Growth and sporulation of Metarrhizium anisopliae and Beauveria bassiana on media containing various peptone sources. J. Invert. Pathol. 25: 301-305.

352

Batra, L. R., and S. W. Batra. 1978. Termite-fungus mutualism, pp. 117-163. In L. R. Batra (ed.), Insect-fungus symbiosis. Wiley and Sons. 276 pp.

Bevan, C. W., A. J. Birch, and H. Caswell. 1961. An insect repellent from black cocktail ants. J. Chem. Soc. p. 488.

Blum, M. S. 1981. Chemical defenses of arthropods. Academic Press, N.Y., NY. 538 pp.

Blum, M. S., F. Padovan, and E. Amante. 1968. Alkanones and terpenes in the mandibular glands of Atta species. Comp. Biochem. Physiol. 26: 291-299.

Boemare, N., E. Bonifassi, C. Laumond, and J. Luciani. 1983. Etude experimentale de l'action pathogene du nematode Neoaplectana carpocapsae Weiser. Recherches gnotobiologiques chez l'insecte Galleria mellonela L. Agronomie 3: 407-416.

Boemare, N., C. Laumond, and J. Luciani. 1982. Mise en evidence de la toxicogenese provoquee par le nematode entomophage Neoaplectana carpocapsae Weiser chez l'insecte Galleria mellonela L. C. R. Acad. Sci. Ser. III 295: 543-546.

Bovien, P. 1937. Some types of associations between nematodes and insects. Vidensk. Medd. Dansk. Naturh. Foren. 101: 1-114.

Cole, L. K., M. S. Blum, and R. W. Roncadori. 1975. Antifungal properties of the insect alarm pheromones, citral, 2-heptanone and 4-methyl-3-heptanone. Mycologia 67: 701-708.

Collins, R. P., and K. Kalnins. 1966. Production of carbonyl compounds by several species of endoconidium-forming fungi. Mycologia 58: 622-628.

Crewe, R. M., and M. S. Blum. 1972. Alarm pheromones of the Attini: Their phylogenetic significance. J. Insect Physiol. 18: 31-42.

Decharme, M. 1978. Contribution a la connaissance de Leucocoprinus gongylophorus (Moeller). Heim, champignon symbiote de la fourmi Acromyrmex octospinosus Reich. Dea de cytologie et morphogenese vegetales. Univ. Paris. 68 pp. 12 planches.

Decharme, M., and M. Issaly. 1980. Contribution a l'etude du champignon symbiote de quelques fourmis de la tribu Attini. Crytog. Mycol. I: 1-18.

Emerson, A. E. 1956. Regenerative behavior and social homeostasis of termites. Ecology 37: 248-258.

Evans, H. C. 1974. Natural control of arthropods with special reference to ants (Formicidae) by fungi in the tropical high forest of Ghana. J. Appl. Ecol. 11: 37-49.

Evans, H. C., and R. A. Sampson. 1984. Cordyceps species and their anamorphs pathogenic on ants (Formicidae) in tropical forest ecosystems. II. The Camponotus (Formicinae) complex. Trans. Br. Mycol. Soc. 82: 127-150.

353

Febvay, G., and A. Kermarrec. 1981. Morphologie et fonctionnement du filtre infrabuccal chez une attine Acromyrmex octospinosus (Reich) (Formicidae: Attini). Roles de la poche infrabuccale. Int. J. Insect Morphol. Embryol. 10: 165-177.

Febvay, G., M. Decharme, and A. Kermarrec. 1984a. Digestion of chitin by the labial glands of Acromyrmex octospinosus (Reich) (Hymenoptera: Formicidae). J. Can. Zool. 62: 229-234.

Febvay, G., F. Mallet, and A. Kermarrec. 1984b. Attractivite du couvain et comportement des ouvrieres de la fourmi Attine Acromyrmex octospinosus (Reich) (Hymenoptera: Formicidae). Actes coll. Insect. Soc. 1: 79-86. Ed. UIEIS, Presses Univ. Paris 12.

Franzke, C., and V. Thurm. 1970. Zur bildung von Methylketonen aus Fettsauren durch Penicillium roqueforti. 2. Mitt. Quantitative Erfassung der durch P. roqueforti aus Fettsauren gebildeten Methylketone. Nahrung 14: 287-291.

Gehrig, R. F., and S. G. Knight. 1961. Formation of 2-heptanone from caprylic acid by spores of various filamentous fungi. Nature 4808: 1185.

Gilliam, M., S. Taber, and G. V. Richardson. 1983. Hygienic behavior of honey bees in relation to chalkbrood disease. Apidologie 14: 29-39.

Glancey, B. M., R. K. Vander Meer, A. Glover, C. S. Lofgren, and S. B. Vinson. 1981. Filtration of microparticles from liquids ingested by the red imported fire ant, Solenopsis invicta Buren (Hymenoptera: Formicidae). Insect. Soc. 28: 395-401.

Gochnauer, T. A., R. Boch, and N. J. Margetts. 1979. Inhibition of Ascosphaera apis by citral and geraniol. J. Invert. Pathol. 34: 57-61.

Gosswald, K. 1938. Uber den insektentotenden pilz Beauveria bassiana (Bals.) Vuill. bisher bekanntes und eigne Versuche. Arb. Biol. Reichsanstalt, Land Forstwirts. 18: 399-452.

Grasse, P. P. 1959. La reconstruction du nid et les coordinations inter-individuelles chez Bellicositermes natabensis et Cubitermes sp. La theorie de la stigmergie: Essai d'interpretation du comportement des termites constructeurs. Insect. Soc. 6: 41-83.

Gupta, R. C., R. N. Tandon, D. K. Arora, A. N. Shukla, and D. B. Singh. 1979. Dynamics of microfungi in fallowland soils with emphasis on volatile fungistasis. Rev. Ecol. Biol. Sol. 16: 491-497.

Hervey, A., and M. S. R. Nair. 1979. Antibiotic metabolite of a fungus cultivated by gardening ants. Mycologia 77: 1064-1066.

Hervey, A., C. T. Rogerson, and I. Leong. 1977. Studies on fungi cultivated by ants. Brittonia 29: 229-236.

354

Iizuka, T., T. Iwadare, and K. Orito. 1979. Antibacterial activity of myrmicacin and related compounds on pathogenic bacteria in silkworm larvae, Streptococcus faecalis AD-4. J. Fac. Agr. Hokkaido Univ., Hokkaido, Japan. 59: 262-266.

Janet, C. H. 1893. Etude sur les nematodes des glandes pharyngiennes des fourmis. C. R. Acad. Sci. Paris. 117: 700.

Jouvenaz, D. P., E. A. Ellis, and C. S. Lofgren. 1984. Histopathology of the tropical fire ant, Solenopsis geminata, infected with Burenella dimorpha (Microspora: Microsporida). J. Invert. Pathol. 43: 324-332.

Kermarrec, A. 1973. Recherches sur les ennemis du champignon de couche. I. Contribution a l'etude de la zoocenose des composts de champignonniere a Agaricus bisporus Lange. Ann. Zool. Ecol. Anim. 5: 425-464.

Kermarrec, A. 1975. Etude des relations synecologiques entre les nematodes et la fourmi manioc, Acromyrmex octospinosus Reich. Ann. Zool. Ecol. Anim. 7: 27-44.

Kermarrec, A., and H. Mauleon. 1975. Quelques aspects de la pathogenie d'Entomophthora coronata Cost. Kervork pour la fourmi manioc de la Guadeloupe, Acromyrmex octospinosus (Formicidae: Attini). Ann. Parasitol. 50: 351-360.

Laumond, C., H. Mauleon, and A. Kermarrec. 1979. Donnees nouvelles sur le spectre d'hotes et le parasitisme du nematode entomophage Neoaplectana carpocapsae. Entomophaga 24: 13-27.

Lindauer, M. 1977. Super-individual aspects of social function, nutrition, care of offspring, and homeostasis. pp. 92-94. Proc. 8th Int. Cong. IUSSI. Wageningen, The Netherlands. 325 pp.

Locke, M. 1974. The structure and formation of the integument in insects, pp. 124-215. In M. Rockstein (ed.), The physiology of insecta, VI. Academic Press, N.Y., NY.

Lofgren, C. S., W. A. Banks, and B. M. Glancey. 1975. Biology and control of imported fire ants. Annu. Rev. Entomol. 20: 1-30.

Martin, M. M., J. G. MacConnell, and G. R. Gale. 1969. The chemical basis for the attine ant fungus symbiosis: Absence of antibiotics. Ann. Entomol. Soc. Am. 62: 386-388.

Maschwitz, U., C. Koob, and H. Schildknecht. 1970. Ein beitrag zur funktion der metathoracicaldruse der ameisen. J. Insect Physiol. 16: 387-404.

Moser, J. C., R. C. Brownlee, and R. Silverstein. 1968. Alarm pheromones of the ant Atta texana. J. Insect Physiol. 14: 529-535.

Muzarelli, R. A. 1977. Chitin. Pergamon Press, Oxford, England. 309 pp.

Nair, M. S., and A. Hervey. 1979. Structure of lepiochlorin, an antibiotic metabolite of a fungus cultivated by ants. Phytochemistry 18: 325-326.

Parry, K., and E. D. Morgan. 1979. Pheromones of ants: A review. Physiol. Entomol. 4: 161-189.

Ponchet, J. 1982. Realites et perspectives de la lutte biologique contre les maladies des plantes. Agronomie 2: 305-314.

Quinlan, R. J., and J. M. Cherrett. 1978. Studies on the role of the infrabuccal pocket of the leaf-cutting ant Acromyrmex octospinosus (Reich) (Hymenoptera: Formicidae). Insect. Soc. 25: 237-245.

Riba, G., K. Katagiri, and K. Kawakami. 1982. Preliminary studies on the susceptibility of the silkworm, Bombyx mori (Lepidoptera: Bombycidae) to some entomogenous Hyphomycetes. Appl. Entomol. Zool. 17: 238-243.

Riley, R. G., R. M. Silverstein, and J. C. Moser. 1974. Isolation, identification, synthesis, and biological activity of volatile compounds from heads of Atta ants. J. Insect Physiol. 20: 1629-1637.

Schildknecht, H., and K. Koob. 1971. Myrmicacin, das erste Insekten-Herbizid. Angew. Chem. 10: 124-125.

Schildknecht, H., Y. Pasty, B. Reed, F. Dewitt, and K. Koob. 1973. Auxin activity in the symbiosis in leaf-cutting ants and their fungus. Insect Biochem. 3: 439-442.

Shearer, D. A., and R. Bock. 1965. 2-Heptanone in mandibular gland secretion of the honey bee. Nature 206: 530.

Steinhaus, E. A., and G. A. Marshall. 1967. Previously unreported accessions for diagnosis and new records. J. Invert. Pathol. 9: 436-438.

Torregrossa, J. P., G. Febvay, and A. Kermarrec. 1981. Les stades larvaires chez la fourmi attine, Acromyrmex octospinosus (Reich) (Hymenoptera: Formicidae). Colemania 1: 141-147.

Vandel, A. 1930. La production d'intercastes chez la fourmi Pheidole pallidula sous l'influence de parasites du genre Mermis L. (Etude morphologique des individus parasites). Bull. Biol. France-Belgique. 64: 457-494.

Wahab, A. 1962. Untersuchungen uber Nematoden in den Drusen des kopfes der Ameisen (Formicidae). Z. Morphol. Okol. Tiere. 52: 33-92.

Wallis, D. I. 1965. Division of labour in ant colonies. Symp. Zool. Soc. London 14: 97-112.

Weber, N. A. 1956. Treatment of substrate by fungus-growing ants. Anat. Rec. 125: 604-605.

Weber, N. A. 1972. Gardening ants, the attines. Mem. Am. Phil. Soc. Vol. 92, 146 pp.

Wilson, E. O. 1980a. Caste and division of labour in leaf-cutter ants (Hymenoptera: Formicidae: Atta). I. The overall pattern in A. sexdens. Behav. Ecol. Sociobiol. 7: 143-156.

Wilson, E. O. 1980b. Caste and division of labor in leaf-cutter ants (Hymenoptera: Formicidae: Atta). II. The ergonomic optimization of leaf cutting. Behav. Ecol. Sociobiol. 7: 157-165.

356

RESUME

Le systeme fourmi-basidiomycete est soumis experimentale-
ment a diverses pressions parasitaires et pathologiques pour tenter
de definir des voies de controle microbiologique. Les organismes
suivants, preselectionnes in vitro pour leur pathogenie sur un des
partenaires de la symbiose, ont ete utilises: Nematodes mycophages,
Nematodes entomoparasites, Champignons mycoparasites et
Champignons entomopathogenes. Les facteurs a l'origine des
resultats decevants observes sur nids complets sont analyses: ils
sont de natures morphologique, sensorielle, comportementale et
biochimique. Les donnees suivantes sont originales: diminution du
parasitisme par Neoaplectana carpocapsae en fonction du developpement des ouvrieres de Acromyrmex octospinosus; detection des
conidiospores des champignons entomopathogenes par les antennes
(electroantennographies) et analyse du comportement d'hygiene
consecutif; secretion de facteurs biochimiques fongistatiques avec la
salive des ouvieres et par le mycelium du symbiote. L'ensemble de
ces donnees experimentales est discute dans ses aspects coevolutifs
et les difficultes d'une application pratique contre les fourmis
champignonnistes sont soulignees.

29
Chemical Control
and Bait Formulations
for Leaf-Cutting Ants

J. M. Cherrett

In 1971, a postal survey of 27 countries was conducted to determine which methods were most commonly used at that time for control of leaf-cutting ants. The results (Table 1) revealed that "Mirex 450," a pelleted bait, was the control method of choice. The next two most popular techniques were pouring organochlorine insecticides as emulsions down nest holes or pumping them down as dusts. The fourth most widely used technique involved fumigation of nests with methyl bromide.

The survey revealed that two broadly different approaches to the chemical control of leaf-cutting ants are used—direct chemical poisoning and poison baits. A summary of the current status of each of these follows.

DIRECT CHEMICAL POISONING OF THE NEST

<u>Atta</u> nests are large and conspicuous objects with a surface area up to 250 m^2 and a mature nest density rarely exceeding 3/ha. Thus, finding and spot treating them is practical on a large scale which is not commonly the case in ant control. Mariconi (1970) provides detailed instructions for direct chemical poisoning.

Fumigation

Where the toxicant is a gas at ambient temperatures, it is introduced into the main nest holes through pipes; each hole is then sealed with earth. Where it is a volatile liquid, it may be vaporized before being piped into the nest, or be poured directly into the holes before sealing. Good kills can be achieved, but the gases may be dangerous to the operator. The application apparatus has been comprehensively reviewed by Mariconi (1970).

358

TABLE 1. Frequency of citation of methods employed to control leaf-cutting ants. Results of a 1971 questionnaire.[a]

Control method	Number of times method was reported
Mirex bait	15
Aldrin introduced into nest as a powder or emulsion	12
Chlordane introduced into nest as a powder or emulsion	10
Methyl bromide as a nest fumigant	7
Dieldrin introduced into nest as a powder or emulsion	6
Aldrin baits	5
Heptachlor introduced into nest as a powder or emulsion	4
Carbon disulphide as a nest fumigant	3
Benzene hexachloride introduced into nest as a powder	2

[a]Fifteen chemicals cited in data from 17 countries.

Aqueous Solutions

The toxicant is usually incorporated as a suspended wettable powder or as an emulsion and then poured into the nest holes. Where nests are deep, and nests of Atta vollenweideri may reach 5 m (Jonkman 1977), penetration can be poor, especially if the soil is dry. However, the application technique is simple and requires minimal apparatus.

Dusts

Toxicants incorporated in a dust are forcibly pumped into the entrance holes and, being more mobile, penetrate the nest more effectively. This is best done in dry conditions. A hand pump or motorized blower is required.

Smokes

The excellent penetration of fumigant gases and the wide

range of modern insecticides available as dusts or aqueous solutions can be combined if the toxicants are incorporated into smokes using a thermal fogger. Its nozzle is inserted into the main nest entrances and smoke is forced into the tunnels until it emerges from nearby holes which are then sealed up. The process is rapid and effective, but the machinery is expensive and requires maintenance (Kennard 1965; Vilela, Chapter 33). Brussell and Vreden (1967) found smokes the most efficient of the methods they compared. The small quantity of toxicant required and the large areas treatable per application can be seen in Table 2.

Direct nest poisoning has been the traditional chemical control technique for leaf-cutting ants, and large-scale control campaigns have been attempted. Oliveira (1934) reported the elimination of 141,100 nests in Sao Paulo with carbon disulfide, and Kennard (1965) killed 5,000 nests in Guyana with swing fog machines. The difficulties with this approach are: (1) locating nests, which may be in thick bush or on a neighbor's property; (2) overlooking small nests; (3) treating large nests, which is time-consuming and requires special training to obtain complete coverage of irregularly-shaped nests; and (4) expensive and complex machinery may be needed requiring transport and maintenance, often to remote areas. The advantages of direct poisoning are: (1) it is efficient and reliable, even in wet weather, if properly carried out and (2) the chemicals may be inexpensive.

Despite many subsequent advances in leaf-cutting ant control, it is interesting to note that in 1983 the only two pesticides registered in the United States for the control of A. texana were methyl bromide and Durham's Sodium Cyanide Balls (Moser 1984).

THE USE OF POISON BAITS

It is apparent (Table 1) that toxic baits are the control method of choice, and their advantages are many. With baits, it is not essential to find the nest, as bait can be sprinkled on working trails or where damage is being experienced. Nests on a neighbor's property can be destroyed without access, and small inconspicuous nests will be easily killed. The nests can be treated very quickly, and the bait is easy to apply and requires no specialized equipment or training. The concentration of toxicant in baits is usually low (0.45% A.I. in Mirex 450), so that handling it is a relatively safe procedure. The ants then collect it and concentrate it in their nest where it has maximal effect. Accordingly, low dose rates are possible, and Ribeiro and Woessner (1979) used mirex bait at 10 g/m^2 of nest surface (0.04 g/m^2 A.I.)—a sharp contrast with the other chemicals listed in Table 2. Lastly, suitable bait formulations can be fairly specific to the target organism.

TABLE 2. Toxicants used for the direct poisoning of <u>Atta</u> nests. Modified after Proctor (1957), Kennard (1965), Mariconi (1970), and Nogueira et al. (1982).

Toxicant	Formulation	Quantity per m^2	Max. nest area per application
Methyl bromide	gas (B.P. 3.5 C)	$4\ cm^3$	$5\ m^2$
Carbon disulphide	liquid (B.P. 46 C)	$75\ cm^3$	$5\ m^2$
Aldrin 5%	dust	30 g	$3\ m^2$
Chlordane 10%	dust	30 g	$3\ m^2$
Heptachlor 5%	dust	30 g	$3\ m^2$
Aldrin 40%	aqueous solutions	$5\ cm^3 + 500\ cm^3$ water	$2\ m^2$
Chlordane 74%	aqueous solutions	$2\ cm^3 + 500\ cm^3$ water	*
Heptachlor	aqueous solutions	$10\ cm^3 + 500\ cm^3$ water	$2\ m^2$
Aldrin	smoke	$0.17\ g + 6.6\ cm^3$ diesel oil	$17\ m^2$
Heptachlor	smoke	$0.37\ g +$ oil carrier	—

*May also be watered over nest surface in addition to pouring down holes.

Toxic baits also have disadvantages, some being peculiar to leaf-cutting ant control. Successful control depends on the ants picking up bait particles and carrying them back to their nest. Preferences in Atta and Acromyrmex are particularly variable (Littledyke and Cherrett 1975), and so their reactions to a bait are never wholly predictable. Contamination of baits when they are handled or stored near other chemicals with a strong smell may inhibit pickup. If bait particles become wet, they deteriorate as a result of disintegration and leaching. Bacterial and fungal growth is another serious problem in the humid tropics, and Ribeiro and Woessner (1979) considered mirex bait unsuitable in the wet season in Amazonia. Commercial baits can be expensive, and the peasant farmers of Central and South America who suffer most from leaf-cutting ant depredations are often very poor. Although adult leaf-cutting ants will drink oil and sugary solutions (Echols 1966), the brood appear to be fed exclusively on fungal staphylae from the older parts of the garden; consequently, there is no direct route for toxicants from bait to brood (Peregrine and Cherrett 1976). Also, the fungus is metabolically active and may be capable of detoxifying some ant insecticides (Little et al. 1977). Recent research on toxic baits for leaf-cutting ants has attempted to overcome these disadvantages and is best reviewed by examining their components.

Toxicants

Two percent aldrin in wheat flour was the first bait in general use around 1957 (Goncalves 1960). Since then, at least 42 leaf-cutting ant baits have been marketed, with aldrin and heptachlor being the toxicants used in most of them. Some toxicant screening has been carried out in the field (Phillips et al. 1976, 1979; Ambrosi et al. 1979; Anon. 1984; Naccarata and Jaffe, in press). Attempts to slow down the rate of kill of toxicants have included microencapsulation. Although some formulations looked promising, none have been better than mirex; and to date, none have been used in a commercial product. A juvenile hormone analogue (altozar) was tested against A. sexdens in a petri dish bioassay without a fungus garden. There was a substantial mortality of adults, and larval and pupal development were fatally affected (Little et al. 1977). When the toxicant was introduced into intact colonies, however, little effect could be detected; and it was suggested that the presence of the fungus garden protected the brood, possibly by detoxification. In contrast, mirex has proven to be an excellent toxicant for leaf-cutting ants, since the first report of its use by Echols (1965) against A. texana in the United States. Its success can be attributed to its: (1) lack of repellency when added to baits at effective concentrations, (2) slow action so it can be fully incorporated into the fungus gardens (Peregrine and Cherrett 1974) and spread from ant to ant by trophallaxis and mutual grooming (Echols 1966), and (3) chemical stability and persistence that allows it to withstand fungus metabolism. Compared with aldrin, mirex has no real advantage for the small nests of Acromyrmex (Cherrett and Sims 1969); but in the large many-chambered Atta nests, partial kill can often result if social activity is disrupted by a fast-acting poison before the aldrin is well-dispersed. If the queen is living in unaffected fungus gardens, the colony will regenerate. Here mirex seems to have the advantage (Liceras 1966).

The use of mirex against leaf-cutting ants was a spinoff from its discovery and use against fire ants, and the next generation of toxicants may be developed similarly (see Lofgren, Chapter 30). This is because (1) funds have not been made available for large-scale screening of chemicals against leaf-cutting ants and (2) they are less satisfactory as laboratory animals, and field trials on the large, widely-spaced nests of Atta are expensive to conduct.

Attractants

Bait particles are more readily found by scout ants if they are attracted to them from a distance. 'Food odours' such as vegetable oils, crushed plant leaves, and citrus fruit extracts can be detected by the ants in a moving air olfactometer (Littledyke and Cherrett 1978). To improve both attractiveness and species specificity, attempts were made to add pheromones to the baits. Robinson and Cherrett (1974) tried without success to extract brood pheromones but abandoned the idea when the ants' reactions to their own living brood were found to be variable and influenced by the distance of the brood from the nest. Following the pioneering discovery and synthesis of the trail pheromone of A. texana (Moser and Blum 1963; Tumlinson et al. 1971), a good deal of effort was put into the possible use of this pheromone as an attractant in baits (Robinson and Cherrett 1978). In the end it was concluded that, although the pheromone acted as a foraging stimulant, and also attracted ants to bait particles from a short distance, its effect was not additive to that of the food odours normally present in baits. As these were cheaper to produce, the use of trail pheromone was not cost effective (Robinson et al. 1982). The addition of food odours to baits, as well as air tight bait storage for retention of volatile attractants, are valuable features for enhancement of their effectiveness.

Arrestants

Chemicals which attract scout ants to baits do not necessarily stimulate them to pick up the bait and carry it back to their nest. The A. texana trail pheromone, for example, does not induce pick-up of filter paper discs (Robinson and Cherrett 1978). Chemical analyses of the albedo of citrus skins, a favored substrate for leaf-cutting ants, showed that it consisted of a mixture of pick-up inducers and inhibitors. If the inhibitory substances are removed, "pick-up inducers" may act synergistically, making a complex mixture more acceptable than any single chemical (Mudd et al. 1978). Not surprisingly for general feeders, sugars were the most acceptable of the refined fractions. This means that the acceptability of any bait can usually be enhanced by the addition of small quantities of molasses or concentrated orange juice.

Matrix Formulations

The matrix provides the physical structure of the bait and may also provide arrestants and attractants. The matrix selected depends upon the way in which the bait is to be used, the ant species involved, and the marketing strategy to be adopted.

Central formulation for spot treatment by farmers. This has been the commonest strategy; and bagasse, cassava flour, cereal grains, dried citrus pulp, soybean meal, wheat middlings, and wheat flour have all been utilized for leaf-cutting ant baits, while dried grass has been employed for grass-cutting ant baits (Robinson et al. 1980). However, natural products vary in quality and storage conditions are critical if deterioration is to be avoided. Some products have variable particle sizes making mechanical spreading difficult (Lewis 1973). Pelleting ground up bait matrix produces an even-sized product, but the pellets are susceptible to disintegration if they become wet. Synthetic matrices of known composition and properties have been used in experimental baits to overcome some of these problems. Foamed gelatin, urea-formaldehyde resin, and vermiculite onto which attractants and arrestants have been sprayed were tested by Cherrett et al. (1973) and Etheridge and Phillips (1976). Vermiculite was the most satisfactory material, and it has been tested in the field (Phillips et al. 1976).

Formulation by farmers on site. Since baits are bulky to market, consideration has been given to packaging concentrates containing toxicant, attractants, arrestants, and formulating agents in tins. Farmers can dilute the concentrates with water and spray them onto locally available waste product matrices such as sawdust, chopped straw, or dried cocoa pods. A concentrate prepared in this way, and sprayed onto vermiculite, has worked well in the field (Phillips et al. 1976).

Formulation for aerial application. For large scale control campaigns, or when high densities of small Acromyrmex spp. nests are the problem, aerial bait application is desirable and can be cost effective (Cherrett and Merrett 1969; Lewis 1973; Lewis and Norton 1973; Robinson et al. 1980). Its effectiveness was inadvertently demonstrated in the Kisatchie National Forest in central Louisiana where aerial applications of mirex granules against fire ants during 1965 and 1966 killed at least 99% of A. texana colonies (Moser 1984). For aerial application, the ideal matrix is one that can be loaded in dense liquid form and then transformed during application into a light, fluffy, bulky bait which is cheap, acceptable to the ants, and resistant to weathering. Jutsum and Cherrett (1981) described such a bait consisting of polyurethane foam, manufactured at the time of spraying from two precursor liquids, one of which contained the toxicant and attractants. One litre of precursors yielded 351 litres of solid foamed bait. When aldrin was used as a toxicant, good kill of Acromyrmex octospinosus was achieved.

The foamed polyurethane bait showed excellent resistance to weathering and to mold growth, but baits based on natural products deteriorated rapidly and became unacceptable to the ants in damp tropical conditions. Propionic acid or sodium benzoate can be used to suppress mold growth (Cherrett et al. 1973), and coating the bait with a water repellent layer of silicone improved its acceptability when wet (Cherrett and Merrett 1969). More recently, Bottrell (1980) showed that A. sexdens workers could complete cuts in up to 400-gauge polythene sheet and that 26% could cut 1,000 gauge if it was made attractive by spraying it with a 4% wt:vol. methanol extract of dried citrus pulp. Dobson and Sherringham (personal communication) demonstrated that attractants from citrus pulp diffused through sealed polythene bags and induced investigation and cutting by A. sexdens workers. This suggests that toxic baits normally marketed in polythene bags can be left unopened for spot treatment of nests as the bags will be opened by the ants. This has been confirmed in the field by Palmer and Hostachy (personal communication). In field trials in Venezuela, Naccarata and Jaffe (in press) used sealed polythene bags sprayed with synthetic orange juice to obtain excellent control of A. laevigata. As they point out, the value of leaving the bait in a sealed polythene bag is that it remains unaffected by rain until opened by actively-foraging ants. The specificity of the bait is increased as few other animals can open the bags, and the sealed bag reduces the chances of operators becoming contaminated by toxicants. Bags that are not opened by leaf-cutting ants within a reasonable time can be collected and used again.

A wide-range of chemical control techniques are available for limiting the damage caused by leaf-cutting ants; and recently, control programs have been attempted in several countries (Pollard 1982). Technically, mirex bait is the most successful method; but because of its persistence in the environment, its toxicity to some nontarget organisms, and its possible carcinogenic properties, it has been banned in several countries, including the United States. As leaf-cutting ant control is essential to Central and South American agriculture, it is imperative that an effective substitute be developed. In the meantime, however, governments should consider whether the disadvantages of mirex used as a spot treatment for nests really outweigh the danger to the operator using some of the alternatives, such as Durham's Sodium Cyanide Balls.

ACKNOWLEDGMENTS

I am grateful to Mr. H. M. Dobson and Miss C. Sherringham of the University College of North Wales Bangor for allowing me to quote from their unpublished student theses, to Drs. J. R. Palmer and B. Hostachy for permission to quote from their unpublished reports, and to Mr. V. Naccarata and Dr. K. Jaffe for an advanced copy of their manuscript.

REFERENCES CITED

Ambrosi, D., G. Bic, J. Desmoras, G. Gallinelli, and G. Roussel. 1979. 32,861 RP, a new product for the control of sucking insects, cockroaches and ants. Proc. Brit. Crop Protection Conf. - Pest and Diseases. 2: 533-540.

Anon. 1984. Town ant research. Texas Forest Pest Report 1982-1983, Tex. For. Serv. Publ. 136: 23-25.

Bottrell, C. M. 1980. Plant defences against leaf-cutting ant attack. Unpublished Ph.D. Thesis, University of Wales, Great Britain.

Brussell, E. W. van, and G. van Vreden. 1967. Nieuwe methoden ter bestrijding van draagmieren (Atta spp.) in Suriname. Surin. landb. 2: 74-81.

Cherrett, J. M., and M. R. Merrett. 1969. Baits for the control of leaf-cutting ants III. Waterproofing for general broadcasting. Trop. Agric., Trinidad 46: 221-231.

Cherrett, J. M., and B. G. Sims. 1969. Baits for the control of leaf-cutting ants II. Toxicity evaluation. Trop. Agric., Trinidad 46: 211-219.

Cherrett, J. M., D. J. Peregrine, P. Etheridge, A. Mudd, and F. T. Phillips. 1973. Some aspects of the development of toxic baits for the control of leaf-cutting ants, pp. 69-75. Proc. VII Int. Cong. I.U.S.S.I., London, England.

Echols, H. W. 1965. Town ants controlled with mirex baits. U.S. For. Serv. Res. Note SO - 18.

Echols, H. W. 1966. Assimilation and transfer of mirex in colonies of Texas leaf-cutting ants. J. Econ. Entomol. 59: 1336-1338.

Etheridge, P., and F. T. Phillips. 1976. Laboratory evaluation of new insecticides and bait matrices for the control of leaf-cutting ants (Hymenoptera: Formicidae). Bull. Entomol. Res. 66: 569-578.

Goncalves, A. J. L. 1960. O emprego das iscas no combate as formigas cortadeiras. Bol. do Campo, Rio de Janeiro 16: 3-10.

Jonkman, J. C. M. 1977. Biology and ecology of the leaf-cutting ant Atta vollenweideri, Forel 1893 (Hym.: Formicidae) and its impact in Paraguayan pastures. Thesis Universiteits bibliotheck, Leiden, The Netherlands.

Jutsum, A. R., and J. M. Cherrett. 1981. A new matrix for toxic baits for control of the leaf-cutting ant Acromyrmex octospinosus (Reich) (Hymenoptera: Formicidae). Bull. Entomol. Res. 71: 607-616.

Kennard, C. P. 1965. Control of leaf-cutting ants (Atta spp.) by fogging. Exp. Agric. 1: 237-240.

Lewis, T. 1973. Aerial baiting to control leaf-cutting ants (Formicidae: Attini) in Trinidad. II. Field application, nest mortality and the effect on other animals. Bull. Entomol. Res. 63: 275-287.

Lewis, T., and G. A. Norton. 1973. Aerial baiting to control leaf-cutting ants (Formicidae: Attini) in Trinidad. III. Economic implications. Bull. Entomol. Res. 63: 289-303.

Liceras, Z. L. 1966. Cebos contra la hormiga 'coqui' Atta cephalotes L. en Tingo Maria Peru. Revista peruana de entomologia agricola 7: 45-49.

Little, C. H., A. R. Jutsum, and J. M. Cherrett. 1977. Leaf-cutting ant control. The possible use of growth regulating chemicals, pp. 89-90. Proc. 8th Int. Cong. I.U.S.S.I. Wageningen, The Netherlands.

Littledyke, M., and J. M. Cherrett. 1975. Variability in the selection of substrate by the leaf-cutting ants Atta cephalotes (L.) and Acromyrmex octospinosus (Reich) (Formicidae: Attini). Bull. Entomol. Res. 65: 33-47.

Littledyke, M., and J. M. Cherrett. 1978. Olfactory responses of the leaf-cutting ants Atta cephalotes (L.) and Acromyrmex octospinosus (Reich) in the laboratory. Bull. Entomol. Res. 68: 273-282.

Mariconi, F. A. M. 1970. As sauvas. Editora Agronomica, 'Ceres,' Sao Paulo. 167 pp.

Moser, J. C. 1984. Town ant, pp. 47-52. Misc. Publ. Tex. Agric. Expt. Sta. (1553), College Station, TX.

Moser, J. C., and M. S. Blum. 1963. Trail marking substance of the Texas leaf-cutting ant: Source and potency. Science 140: 1228.

Mudd, A., D. J. Peregrine, and J. M. Cherrett. 1978. The chemical basis for the use of citrus pulp as a fungus garden substrate by the leaf-cutting ants Atta cephalotes (L.) and Acromyrmex octospinosus (Reich) (Hymenoptera: Formicidae). Bull. Entomol. Res. 68: 673-685.

Naccarata, V., and K. Jaffe. Formulacion y desarrollo de un cebo atractivo toxico para control de bachacos, Atta spp. (Hymenoptera: Formicidae). Bol. Entomol. Venezolono. (In press).

Nogueira, S. B., J. A. Barrigossi, G. O. da Silva, D. S. dos Martins, and P. R. Nunes. 1982. Novos produtos termonebulizaveis, no combate as especies de sauvas, Atta bisphaerica, Forel 1908, A. laevigata F. Smith, 1858, e A. sexdens rubropilosa Forel, 1908 (Hymenoptera: Formicidae). Revista Seiva 42: 90-93.

Oliveira, F. M. L. 1934. Combate a sauva. Bolm. Agric. S. Paulo, Brasil 35: 541-610.

Peregrine, D. J., and J. M. Cherrett. 1974. A field comparison of the modes of action of aldrin and mirex for controlling colonies of the leaf-cutting ants Atta cephalotes (L.) and Acromyrmex octospinosus (Reich). Bull. Entomol. Res. 63: 609-618.

Peregrine, D. J., and J. M. Cherrett. 1976. Toxicant spread in laboratory colonies of the leaf-cutting ant. Ann. Appl. Biol. 84: 128-133.

Phillips, F. T., P. Etheridge, and A. P. Martin. 1979. Further laboratory and field evaluations of experimental baits to control leaf-cutting ants (Hymenoptera: Formicidae) in Brazil. Bull. Entomol. Res. 69: 309-316.

Phillips, F. T., P. Etheridge, and G. Scott. 1976. Formulation and field evaluation of experimental baits for the control of leaf-cutting ants (Hymenoptera: Formicidae) in Brazil. Bull. Entomol. Res. 66: 579-585.

Pollard, G. V. 1982. A review of the distribution, economic importance and control of leaf-cutting ants in the Caribbean region with an analysis of current control programmes, pp. 43-61. In C. W. D. Brathwaite and G. V. Pollard (eds.), Urgent plant pest and disease problems in the Caribbean. IICA, Misc. Publ. 378.

Proctor, J. H. 1957. Coushi ant control with chlordane and aldrin. British Guiana Dept. of Agric., Farmer's Leaflet 5. 2 pp.

Ribeiro, G. T., and R. A. Woessner. 1979. Teste de eficiencia com seis (6) sauvicidas no controle de sauvas (Atta spp.) na Jari, Para, Brasil. Anais da Soc. Entomol. Brasil. 8: 77-84.

Robinson, S. W., and J. M. Cherrett. 1974. Laboratory investigations to evaluate the possible use of brood pheromones of the leaf-cutting ant Atta cephalotes (L.) (Formicidae: Attini) as a component in an attractive bait. Bull. Entomol. Res. 63: 519-529.

Robinson, S. W., and J. M. Cherrett. 1978. The possible use of methyl 4-methylpyrrole - 2 - carboxylate, an ant trail pheromone, as a component of an improved bait for leaf-cutting ant (Hymenoptera: Formicidae) control. Bull. Entomol. Res. 68: 159-170.

Robinson, S. W., A. Aranda, L. Cabello, and H. Fowler. 1980. Locally produced toxic baits for leaf-cutting ants for Latin America; Paraguay, a case study. Turrialba 30: 71-76.

Robinson, S. W., A. R. Jutsum, J. M. Cherrett, and R. J. Quinlan. 1982. Field evaluation of methyl 4-methylpyrrole - 2 - carboxylate, an ant trail pheromone, as a component of baits for leaf-cutting ant (Hymenoptera: Formicidae) control. Bull. Entomol. Res. 72: 345-356.

Tumlinson, J. H., R. M. Silverstein, J. C. Moser, R. G. Brownlee, and J. M. Ruth. 1971. Identification of the trail pheromone of the leaf-cutting ant, Atta texana. Nature 234: 348-349.

30
The Search
for Chemical Bait Toxicants

C. S. Lofgren

Control of pest ant species relies characteristically on toxic baits which consist of a toxicant combined with an attractive food material and, if needed, a granular carrier for ease of dispersal in outdoors or large-scale control programs. The effectiveness of this approach depends upon the availability of a delayed-action toxicant so that foraging ants are not killed before they carry it back to the nest and distribute it to all colony members. The ultimate transfer of the toxicant to the queen is of paramount importance since her continued egg production will soon result in colony growth even though 85% or possibly more of the original workers are killed (Williams et al. 1980).

The first studies with bait toxicants for fire ant control were conducted by Travis (1939) against Solenopsis geminata. He used thallium sulfate or thallium acetate in syrup baits. The former compound was toxic in cage tests but not in the field while the latter appeared effective in both types of tests. Results in the latter tests were reported to be erratic, probably due to weather conditions. Green (1952) tested a bait containing thallium sulfate against the black imported fire ant, S. richteri, and found it effective against captive colonies but ineffective when scattered around mounds in the field.

With the advent of the red and black imported fire ants (IFA), S. invicta and S. richteri, as major pest species and the promulgation of the Federal-State Imported Fire Ant Control Program in 1957, a concerted drive was made by the United States Department of Agriculture (USDA) to identify effective toxicants and bait formulations. The procedures used to identify toxicants were described first by Stringer et al. (1964). Various modifications of these techniques were made over the next 20 years; the most recent description was published by Williams (1983) who described three steps in the evaluations. These included primary screening tests with individual workers, secondary tests with queen right laboratory colonies, and field tests with natural infestations.

Basically the primary screening tests involved allowing groups of 20 worker ants in test chambers to feed for 24 hours on soybean oil or sugar-water solutions of the candidate toxicants. After this they were transferred to clean chambers and given neat food; mortality was recorded at regular intervals for 14 days. In the initial tests, field-collected ants were used; however, later, when colony laboratory-rearing procedures were perfected, workers from these colonies were utilized. Any toxic chemicals that produced delayed toxicity (<15% mortality after 24 hours) over a tenfold dosage range were re-evaluated as baits with queen right colonies. In these tests, the chemical was dissolved in soybean oil and fed directly to the ants from micropipettes or following absorption on a granular carrier (corncob grits or pregel defatted corn grits). Observations of worker and brood mortality and effects on the queen were recorded over a period of one to several months, depending upon the severity of the toxicological response. Toxicants that killed the colonies directly or inhibited development and reproduction were next formulated in soybean oil and a carrier for distribution on small field plots. Effectiveness of the baits was determined by comparing the total number of active nests before and after treatment or by a more complex method that involved rating colony size and presence or absence of brood.

Toxicants that are effective in all of these stages of development are considered candidates for commercial development; however, other factors could influence their successful development. Accordingly, we have determined several characteristics that are essential for a particular chemical to be a suitable ant bait toxicant. The toxicants must:

a) Be soluble in vegetable oils,

b) Lack repellency to the ants,

c) Exhibit delayed toxicity over greater than a tenfold range of concentrations,

d) Be transferred readily from one ant to another during trophallaxis,

e) Be rapidly biodegraded,

f) Exhibit low mammalian toxicity in the formulated bait, and

g) Be environmentally acceptable.

From 1958 to 1977, over 3200 chemicals were evaluated for activity as bait toxicants (Banks et al. 1977). Only one of these, mirex, was eventually developed commercially as an ant bait. These data vividly illustrate the difficulty in discovering and developing delayed-action toxicants. When mirex registrations were cancelled by the Environmental Protection Agency (EPA) in 1977, we had no suitable alternative so we were faced with the problem of developing a comprehensive program to find other chemicals. In meetings with our own staff and others within USDA, we reviewed all possible approaches to obtaining delayed-action chemicals or formulations. It was recognized at this point that chemicals other than those that caused direct toxicity needed to be investigated since inhibition of egg production and/or larval development could also cause death of the colony. As a result of our deliberations, the following potential approaches were considered for investigation:

a) Controlled release technology,

b) Inhibition of enzymes that detoxify insecticides,

c) Evaluation of insect growth regulators,

d) Evaluation of chemicals that inhibit reproduction by the queen,

e) Synthesis of chemicals with delayed-action, and

f) Evaluation of conventional chemicals obtained from industry.

All of these approaches have been studied since 1977 through our own research effort and cooperative studies with other USDA laboratories, private research laboratories, university scientists, and private industry. Despite some disappointments, several of the projects were successful and provide us today with an array of potential chemical baits for IFA control. A review of the research on the different projects follows.

CONTROLLED RELEASE TECHNOLOGY

Vander Meer et al. (1980) suggested four possible controlled release techniques: A) matrix-bound toxicants, B) microencapsulation, C) modification of toxicant structure (protoxicants), and D) attachment of insecticides to polymers (pendent toxicants). Methods A and B proved impractical for two reasons. All of the formulations tested leaked toxicant, thus, the food attractant always became contaminated. The second problem involved the unique structure of the pharynx of IFA workers that filters out solid particles. The filtration mechanism involves a series of ridges and hairs at the entrance to and in the buccal tube. During feeding, food is taken into the infrabuccal pocket and compressed. The liquid is forced through the pharynx while the particles are retained by the filtering hairs and expelled with the infrabuccal pellet. Tests conducted with fluorescent latex particles revealed almost all particles 0.88 microns in diameter were retained in the infrabuccal pocket (Glancey et al. 1981). This diameter was well below practical technology for microencapsulation at that time.

Method C listed above is defined as the chemical alteration of a known toxicant to a relatively nontoxic material that is structured so that the toxicant can be released in the ant by an enzymatic or hydrolytic mechanism. This technique is used in the pharmaceutical industry. One of the major problems with this technique for ant toxicants is the lack of toxicants with functional groups available for modification. Research on this method was conducted in collaboration with the USDA Insect Physiology Laboratory, Beltsville, Maryland, who prepared modified structures of trichlorfon and sodium fluoroacetate. Results of these tests (Kochansky et al. 1979) showed some delay in toxicity with caprate and sterol esters, but the range of dosages over which delay occurred was very narrow.

A study of the pendent toxicant method (D) was conducted with the Southern Research Institute at Birmingham, Alabama. With this method, numerous molecules of the toxicant are attached to a polymer backbone. It differs from the protoxicant approach in that the high-molecular weight pendent toxicants do not pass through membranes, and thus, are less readily metabolized. Two toxicants were utilized in the study, trichlorfon and 1H,1H-pentadecafluoro-octanol-1-ol. Pendent toxicant formulations of both compounds were prepared; however, the materials were not soluble in soybean oil. Various methods (sonication, surfactants) were used to suspend the materials so that they would be ingested by the ants, but no increased delay in toxicity was noted in the bioassays. Fortunately, it was the use of the fluorinated surfactants in these studies that led to the discovery that some fluoroaliphatic sulfones were good delayed-action toxicants (see section on Delayed-Action Toxicants). Further studies with protoxicants and pendent toxicants seem warranted but the research is expensive and time-consuming.

ENZYME INHIBITION

SKF-525A is an inhibitor of mixed-function oxidases that are involved in the conversion of the sulfur analogues of some fast-acting organophosphate insecticides to the oxygen analogue (e.g., malathion). The inhibitor was combined at a 10:1 ratio with five different organophosphates and tested in our primary screening test. In every case, the SKF-525A reduced the toxic action of the insecticides to an unacceptable level. No further research was conducted with this approach.

INSECT GROWTH REGULATORS

Since the effects of insect growth regulators (IGRs) are exhibited on caste and larval development rather than direct toxicity (Vinson and Robeau 1974), good colony rearing techniques (Banks et al. 1981b) and the availability of large numbers of colonies is imperative. All active IGRs cause a surge in the development of sexual forms within 2 to 4 weeks after the colony is exposed. A small amount of brood may appear in later months but the colony eventually dies from lack of adult workers. Less active compounds may produce similar temporary effects but the colony almost always recovers. Sterility of queens from exposure to methoprene as reported for the Pharaoh ant, Monomorium pharaonis, has not been observed (Edwards 1975).

Banks et al. (1978) described laboratory tests with 26 juvenile hormone mimics against queen right colonies and found that AI3-36206 (1-(8-methoxy-4,8-dimethylnonyl)-4-(1-methylethyl)benzene) was the most effective. Since then, better activity has been obtained with several other IGRs (see Banks, Chapter 32). The most active of these compounds (fenoxycarb) gives good control of ants in the field; however, the slow death of the colonies may cause concern because it appears as though the colonies will recover and incipient colonies may appear before all the older colonies die. Repeated treatments every 6 to 12 months and an educational program for the users may be the best answer to this problem. Development of a commercial bait formulation (Logic®) is in progress, and a request for registration from EPA has been submitted by the developer, Maag Agro-Chemicals, Vero Beach, Florida. (Registration was granted in late 1985.)

Chitin inhibitors which are also classified as IGRs have been tested; but the best compounds, such as dimilin, are extremely insoluble and, thus, a fair evaluation on their effectiveness has never been obtained.

REPRODUCTIVE INHIBITORS

One of our early screening programs involved the evaluation of chemicals that caused sterility in insects. As with the IGRs, it was necessary to conduct primary screening tests with these compounds on queen right colonies. Several hundred materials were tested with little success (Banks, unpublished data). Our one success in this area came from an unexpected source when Merck and Company, Rahway, New Jersey, submitted some novel compounds obtained from the soil microorganism, Streptomyces avermitilis. In the initial primary screening tests, some mortality of workers was noted; but it was not sufficiently delayed to warrant further tests. However, because these compounds were good nematicides and were from a unique source, we also tested one of them, avermectin B_1, against queen right colonies at a concentration of 1%. Limited worker mortality occurred but, much to our surprise, no brood was found in the colonies after 4 weeks, even though the queen was alive. We next tested successively lower concentrations and found that brood production could be stopped at concentrations in soybean oil as low as 0.0025%. In no instance did brood production resume. In field tests, we obtained similar results. Worker brood was found in only 8 of 928 colonies that fed on baits that were applied at rates of AI ranging from 0.0077 to 7.41 g/ha (Lofgren and Williams 1982). As with the IGRs, total colony death was slow since most workers present at the time of exposure died slowly of "natural" causes.

Commercial development of a formulation of avermectin B_1 (Affirm®) containing 0.011% AI is underway. It will be applied at the rate of 1 lb/acre or 120 mg AI/ha.

Toxicological studies of the avermectins indicate that they are gamma aminobutyric acid (GABA) agonists. Glancey et al. (1982) found that avermectin B_1 caused irreversible damage to the ovaries of IFA queens which resulted in complete sterility.

SYNTHESIS OF TOXICANTS

The only synthesis program for IFA toxicants was undertaken by the Chemistry Department of Mississippi State University. Their syntheses were based on data from the USDA screening program. Fisher et al. (1980) reported that alkoxycarbonylphenyl compounds showed promise but the toxic delay was inadequate. In a second study, Fisher et al. (1983) synthesized and tested 21 diethyl aryl phosphorothionates. Compounds with either a bromine or ester substituent were the better toxicants. While their studies were partially successful, they could not be continued because of a lack of funds.

CONVENTIONAL DELAYED-ACTION TOXICANTS

Three new classes of chemicals were discovered that exhibited good delayed toxicity: amidinohydrazones, phenylenediamines, and fluorinated sulfones. Williams et al. (1980) published the results of tests with nine of the amidinohydrazones. The most effective of these compounds was American Cyanamid AC 217,300 (tetrahydro-5, 5-dimethyl-2(1H)-pyrimidinone, [3-[4-(trifluoromethyl)phenyl]-1-[2-[4-(trifluoromethyl)phenyl]ethynyl]-2-propenylidene]hydrazone). In preliminary screening tests, the chemical in soybean oil gave delayed kill over more than a tenfold dosage range; however, the compound had low solubility in soybean oil (<1%). This was eventually overcome by using oleic acid as a cosolvent. In secondary tests, concentrations of 2.5 to 10% in the oil gave either complete colony mortality or the queen and most of the workers were killed. Later studies showed queen kill and mortality at concentrations as low as 0.1%. The effects of this compound on the queen are its most striking property. Field studies with baits containing AC 217,300 were effective in field tests (Banks et al. 1981a; Harlan et al. 1981). The most effective concentration in the soybean oil baits was 2.5%. A bait (Amdro) containing AC 217,300 was registered for fire ant control in August 1980.

The second promising toxicant was a phenylenediamine provided by Eli-Lilly and Company, Greenfield, Indiana. This compound, EL-468 (N-[2-amino-3-nitro-5-(trifluoromethyl)phenyl]-2,2,3,3-tetrafluoropropanamide), gave delayed kill at 0.1 and 1.0%

concentrations in primary screening tests; in secondary tests, it killed laboratory colonies at concentrations in soybean oil of 1, 2.5, and 5%. In field tests, baits with 2.5 and 5.0% concentrations in the soybean oil gave 86 to 91% control (Williams and Lofgren 1981). EL-468 was on the verge of commercialization when toxicological studies revealed possible teratogenic effects. It was then withdrawn and no further research or development has taken place.

A third group of new toxicants are the fluoroaliphatic sulfones. These toxicants were discovered serendipitously when they were used as surfactants to suspend pendent toxicant formulations. Over 300 compounds of this type have been screened and many of them have shown excellent delayed toxicity (Vander Meer et al. 1985). The most consistent results in field tests have been obtained with AI3-29757 (N-ethyl heptadecafluoroctyl sulfonamide). This compound is not soluble above 2% and it is somewhat repellent to the ants. Commercial development is being undertaken by Griffin Corporation, Valdosta, Georgia. Little is known at this point about its mammalian toxicology. Biodegradability might be a problem also.

SUMMARY

Delayed-action toxicants are difficult to obtain because this type of toxicity runs counter to standard synthesis programs that seek chemicals that kill very fast. However, the search for delayed toxicants or compounds that affect reproduction over the past 8 years has been rewarding. It is possible that four chemicals will be registered in bait formulations for IFA control by the end of 1986. The leads obtained, especially with IGRs and inhibitors of reproduction, suggest that many other similar compounds await discovery and development. The synthesis program described was very limited in scope and more intensive research in this area should be productive. Finally, the controlled release technique, while unsuccessful, did lead to valuable information on ant morphology and formulation techniques.

REFERENCES CITED

Banks, W. A., H. L. Collins, D. F. Williams, C. E. Stringer, C. S. Lofgren, D. P. Harlan, and C. L. Mangum. 1981a. Field trials with AC-217,300, a new amidinohydrazone bait toxicant for control of the red imported fire ant. Southwest. Entomol. 6: 158-164.

Banks, W. A., C. S. Lofgren, D. P. Jouvenaz, C. E. Stringer, P. M. Bishop, D. F. Williams, D. P. Wojcik, and B. M. Glancey. 1981b. Techniques for collecting, rearing, and handling imported fire ants. U.S. Dept. Agric., Sci. Ed. Admin., AAT-S-21: 1-9.

376

Banks, W. A., C. S. Lofgren, and J. K. Plumley. 1978. Red imported fire ants: Effects of insect growth regulators on caste formation and colony growth and survival. J. Econ. Entomol. 71: 75-78.

Banks, W. A., C. S. Lofgren, C. E. Stringer, and R. Levy. 1977. Laboratory and field evaluation of several organochlorine and organophosphorous compounds for control of imported fire ants. USDA, ARS-S-169. 13 pp.

Edwards, J. P. 1975. The effects of a juvenile hormone analogue on laboratory colonies of Pharaoh's ants, Monomorium pharaonis (L.) (Hymenoptera: Formicidae). Bull. Entomol. Res. 65: 75-80.

Fisher, T. H., W. E. McHenry, E. A. Alley, and H. W. Chambers. 1980. Toxicity of O,O-diethyl O-carboalkoxyphenyl phosphorothionates to the imported fire ant. J. Agric. Food Chem. 28: 731-735.

Fisher, T. H., W. E. McHenry, E. G. Alley, C. S. Lofgren, and D. F. Williams. 1983. Some phosphorothionate imported fire ant toxicants with delayed kill. J. Agric. Food Chem. 31: 730-733.

Glancey, B. M., C. S. Lofgren, and D. F. Williams. 1982. Avermectin B_1a: Effects on the ovaries of red imported fire ant queens. J. Med. Entomol. 19: 741-745.

Glancey, B. M., R. K. Vander Meer, A. Glover, C. S. Lofgren, and S. B. Vinson. 1981. Filtration of microparticles from liquids ingested by the red imported fire ant, Solenopsis invicta Buren (Hymenoptera: Formicidae). Insect. Soc. 28: 395-401.

Green, H. B. 1952. Biology and control of the imported fire ant in Mississippi. J. Econ. Entomol. 45: 593-597.

Harlan, D. P., W. A. Banks, H. L. Collins, and C. E. Stringer. 1981. Large area tests of AC 217,300 bait for control of imported fire ants in Alabama, Louisiana and Texas. Southwest. Entomol. 6: 150-157.

Kochansky, J. P., W. E. Robbins, C. S. Lofgren, and D. F. Williams. 1979. Design of some delayed-action toxicants for baits to control red imported fire ants. J. Econ. Entomol. 72: 655-658.

Lofgren, C. S., and D. F. Williams. 1982. Avermectin B_1a, a highly potent inhibitor of reproduction by queens of the red imported fire ant. J. Econ. Entomol. 75: 798-803.

Stringer, C. E., C. S. Lofgren, and F. J. Bartlett. 1964. Imported fire ant toxic bait studies: Evaluation of toxicants. J. Econ. Entomol. 57: 941-945.

Travis, B. V. 1939. Poisoned-bait tests against the fire ant, with special reference to thallium sulfate and thallium acetate. J. Econ. Entomol. 32: 706-713.

Vander Meer, R. K., C. S. Lofgren, D. H. Lewis, and W. E. Meyers. 1980. Controlled release formulations and control of the imported fire ant: What are the possibilities?, pp. 251-256. In

R. Blake (ed.), Controlled release of bioactive materials. Academic Press, Inc., New York. 473 pp.

Vander Meer, R. K., C. S. Lofgren, and D. F. Williams. 1985. Fluoroaliphatic sulfones: A new class of delayed-action insecticides. J. Econ. Entomol. 78: 1190-1197.

Vinson, S. B., and R. Robeau. 1974. Insect growth regulator: Effects on colonies of the imported fire ant. J. Econ. Entomol. 67: 584-587.

Williams, D. F. 1983. The development of toxic baits for the control of the imported fire ant. Fla. Entomol. 66: 162-172.

Williams, D. F., and C. S. Lofgren. 1981. Eli Lilly EL-468, a new bait toxicant for control of the red imported fire ant. Fla. Entomol. 64: 472-477.

Williams, D. F., C. S. Lofgren, W. A. Banks, C. E. Stringer, and J. K. Plumley. 1980. Laboratory studies with 9 amidinohydrazones, a promising new class of bait toxicants for control of red imported fire ants. J. Econ. Entomol. 73: 798-802.

31
Chemical Baits: Specificity
and Effects on Other Ant Species

D. F. Williams

Chemical baits have been used for many years to control pest ant species such as fire ants and leaf-cutting ants (see Cherrett, Chapter 29; Lofgren, Chapter 30). Prior to 1978, the most effective toxic baits contained the toxicant mirex which was developed in the early 1960s for control of the red and black imported fire ants (IFA), Solenopsis invicta and S. richteri (Lofgren et al. 1964). Because of its effectiveness (usually >95%), it replaced residual applications of heptachlor in 1962 as the preferred agent for IFA control. This chemical was also effective against leaf-cutting ants and is still used in some countries to control this pest. However, because several studies in the early 1970s revealed mirex residues in fish and wildlife (Lowe et al. 1971; Baetcke et al. 1972; Borthwick et al. 1973), the U.S. Environmental Protection Agency (EPA) in April 1973 filed notice in the Federal Register for a public hearing to determine if registrations of mirex should be cancelled. After several years of hearings, these registrations were withdrawn on December 31, 1977.

As reported by Lofgren (Chapter 30), the loss of mirex necessitated a concerted effort to discover and develop other chemical agents for IFA control, particularly ones that were environmentally acceptable. Two chemicals developed through this research were registered by the EPA, AC 217300 (tetrahydro-5,5-dimethyl-2(1H)-pyrimidinone [3-[4-(trifluoromethyl)phenyl]-1-[2-[4-(trifluoromethyl)phenyl]ethynyl]-2-propenylidene]hydrazone), the active component in Amdro® and Stauffer MV-678 (1-(8-methoxy-4,8-dimethylnonyl)-4-(methylethyl)benzene), the active ingredient in Pro-Drone®. Two additional chemicals currently being considered for registration are avermectin B_1, which causes irreversible sterility of the queen, and fenoxycarb (ISO proposed), which acts as an insect growth regulator (see Lofgren, Chapter 30).

Although the effects that the five bait toxicants just described exert on an IFA colony may differ, all of the chemicals use the same basic bait formulation, that is, the chemical is dissolved in a food attractant (usually soybean oil) and applied to a corn grit carrier. A major problem with this system is that large numbers of nontarget insects, especially beneficial ant species, feed on the bait and are affected by the toxicants. For example, Mr. H. T. Vanderford, an entomologist working for the Georgia Department of Agriculture, made numerous unpublished observations on nontarget ant species following applications of mirex bait from 1962 to 1967 in Georgia. He found high mortality (>75%) of 17 of the 37 species observed, moderate reductions of 13 others, and no effect on 7 species. In addition to these 37 species, more than a dozen other species were observed feeding on the bait; but he was unable to monitor the effects, if any. Clearly, these observations indicate that a very large number of other ant species are attracted to and feed on granular soybean oil baits. Markin et al. (1974) studied the effects of mirex bait on 14 common ant species in Louisiana and found that those species most affected were classified as oil-feeding with the exception of a Pheidole spp. which, although oil-loving, was not affected by the treatment. Following aerial applications of mirex bait in Texas, Summerlin et al. (1977) monitored 14 species of ants and found that omnivorous and highly predacious ants were the first to be affected by the bait and were eliminated within two weeks. After eight weeks, only two species of ants could be found in the study plot and both contained residues of mirex. Thus, mirex, a broad spectrum insecticide, was toxic to most of the ants that ingested it (Markin et al. 1972).

Although Amdro does not present the residue problem of mirex, it nevertheless still affects many nontarget ant species. Edmunson (1981) indicated in both laboratory and field studies that Amdro was relatively non-selective in its effects against several other ant species and that overall ant activity in the field was reduced 80 to 100% following treatments. A redeeming feature, however, was that all of the species affected, except one, recovered to pretreatment population levels within one year following treatment. On the other hand, Apperson et al. (1984) did not observe any deleterious effects from Amdro on the nontarget ant species he studied. The differences between the two studies might be explained by the fact that the study area used by Apperson et al. was heavily infested with S. invicta and this species generally outcompetes other ant species; thus, few other ant species were collected. Edmunson, on the other hand, wanted to observe the maximum effects of Amdro so he conducted a study in an area free of S. invicta, although some colonies were located nearby. Therefore, he not only had more species but greater numbers on which to evaluate the effects of Amdro. Stimac (personal communication) also indicated that Amdro exhibited deleterious effects on nontarget ant species in studies he conducted in a pasture near Gainesville,

Florida.

Another problem with baits such as mirex and Amdro is the ability of the most aggressive and highly reproductive ant species to quickly reinfest areas following bait treatments. This is especially true if all competitors, even those offering a small degree of competition, are also eliminated by the bait treatment. Reinfestation of treated areas by S. invicta is well documented (Lofgren et al. 1964; Lofgren and Weidhaas 1972; Markin et al. 1974; Summerlin et al. 1976, 1977; Brown 1980; Apperson et al. 1984). Because S. invicta colonies produce large numbers of sexuals (3,000 to 5,000/ year), newly-mated queens quickly inundate bait-treated areas. If these queens are free from competition from other ant species or conspecific colonies, then large numbers of incipient colonies can be established. In areas heavily infested with S. invicta, workers from the conspecific colonies are the most significant mortality factor for newly-mated queens (Whitcomb et al. 1973). In contrast, in lightly infested areas, Conomyrma insana (Buckley) is the significant mortality factor for newly-mated queens of S. invicta. Whitcomb et al. (1973) also indicated that 11 other species of ants were seen attacking these queens. Thus, it is obvious that predation by already established ant species plays a major role in determining the success of colony-founding queens.

Studies of the effects of Pro-Drone, fenoxycarb, and avermectin B_1 on nontarget insects, especially ants, have been or are currently underway; but at the present time, no published information is available. L. Lemke (personal communication) stated that in her studies in South Carolina Pro-Drone caused no deleterious effects on nontarget ants. However, no pretreatment counts were recorded, which may have influenced the interpretation of the results. S. A. Phillips (personal communication) indicated that Pro-Drone applied over a large area in Texas had no effect on nontarget ants and little effect on target ants. In contrast, S. B. Vinson (personal communication) found some detrimental effects on nontarget ant species following Pro-Drone applications to IFA populations in Texas. Studies presently underway by D. F. Williams et al. (unpublished data) indicate that both fenoxycarb and avermectin B_1 reduce nontarget ant populations; however, all affected species are returning to pretreatment population levels after one year.

Although insect growth regulators and other chemicals with different modes of action may have lesser effects on nontarget arthropods or affect fewer species than the stomach poisons and neurotoxins, they probably will still have deleterious effects on many nontarget organisms. This potential problem is particularly true if they are attracted to and feed on the soybean oil in the bait. Unfortunately, little hope exists at this time of finding a replacement for soybean oil because (1) it is an excellent attractant for IFA, (2) it is reasonably priced and readily available, (3) most chemical control agents easily dissolve in it, and (4) it is readily absorbed on carriers, making an excellent flowable bait.

If the problem of non-selectivity is to be overcome, all potential areas of research should be explored. For example, size and texture of bait particles, timing of applications, and rates of application may offer some promise in making baits more specific to S. invicta. The ant species that are the most aggressive foragers tend to dominate baits soon after they are applied; therefore, larger particles may not be readily picked up by fast foraging nontarget ant species giving S. invicta a chance to displace them at the bait. Applying the baits at optimum foraging and feeding times for S. invicta would also lessen the chances of other ant species feeding on the bait. The rate of application may play a role in selectivity since aggressive foragers such as S. invicta would collect most, if not all, of the bait applied at optimal rates; but amounts above optimal would allow other ant species greater access to the bait.

The previously mentioned areas of research deal with the improvement of the fire ant bait in its present form; i.e., the soybean oil and toxicant applied to a granular carrier. Four other areas of bait toxicant research which do not include the soybean oil attractant offer promise. Examples of these are: (1) pheromones, (2) phagostimulants, (3) biological control agents, and (4) carriers and formulations. Research in each of these categories is underway, but much more needs to be done.

The use of pheromones to increase the specificity of baits to the target insect appears very promising. Elucidation of portions of the queen and trail pheromone complexes have been reported (Rocca et al. 1983a, 1983b; Vander Meer et al. 1981), and Vander Meer (1983) presented an indepth review of the pheromones found in S. invicta and the role they play in this species' behavior. Some research on baits in combination with trail pheromones to control leaf-cutting ants has been reported (see Kermarrec, Chapter 29). While some of the baits appeared more attractive, they did not surpass the attractiveness of food odors already in the bait. Other attractants that look very interesting include the colony odors involved in nestmate recognition. These may involve cuticular hydrocarbons which make up 65 to 75% of the ant's cuticular lipids (Lok et al. 1975). These fire ant species "recognition" chemicals, when combined with toxicants, may aid in developing species specific baits. Vander Meer (1983) indicated that the four species of Solenopsis, S. invicta, S. geminata, S. xyloni, and S. richteri, can all be identified by their cuticular hydrocarbon patterns. Also of special interest is the queen recognition pheromone which is discussed by Glancey in Chapter 19. Surrogate queens (rubber septa, wood sticks) treated with these compounds are readily carried to the nest. Thus, the pheromone may offer a means of enticing foragers to collect and carry bait particles back to their nestmates.

Phagostimulants, plant chemicals which elicit feeding responses, are another area of research that should be pursued. Most insects are attracted to specific food substances and fire ants and leaf-cutting ants are no exception. For example, fire ant workers are highly attracted to, and feed on, the calyx of okra flowers (Vander Meer, unpublished data). If the attractants or phagostimulants found in okra could be incorporated into baits, then the possibility of enhancing their attractiveness to fire ants would increase.

Biological agents can be very species specific in their effects on certain organisms. If a pathogen (bacteria, virus, or fungus) was discovered that exerted an effect on one or more stages of only the target insect, then introducing this infectious agent into an IFA population would be a species-specific control measure. Also, pathogens may be successfully utilized as stress agents rather than outright mortality factors. For example, if pathogens can be used to stress a colony, other species of ants may then be able to successfully compete for resources and displace the treated colony. Jouvenaz (Chapter 27) presents a review of the status of diseases of IFA. None of these diseases seem to offer a solution; however, others undoubtedly remain to be discovered. The signing of a cooperative agreement between USDA, ARS, Insects Affecting Man and Animals Research Laboratory, Gainesville, Florida, and EMBRAPA—the Empresa Brasileira de Pesquisa Agropecuaria of the Brazilian Ministry of Agriculture, has provided the basis for an extensive search for other fire ant diseases and parasites. If a suitable IFA pathogen could be mass-produced, baits inoculated with this organism could be formulated and applied to IFA populations, just as the current chemical baits are applied.

Carrier and formulation research also offers some promise in bait specificity. The bait carrier and formulation used against IFA is a pregel defatted corn grit containing 30% once-refined soybean oil in which the chemical toxicant is dissolved. The problem with this formulation, as previously mentioned, is the attractiveness of the soybean oil to other ants. Recent laboratory and field studies by Williams et al. (unpublished data) have shown good control of IFA with a bait composed only of fenoxycarb and the carrier, pregel defatted corn grits. The carrier was immersed in a 2.5% acetone solution of the technical chemical for 30 minutes, removed, and spread out in a shallow pan to air dry under a fume hood for 24 hours. The bait was then ready for use. Apparently, enough residual oil remained in the defatted corn grits to attract foraging workers of the fire ant. Also, technical fenoxycarb has a low repellency to the IFA workers. Studies are underway to determine if, in fact, this formulation with greatly reduced oil content lessens the effect of fenoxycarb on other ant species.

Food flow studies of Howard and Tschinkel (1981), Sorensen

and Vinson (1981), and Sorensen et al. (1983) indicate that we may target baits towards queens and larvae, bypassing workers, by utilizing protein baits. Substituting a protein bait for the soybean oil may allow us to utilize otherwise unusable toxicants.

Another approach to IFA baits has been the use of a "natural" component of their diet, insect pupae. House fly (Musca domestica) and eye gnat (Hippelates pusio) pupae are not only very attractive to fire ants but also make a very flowable carrier which is easily dispersed with application equipment. The acetone-immersing method described previously can be used to apply the technical chemical to the pupae. The treated pupae then become an attractant-carrier with the toxicant trapped inside until the IFA feed on it. In previous studies, Williams et al. (unpublished data) have shown very good control with house fly pupae treated with fenoxycarb (Table 1). Control was as good as that achieved with the Amdro standard, and the amount of active ingredient applied was much smaller. In laboratory studies, some nontarget ant species were found not to feed on pupae, thus imparting some bait specificity.

In conclusion, the area of bait specificity is one in which little interest has been shown. Most control programs using insecticides, including bait techniques, have been developed with little regard to the impact these chemicals might have on the environment. The majority of available insecticides are broad spectrum because developmental costs are extremely high, and highly specific chemicals can quickly become very unprofitable, especially in the case of small markets (Zeck 1985). Nevertheless, if we are to develop programs for better management of insect populations, such as the IFAs and the leaf-cutting ants, we must minimize the impact of these programs on nontarget species. The future goal of pest management for fire ants and leaf-cutting ants should be to provide long-term population suppression. For us to accomplish this goal, we must utilize any method or combination of methods that reduce populations of the pest species while minimizing the effects on beneficial and other nontarget species. Thus, we are seeking management schemes which offer more bait selectivity, defined by Bartlett (1964) as the capacity of a treatment to spare natural enemies while destroying pests. To further illustrate this principle, Ripper et al. (1956) have divided selectivity into two types, ecological and physiological. Ecological selectivity is obtained by manipulating the amount of chemical toxicant reaching the target species with little or no effects on the nontarget species. Physiological selectivity would occur by making the chemical itself more selective by designing the molecular structure of the toxicant so that it is more toxic to the target species than it is to nontarget species (Zeck 1985). Both of these methods have great potential, especially in view of our rapidly increasing knowledge of ant ecology, physiology, and biochemistry.

TABLE 1. Control of IFA with fenoxycarb and AC 217300 (Amdro) on house fly pupae or pregel defatted corn grits. Alachua County, FL, April 1982. (Avg. of three 1-acre plots.)

Treatment	g AI/ acre	Pretreatment counts		% reduction in PI after wks indicated[d]		
		No. mounds[c]	PI[e]	6	14	18
AC 217300[a]	0.81	52	1203	20	53	62
Fenoxycarb[a]	0.66	49	1117	78	92	85
AC 217300 (std.)[b]	4.35	48	1065	69	92	89
Fenoxycarb (std.)[b]	3.50	45	1005	79	95	94
Untreated check	—	52	1132	28	33	32

[a]House fly pupae baits were prepared by dissolving the test chemical in acetone (2.5% AC 217300 and 3.0% fenoxycarb), immersing the pupae in this solution for 30 minutes, and air drying under a fume hood for 24 hours.
[b]Standard (std.) baits consisted of 70% pregel defatted corn grits impregnated with 30% of the soybean oil-toxicant solution.
[c]No. mounds per 1/2-acre circle within center of each 1-acre square plot.
[d]Corrected for check mortality by Abbott's formula.
[e]Population index (PI). See Banks, Chapter 32, for method of calculation.

REFERENCES CITED

Apperson, C. S., R. B. Leidy, and E. E. Powell. 1984. Effects of Amdro on the red imported fire ant (Hymenoptera: Formicidae) and some nontarget ant species and persistence of Amdro on a pasture habitat in North Carolina. J. Econ. Entomol. 77: 1012-1018.

Baetcke, K. P., J. D. Cain, and W. E. Poe. 1972. Residues in fish, wildlife, and estuaries: Mirex and DDT residues in wildlife and miscellaneous samples in Mississippi—1970. Pestic. Monit. J. 6: 14-22.

Bartlett, B. R. 1964. Interaction of chemical and biological control, Ch. 17. In P. Debach (ed.), Biological control of insect pests and weeds. Reinhold Publ. Co., N.Y., NY.

Borthwick, P. W., T. W. Duke, A. J. Wilson, Jr., J. I. Lowe, J. M. Patrick, Jr., and J. C. Oberheu. 1973. Residues in fish, wildlife, and estuaries: Accumulation and movement of mirex in selected estuaries of South Carolina, 1969-71. Pestic. Monit. J. 7: 6-26.

Brown, R. E. 1980. The imported fire ant in Florida. Proc. Tall Timbers Conf. Ecol. Anim. Control Habitat Manage. 7: 15-21.

Edmunson, M. B. 1981. The effect of Amdro on nontarget ant species associated with Solenopsis invicta Buren in Florida. M.S. Thesis, University of Florida, Gainesville, FL. 112 pp.

Howard, D. F., and W. R. Tschinkel. 1981. Internal distribution of liquid foods in isolated workers of the fire ant, Solenopsis invicta. J. Insect Physiol. 27: 67-74.

Lofgren, C. S., and D. E. Weidhaas. 1972. On the eradication of imported fire ants: A theoretical appraisal. Bull. Entomol. Soc. Am. 18: 17-20.

Lofgren, C. S., F. J. Bartlett, C. E. Stringer, and W. A. Banks. 1964. Imported fire ant toxic bait studies: Further tests with granulated mirex-soybean oil bait. J. Econ. Entomol. 57: 695-698.

Lok, J. B., E. W. Cupp, and B. J. Blomquist. 1975. Cuticular lipids of the imported fire ants, Solenopsis invicta and Solenopsis richteri. Insectic. Biochem. 5: 821-829.

Lowe, J. I., P. R. Parrish, A. J. Wilson, Jr., P. D. Wilson, and T. W. Duke. 1971. Effects of mirex on selected estuarine organisms. Trans. North Am. Wildl. Nat. Res. Conf. 36: 171-186.

Markin, G. P., J. H. Ford, J. C. Hawthorne, J. H. Spence, J. Davis, H. L. Collins, and D. C. Loftis. 1972. The insecticide mirex and techniques for its monitoring. USDA, APHIS, 81-3 series, November 1972. 19 pp.

Markin, G. P., J. O'Neal, and H. L. Collins. 1974. Effects of mirex on the general ant fauna of a treated area in Louisiana. Environ. Entomol. 3: 895-898.

Ripper, W. E. 1956. Effects of pesticides on balance of arthropod populations. Annu. Rev. Entomol. 1: 403-438.

Rocca, J. R., J. H. Tumlinson, B. M. Glancey, and C. S. Lofgren. 1983a. The queen recognition pheromone of Solenopsis invicta, preparation of (E)-6-(1-pentenyl)-2H-pyran-2-one. Tetrahedron Lett. 24(18): 1889-1892.

Rocca, J. R., J. H. Tumlinson, B. M. Glancey, and C. S. Lofgren. 1983b. Synthesis and stereochemistry of tetrahydro-3,5-dimethyl-6-(1-methylbutyl)-2H-pyran-2-one, a component of the queen recognition pheromone of Solenopsis invicta. Tetrahedron Lett. 24(18): 1893-1896.

Sorensen, A. A., and S. B. Vinson. 1981. Quantitative food distribution studies within laboratory colonies of the imported fire ant, Solenopsis invicta Buren. Insect. Soc. 28: 129-160.

Sorensen, A. A., T. M. Busch, and S. B. Vinson. 1983. Behaviour of worker subcastes in the fire ant, Solenopsis invicta, in response to proteinaceous food. Physiol. Entomol. 8: 83-92.

Summerlin, J. W., A. C. F. Hung, and S. B. Vinson. 1977. Residues in nontarget ants, species simplification, and recovery of populations following aerial applications of mirex. Environ. Entomol. 6: 193-197.

Summerlin, J. W., J. K. Olson, and J. O. Fick. 1976. Red imported fire ant: Levels of infestation in different land management areas of the Texas coastal prairies and an appraisal of the control program in Fort Bend County, Texas. J. Econ. Entomol. 69: 73-78.

Vander Meer, R. K. 1983. Semiochemicals and the red imported fire ant (Solenopsis invicta Buren) (Hymenoptera: Formicidae). Fla. Entomol. 66: 139-161.

Vander Meer, R. K., F. D. Williams, and C. S. Lofgren. 1981. Hydrocarbon components of the trail pheromone of the red imported fire ant, Solenopsis invicta. Tetrahedron Lett. 22: 1651-1654.

Whitcomb, W. H., A. Bhatkar, and J. C. Nickerson. 1973. Predators of Solenopsis invicta queens prior to colony establishment. Environ. Entomol. 2: 1101-1103.

Zeck, W. M. 1985. The future of narrow versus broad-spectrum insecticides, Ch. 23. In J. L. Hilton (ed.), Agricultural chemicals of the future. Rowman and Allanheld Publ., Totowa, NJ.

32
Insect Growth Regulators for Control of the Imported Fire Ant

W. A. Banks

Studies with insect growth regulators (IGR) and the red imported fire ant (RIFA), Solenopsis invicta Buren, began in the early 1970s when Cupp and O'Neal (1973) and Troisi and Riddiford (1974) found that methoprene and hydroprene (ingested or by contact) interfered with maturation of developing larvae, normal metamorphosis, and caused worker mortality. However, Troisi and Riddiford (1974) suggested that, because of lack of persistence in a colony, hormonal growth regulators did not appear to be suitable for control of RIFAs.

Subsequently, Vinson et al. (1974) demonstrated that a number of IGRs were active against pharate reproductive pupae of RIFAs at 0.1 ug/ant or less when topically applied. In other tests, small laboratory colonies fed soybean oil or egg yolk baits containing certain IGRs decreased or stopped egg production, reduced or stopped larval and pupal production, and ultimately died from effects of the chemical (Vinson and Robeau 1974). The same IGRs were more active by contact or fumigation than by ingestion. In these and other studies (Robeau and Vinson 1976), IGRs were shown to be strongly active in shifting caste differentiation in RIFAs. IGR introduction into a colony producing only minor workers stimulated production of major workers, intercastes, and alate queens.

Further studies (Texas Agricultural Experiment Station 1980) showed that although soybean oil baits containing an IGR were very effective in eliminating laboratory colonies of RIFAs, they gave very erratic results against field populations. This was attributed in part to the fact that the ants very rapidly metabolized and/or excreted two of the IGRs found to be most effective, methoprene and Stauffer R-20458 (1-(4'-ethylphenoxy)-6,7-epoxy-3,7-dimethyl-2-octene) (Wendel and Vinson 1978; Bigley and Vinson 1979). They concluded that for effective control the IGR must be more resistant to breakdown or it must be formulated in a way that prevented or retarded breakdown. Also, new techniques were needed to introduce the chemicals into the colonies (Texas Agricultural Experiment Station 1980). Subsequently, Bigley and Vinson (1979) demonstrated that piperonyl butoxide or DEF (S,S,S-tributyl phosphorothioate) retarded breakdown of methoprene by RIFAs.

Banks et al. (1978) demonstrated in laboratory tests that 3 of 26 IGRs administered in peanut butter baits were highly active, eliminating 65 to 75% of the treated colonies. After modification of test procedures in 1977, an additional 29 IGRs were tested in soybean oil bait and 4 additional IGRs were equal or superior in activity to the 3 found in earlier tests (Banks et al. 1983).

In field tests, it was found that one of the more active IGRs, Stauffer MV-678 (1-(8-methoxy-4,8-dimethylnonyl)-4-(methylethyl) benzene), killed up to 76% of treated field colonies in small plots and left most of the surviving colonies without worker brood 26 to 40 weeks following treatment with 4.75 g AI/ha. On larger plots, two applications of a granular soybean oil bait (11.85 g AI/ha) applied with aircraft at 6-month intervals eliminated 89.5% of active colonies and reduced the population index by 95.8% (Banks et al. 1983). Small plot tests with CIBA-GEIGY CGA-38531 (1-(3-ethoxybutoxy)-4-phenoxybenzene), MAAG Agrochemicals RO 13-5223 (ethyl[2-(p-phenoxyphenoxy)ethyl]carbamate), and Montedison JH-286 (1,[5-chloropent-4-ynyl)oxy]-4-phenoxybenzene (1)) showed that these chemicals killed 61.6 to 82.1, 49.8 to 84.8, and 72.9 to 90.1% of active nests and reduced population indexes by 81.3 to 96.2, 82.4 to 97.8, and 97.8 to 98.4%, respectively (Banks and Harlan 1982; Banks et al. 1983).

Continuing laboratory and field studies have identified additional IGRs with excellent activity against RIFAs and substantiated the results of earlier IGR tests. The results of some of this work are reported here.

MATERIALS AND METHODS

Laboratory Studies

Procedures for laboratory evaluation of IGRs were standardized about 1977. The candidate chemicals are dissolved in once-refined soybean oil and the solutions are fed to the test colonies in micropipets. Tests are conducted against laboratory-reared RIFA colonies (Banks et al. 1981) that consist of a queen, 10 to 30 ml of brood (eggs, larvae, and pupae), and 40 to 60 thousand worker ants. All chemicals are initially tested against three colonies at 10 mg/colony (0.5 ml of 2.0% solution). Check colonies are given an equivalent volume of neat soybean oil. After treatment, all colonies are returned to the normal diet and held in the laboratory at 28 ± 2°C for observation. Each colony is examined biweekly through 16 weeks post-treatment and monthly thereafter until the colony dies or recovers from obvious IGR effects. Each colony is rated pretreatment and at each post-treatment interval based on the estimated number of worker ants and quantity of worker brood according to the following colony index scale:

Estimated number of worker ants				Estimated quantity of worker brood (ml)		
	Rating	Value			Rating	Value
<100	1	1		0	A	1
101–5000	2	2		1–5	B	5
5001–20000	3	3		5–10	C	10
20001–35000	4	4		10–20	D	15
35001–50000	5	5		20–30	E	20
>50000	6	6		>30	F	25

The colony index is derived for each colony by multiplication of the assigned numerical value for worker numbers times the numerical value for quantity of brood; e.g., a colony with a rating of 6E would have a colony index of 120 (6 x 20). The effectiveness of a chemical against a colony is determined by comparison of the pre- and post-treatment colony indices.

Field Studies

Those chemicals that cause death of laboratory colonies are formulated into granular baits and tested against natural infestations of the RIFA according to procedures described by Banks et al. (1983). Baits are prepared by dissolving the IGR in once-refined soybean oil and spraying the solution (30% by weight) onto 8 to 30 mesh pregel defatted corn grits as they are tumbled in a mixer. The baits are applied with a tractor-mounted granular applicator to small plots (0.25 to 1.0 ha) or with fixed-wing aircraft to large plots (40 to 350 ha) located in nongrazed pasture and on grass-sod military stage fields and airports. Circular subplots (0.1 to 0.2 ha) are established within each larger plot for pre- and post-treatment evaluation of the ant populations.

Since the IGRs are essentially nontoxic to the worker ants and the queen and express their effects primarily through suppression of brood production, evaluation of these effects is sometimes very difficult. Similar problems were experienced with some of the other materials evaluated for RIFA control, and a population index system was devised for evaluation of these materials. The population index system developed by Harlan et al. (1981) for work with American Cyanamid AC-217,300 (Amdro®) and modified by Lofgren and Williams (1982) for work with avermectin was also adopted for evaluation of IGRs. With this method the entire area within each subplot is searched carefully before treatment and at predetermined intervals post-treatment, and each active nest found is opened with a spade and the contents carefully examined. Each nest is then assigned a rating of from 1 to 10 based on the estimated number of worker ants and the presence or absence of worker brood. This rating is then used to calculate a population index using the following scale:

Number of worker ants	Without worker brood		With worker brood	
	field rating	nest index	field rating	nest index
<100	1	1	6	5
100-1000	2	2	7	10
1000-10000	3	3	8	15
10000-50000	4	4	9	20
>50000	5	5	10	25

The population index for each subplot can be expressed mathematically as follows:

$$\text{Population index (PI)} = \sum_{K=1}^{25} K(N_k)$$

where N_k = the number of ant colonies on a given subplot with a nest index of K, where $25 \geq K \geq 1$.

The population index for all plots or treatments is obtained by summation of the indices for all subplots within the unit. Effectiveness of the treatment is determined by comparison of the pre- and post-treatment population indices.

RESULTS AND DISCUSSIONS

Laboratory Studies

Twenty-six IGRs in addition to those reported by Banks et al. (1978) and Banks et al. (1983) were evaluated in the laboratory from 1982 to 1984. Two chemicals (Table 1) were highly effective in suppressing worker brood production, Sumitomo S-4496 (propionaldehyde oxime O-2-(4-phenoxyphenoxy) ethyl ether) and S-4624 (propionaldehyde oxime O-2-(4-phenoxyphenoxy) propyl ether). The effects of S-4624 were slightly slower than those of S-4496; however, both reduced colony indices by greater than 95% within 8 weeks after treatment and all colonies had died by 48 weeks after treatment. The effects of these compounds on laboratory colonies were similar to those noted with other compounds we have found to be effective.

TABLE 1. Effects of Sumitomo S-4496 and S-4624 on laboratory colonies of red imported fire ants.

Chemical and dosage (AI)	Pretreatment colony index[a]	Percent reduction in colony index after indicated weeks					
		4	8	12	16	20	24
				Test 1			
S-4496 10 mg	104.2±42.7	95.5±1.5	96.2±1.0	98.0±1.1	98.2±1.0	99.3±0.8	99.8±0.4
S-4624 10 mg	129.2±24.6	67.5±20.9	96.5±0.8	97.7±0.8	98.0±0.6	98.7±0.5	97.3±3.6
Check[b]	114.2±32.3	33.3±26.6	7.7±22.5	-6.5±50.3	14.7±61.4	28.0±54.0	32.0±60.6

[a]See text for method of calculating colony index; mean and standard deviation (n=6).
[b]The large standard deviation from weeks 12 to 24 was due to the death of the queen in two of the four colonies.

Figure 1 outlines the general effects observed in all of our tests with active IGRs. Although the two compounds involved in these studies were accepted very well by the ant colonies, some chemicals are repellent to the ants and are either totally rejected or consumed in ineffective quantities. Even if an IGR is readily accepted by the ants, it may produce no effect or only transient effects.

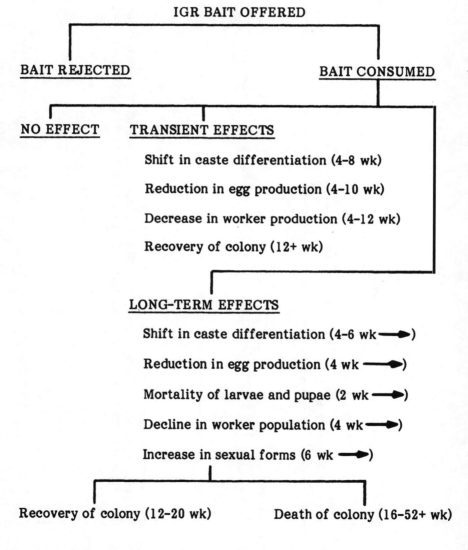

FIGURE 1. Effects of insect growth regulators administered to imported fire ant colonies in oil-based baits.

No detailed physiological studies have been done to determine the mode of action of any IGRs on RIFAs; however, the following changes have been consistently observed with all the active IGRs. As a general rule, they are not lethal to the queen or worker ants, although death of either or both may be accelerated because of general breakdown in maintenance and social organization of the colony. The most active IGRs are effective in (1) shifting caste differentiation from worker to sexual form, (2) causing deformities and death of many developing larvae, and (3) reducing or stopping egg production by the queen. The net result is that no new workers are produced and the colony declines in size and vigor as the existing worker force ages and dies. This decline may continue to the death of the entire colony or it may be reversed if the IGR is eliminated by metabolism and/or excretion before the colony declines below a critical point that has not yet been determined.

Dosage level does not appear to be a critical factor in the ability of a colony to overcome the effects of an IGR. We found that 1 of 2 colonies treated with 120 mg each of JH-25 (E-1-[(7-ethoxy-3,7-dimethyl-2-octenyl)oxy]-4-ethylbenzene) survived treatment even though 3 of 5 treated at 20 mg each died (Banks et al. 1978). Similarly, 2 of 4 colonies treated at 5 mg each with CIBA-GEIGY CGA-38531 survived although 2 of 2 treated at 1 mg each and 6 of 6 treated at 2 mg each died (Banks and Harlan 1982). We have postulated (Banks and Schwarz 1980; Banks and Harlan 1982) that colony makeup and quality and quantity of food available to the colony just prior to treatment may strongly influence long-term effects of an IGR on the ant colony.

Glancey et al. (1973) found that large workers in RIFA colonies serve as repletes, retaining oily solutions in the gastral crop for up to 18 months. We found that red dye used as a tracer in oil solutions of IGRs was retained by some workers for several months after treatment. Traces of dye appeared in sexual larvae 3 to 5 months after treatment. This supported the hypothesis that IGRs were retained for extended periods of time by some workers and slowly released into the colony food supply. Longevity of such stored material, theoretically, would be proportional to the number of repletes and their capacity for storage at the time of treatment. The long-term retention concept is further supported by the fact that queens removed from treated colonies at 12 weeks post-treatment resumed worker brood production when placed with untreated workers even though they had produced no workers since the fourth week post-treatment. Sister test colonies in which the queens remained with treated workers did not produce worker brood to the time of colony death (20 to 24 weeks post-treatment).

Retention of IGRs does not, however, appear to be as simplistic as we originally envisioned. As we noted in the introduction to this paper, Wendel and Vinson (1978) and Bigley and Vinson (1979) showed, respectively, that Stauffer R-20458 and methoprene were very rapidly metabolized and excreted by RIFA colonies. Preliminary data suggests that MAAG RO 13-5223 may also be metabolized relatively quickly by RIFAs. The level of parent compound declined by about 45% and the level of metabolites rose 1 to 72 hours after the ants fed on oil bait containing the compound. In companion studies, we found that radioactivity in a colony dosed with ^{14}C labelled RO 13-5223 declined from an average 2220 dpm/ant on day 2 to background levels by day 56. Highly radioactive detritus removed from the rearing tray during the first week after treatment contained only metabolites of the IGR. Obviously further studies are needed to determine if the IGR is indeed stored in the RIFA colony and, if so, in what form, by whom, and for how long.

Field Studies

Field studies with S-4496 and S-4624 were conducted on 0.3-ha plots and with MAAG RO 13-5223 on aerially-treated 85-ha plots. In the small plot tests, Sumitomo S-4496 was more effective than S-4624, causing population reductions of 99.6 and 99.3% and eliminating 94.3 and 91.7% of active nests at 9.06 and 18.12 g AI/ha, respectively. S-4624 was slightly less effective, causing population reductions of 90.8 and 97.4% and eliminating 82.4 and 82.1% of active nests at 9.13 and 18.26 g AI/ha respectively (Table 2).

In the large area tests, MAAG RO 13-5223 formulated as a pregel defatted corn grit bait eliminated 94.0% of the active nests and reduced the population index by 99.5% after 12 weeks posttreatment (Table 3). The corncob grit formulation was less effective providing 89.1% elimination of active nests and 98.0% reduction in the population index.

The reduction in population indices has proven to be the best indicator of IGR effectiveness in field studies, since colony mortality is usually long-term. However, confounding reinfestation of small to medium plots sometimes occurs before kill is complete.

The effects of IGRs on RIFAs in both laboratory and field have been dramatic and show that these chemicals can be effectively used in RIFA population management programs. Available information indicates that IGRs are nonpersistent, which eliminates one of the environmental problems encountered with previous RIFA control methods. Development and registration of formulations containing IGRs for RIFA control have been relatively slow. The U.S. Environmental Protection Agency (EPA) granted registration in 1983 to a pregel defatted corn grit formulation containing Stauffer MV-678 (Prodrone). However, it has not gained wide acceptance because of its erratic performance. EPA registration of a pregel defatted corn grit bait containing MAAG RO 13-5223 (Logic®) is pending. Further studies with CIBA-GEIGY CGA-38531 and

TABLE 2. Effects of Sumitomo S-4496 and S-4624 on field populations of red imported fire ants.

| Formulation | Application rate | | Pretreatment | | Percent reduction after 12 wks | |
	Bait (kg/ha)	AI (g/ha)	Number active nests	Population index[a]	Number active nests	Population index
S-4496 1.0% Bait	0.905	9.05	53	1090	94.3	99.6
	1.81	18.12	48	1030	91.7	99.3
S-4624 1.0% Bait	0.913	9.13	51	1122	82.4	90.8
	1.83	18.26	56	1164	82.1	97.4
Untreated Check	—	—	47	980	14.9	21.9

[a]Population index determined by modification of system of Lofgren and Williams (1982). See text for description of system.

TABLE 3. Effectiveness of the IGR MAAG RO 13-5223 against populations of red imported fire ants in large field plots. Brunswick, GA – 1983.

| Application rate | | Number active nests | | Population index | | Mean percentage reduction | |
Bait kg/ha	AI g/ha	Pre-treatment	Post-treatment[a]	Pre-treatment	Post-treatment	Number nests	Population index
Pregel defatted corn grit bait[b]							
1.12	11.2	265	16	6074	29	94.0	99.5
Expanded corncob grit bait[b]							
1.12	11.2	274	30	5743	112	87.9	98.0
Untreated Check							
—	—	260	202	5665	4030	22.3	28.9

[a]Post-treatment evaluations made at 12 weeks.
[b]Formulations comprised of 70% pregel defatted corn grit, 29% once-refined soybean oil, 1.0% RO 13-5223 or 75% expanded corncob grits, 24% once-refined soybean oil, 1.0% RO 13-5223.

Montedison JH-286 are waiting decisions of the manufacturers on future development and registration of these materials. Studies are progressing with the Sumitomo compounds and they may eventually receive EPA registration.

Research is continuing to identify other IGRs that may be effective against the RIFAs and to develop improved formulations or application techniques for maximum field results.

REFERENCES CITED

Banks, W. A., and D. P. Harlan. 1982. Tests with the insect growth regulator, CIBA-GEIGY CGA-38531, against laboratory and field colonies of red imported fire ants. J. Ga. Entomol. Soc. 17: 462-466.

Banks, W. A., and M. Schwarz. 1980. The effects of insect growth regulators on laboratory and field colonies of red imported fire ants. Proc. Tall Timbers Conf. Ecol. Anim. Habitat Manage. 7: 95-105.

Banks, W. A., C. S. Lofgren, D. P. Jouvenaz, C. E. Stringer, P. M. Bishop, D. F. Williams, D. P. Wojcik, and B. M. Glancey. 1981. Techniques for collecting, rearing, and handling imported fire ants. USDA, SEA, Adv. Agric. Tech. S-21, 9 pp.

Banks, W. A., C. S. Lofgren, and J. K. Plumley. 1978. Red imported fire ants: Effects of insect growth regulators on caste formation, and colony growth and survival. J. Econ. Entomol. 71: 75-78.

Banks, W. A., L. R. Miles, and D. P. Harlan. 1983. The effects of insect growth regulators and their potential as control agents for imported fire ants. Fla. Entomol. 66: 172-181.

Bigley, W. S., and S. B. Vinson. 1979. Degradation of ^{14}C-methoprene in the imported fire ant, Solenopsis invicta. Pestic. Biochem. Physiol. 10: 1-13.

Cupp, E. W., and J. O'Neal. 1973. The morphogenetic effect of two juvenile hormone analogues on larvae of imported fire ants. Environ. Entomol. 2: 191-194.

Glancey, B. M., C. E. Stringer, Jr., C. H. Craig, P. M. Bishop, and B. B. Martin. 1973. Evidence of a replete caste in the fire ant, Solenopsis invicta. Ann. Entomol. Soc. Am. 66: 233-234.

Harlan, D. P., W. A. Banks, H. L. Collins, and C. E. Stringer. 1981. Large area tests of AC-217,300 bait for control of imported fire ants in Alabama, Louisiana, and Texas. Southwest. Entomol. 6: 150-157.

Lofgren, C. S., and D. F. Williams. 1982. Avermectin B_1a, a highly potent inhibitor of reproduction by queens of the red imported fire ant. J. Econ. Entomol. 75: 798-803.

Robeau, R. M., and S. B. Vinson. 1976. Effects of juvenile hormone analogues on caste differentiation in the imported fire ant, Solenopsis invicta. J. Georgia Entomol. Soc. 11: 198-203.

Texas Agricultural Experiment Station. 1980. The imported fire ant program: A search for new control methods. 28 pp.

Troisi, S. J., and L. M. Riddiford. 1974. Juvenile hormone effects on metamorphosis and reproduction of the fire ant, Solenopsis invicta. Environ. Entomol. 3: 112-116.

Vinson, S. B., and R. Robeau. 1974. Insect growth regulator effects on colonies of the imported fire ant. J. Econ. Entomol. 67: 584-587.

Vinson, S. B., R. Robeau, and L. Dzuik. 1974. Bioassay and activity of several insect growth regulator analogues on the imported fire ant. J. Econ. Entomol. 67: 325-328.

Wendel, L. E., and S. B. Vinson. 1978. Distribution and metabolism of a juvenile hormone analogue within colonies of the red imported fire ant. J. Econ. Entomol. 71: 561-565.

33
Status of Leaf-Cutting Ant Control in Forest Plantations in Brazil

E. F. Vilela

Leaf-cutting ants are considered to be one of the most serious polyphagous insect pests in the Neotropics (Howse and Bradshaw 1977). They damage a wide range of plant species by defoliation while gathering substrate on which to cultivate their fungus. They are very well-adapted for attacking monocultures and cause extensive damage to crops such as eucalyptus, cotton, cassava, sugarcane, citrus, cocoa, and pasture grass (Gonzales 1976). According to Simoes et al. (1981), leaf-cutting ants are a limiting factor in the establishment of eucalyptus plantations in Brazilian "cerrado," (typical tropical savanna that has been planted with such crops and which is heavily infested with leaf-cutting ants). Clark (1972) calculated that at least 5% of the total costs of establishing a eucalyptus plantation are for ant control. As the Brazilian government is now placing emphasis on the production of alternative types of fuel from renewable resources, eucalyptus plantations have gained an even more important role in the economy.

Efforts to prevent leaf-cutting ant infestations go back to the sixteenth century. The methods of control have been various, ranging from the use of explosives to blow up the nest to blowing poisonous gases into the nest. Current control methods have reached a satisfactory standard but can be dangerous to man and the environment; thus, improvements are still needed. Toxic baits are considered to be the safest method and sometimes give the best results, but they unquestionably need to be improved. Applications of liquid formulations of insecticides are no longer used.

In an attempt to evaluate the present leaf-cutting ant control situation and to predict future problems likely to arise in Brazil, a survey was carried out among companies in the State of Minas Gerais, in which there are the largest number of reforestation programs in all of the Brazilian States. This paper presents the results of this survey, a review of techniques used for ant control, and an analysis of the benefits and problems associated with their use.

399

400

THE SURVEY DATA

Questionnaires were sent to forestation and reforestation companies through the Forest Research Society of the University of Vicosa. Five of these questionnaires have been returned. The company areas varied from 45,000 to 150,000 ha.

Eucalyptus spp. are the primary trees grown on the plantations. Most of the companies have a permanent staff to take care of their pest problems, while others have agronomists and foresters to deal with the many silvicultural problems that occur. As members of the Forest Research Society of the University of Vicosa, they can seek advice about their pest problems from the university research staff.

The number of Atta spp. nests in "cerrado" was between 16 to 26/ha in the areas being cleared for the first planting, and 5 to 10/ha in the planted areas, including both small- and large-sized nests. The companies estimated that the control of these pests accounts for more than 75% of the total time spent in pest control and for more than 75% of the costs involved. The control of ants is continuous throughout the year, and the entire area must be patrolled so that all nests can be located. Leaf-cutting ants are the main and the most persistent pest problem according to the companies surveyed.

METHODS OF CONTROL

Mechanical

By searching for entrance holes, it is possible to find new leaf-cutting ant nests in an area approximately 90 days after their mating swarms. At this time, each nest has only one chamber. One can easily destroy the ant colony by digging it up with a hoe.

Dusts

Many insecticides, in dust form, have been used against leaf-cutting ants in Brazil. Currently, only aldrin and heptachlor are used, with limited success, to control small nests.

In large colonies, the ants can quickly modify their ventilation system within the nest to seal off affected zones. It sometimes happens with the application of dusts in large nests that the ants abandon the nest and establish another. The numerous chambers in the nest provide potential dead spaces that the toxicants do not reach and this may lead to control failures (Howse and Bradshaw 1977).

Gas (fumigants)

Methyl bromide, in spite of being the most expensive method, is still used with success to control leaf-cutting ants. In 1980, a private forestry company in the State of Minas Gerais controlled 4.8 million leaf-cutting ant nests, 25% of these with methyl bromide (Fardin, personal communication).

Baits

Baits are the most commonly used control technique for leaf-cutting ants in Brazil. The majority of baits make use of the stomach poison, mirex (dodecachlor). The two most serious weaknesses attributed to the baiting method are degradation of the bait particles (formulation) under field conditions and poor attractiveness of the bait to the ants. Both have been reported to be responsible for failures with baits. Another inconvenience of baits is their dependence on suitable weather conditions which can be mitigated with the use of highly attractive baits which the forager ants readily pick up and carry back to the nest. It is necessary also to use the smallest possible quantities of insecticides in the baits in order to avoid undesirable side-effects. This implies that a high proportion of bait applied in the field must be collected by the foraging ants in a short span of time.

A potential method of increasing bait attractiveness is through the use of pheromones. Since the first reported successful use of pheromones for insect control by Gaston et al. (1967), interest has been growing steadily in this approach. Some pheromones of leaf-cutting ants have been identified and the possibility of using them as bait components to increase attractiveness has been suggested by Moser (1967) and Lewis (1972), and evaluated by Robinson and Cherrett (1973, 1974, and 1978). Cross et al. (1979) made an attempt to enhance bait pickup by Atta sexdens rubropilosa by using their trail pheromone components, 3E25DMP (3-ethyl-2,5-dimethyl-pyrazine) and 2E25DMP (2-ethyl-2,5-dimethylpyrazine). They concluded that Paraguayan field colonies of this species did not preferentially pickup baits impregnated with the pheromone mixture.

Robinson et al. (1982) found in the laboratory that the A. texana and A. cephalotes trail pheromone, M4MP2C (methyl 4-methylpyrrole-2-carboxylate), acted as an attractant to leaf-cutting ants when added to soybean baits. It increased the percent pick-up of pheromone-impregnated bait as compared with the plain bait; however, subsequent field trials were disappointing. Vilela (1983) evaluated leaf-cutting ant pheromones for increasing bait attractiveness. He concluded that the addition of trail pheromones and also (2)-9-nonadecene, isolated or in mixtures, to food baits did not increase bait pickup in field trials with A. sexdens rubropilosa. In spite of the poor results obtained with pheromones in baits, it is still hoped that progress can be made in this field. Until now, little has been known about the behavior of leaf-cutting ants toward the different chemicals they use for communication. Therefore, in the long run, behavioral control agents may play a role in better and safer methods of ant control.

The use of aerial applications of toxic baits, as reported by Lewis and Phillips (1973), has never been considered by Brazilian forestry companies involved in leaf-cutting ant control.

Phillips et al. (1976) field-tested many new non-chlorinated hydrocarbon bait toxicants as alternatives to mirex. Some of these gave control better or equal to mirex; but, to date, they have not been marketed as baits. For instance, Loeck and Nakano (1984) found that diflubenzuron had an effect similar to that of the dodecachlor, controlling 100% of small leaf-cutting ant nests in the laboratory and field. Nogueira (in press) has tested avermectin and Amdro® on citrus pulp baits against large field nests of Atta and Acromyrmex. The results were promising, although avermectin baits took longer to kill the large nest populations than did mirex baits. Also, copper oxychloride has been tested as a bait toxicant against Atta nests and preliminary results have shown that it can be as effective as mirex (Loeck 1982).

Since baits may lose their attractiveness and/or may deteriorate in the field, a new technique to spread the baits has been practiced in Brazil. It consists of sheltering the bait particles in plastic cups and scattering these throughout the area. According to studies carried out by one of the companies surveyed, the most efficient way to utilize these "bait cups" in the field is to survey 10% of the area to be baited and determine the number of ant nests per hectare. This figure is the number of "bait cups" to be used per hectare in the whole area. These cups can remain for as long as six months in good condition in the open field. The results with this technique, however, have not been consistently good. Their effectiveness depends on the ant infestation levels. If they are high, satisfactory final control is achieved. On the negative side, however, when the "bait cups" are in the environment a long time, they can be picked up by other organisms with attendant dangers to wildlife and domestic animals.

Dried citrus pulp has been the most commonly used bait matrix in Brazil. It is now becoming expensive due to its multiple uses; therefore, a new bait matrix is needed. Jutsum and Cherrett (1981) have developed a light-weight matrix of polyurethane that is prepared from dense precursors into which suitable attractants and toxicants can be incorporated; however, it has not as yet been used commercially.

Thermal-fogs

Since the bait toxicant method has not yielded consistently good results on large farms, a new technique to control leaf-cutting ants was introduced about 9 to 10 years ago (Couto et al. 1977). It involves a Swing-Fog® machine similar to one used unsuccessfully in the past to control leaf-cutting ants but which has now been redesigned to incorporate a burner. The insecticide is first heat-nebulized in the burner and then added to the smoke which is injected into an entrance hole of the nest. At the time that the toxic fog is introduced into one nest entrance, all other entrances must be sealed in order to ensure good penetration.

Aldrin (Atafog® 20%) and heptaclor (Arbinex® 20%) have given the highest percentage of nest mortality in tests carried out by the Department of Entomology, University of Vicosa. Fungicides have also been field-tested in the thermal fog system as well as in baits, but the results have not been satisfactory. Three or, at a minimum, two men are required to apply the fogs in the field, one to carry the equipment and the other(s) to seal the nest entrances.

The fogging technique is being used on an increasing scale, mainly by companies involved in forestation and reforestation projects. The firms are now creating their own teams to apply the thermal-fogs since the technique has been proven to be successful. Prior to this, leaf-cutting ant control was sub-contracted to specialized firms. We have noticed that there is a general belief among users that this method can solve the leaf-cutting ant problem. While thermal-fogging and baits admittedly give the best control, some failures still occur, although they are less common with the thermal-fogging system.

When failures do occur with the thermal-fogging method, they are usually caused by operator inexperience or by difficulties in delivering the amount of insecticide needed to kill the entire nest population. Transport of the equipment in the field, maintenance problems, and noise production have also been considered as inconveniences. Locating the nests is another problem in control, but one which also applies to some extent to the baiting method.

In an effort to make the thermal-fogging method more effective, Cruz and Nogueira (personal communication) adapted the exhaust system of a motorcycle to apply insecticide into the nest. These workers reported that the efficiency, in terms of labor, was greater than that of thermal-fogging using conventional equipment.

Biological Control

Studies in the past with biocontrol of leaf-cutting ants using predator ants have been unsuccessful (Mariconi 1970). A different attempt has been undertaken at the University of Sao Paulo at

Piracicaba by Berti Filho (personal communication) using Canthon virens (Coleoptera: Scarabaeidae). This beetle attacks new queens by cutting off their heads, laying eggs in their gasters, and then burying them.

Another attempt at using the fungi Metarrhizium anisopliae and Beauveria bassiana directly against leaf-cutting ants has been tried by Diehl-Fleig (unpublished data) in the State of Rio Grande do Sul. The results are still inconclusive. Baits impregnated with these fungi have also been tested on many occasions but have been unsuccessful.

Flechtmann (1981) tested the potential of the mite Pyemotes tritici (Prostigmata: Tarsonemoidea: Pyemotidae), an effective parasite of stored grain insects, for the control of leaf-cutting ants. Mites were introduced into the nest by exposing mite cultures at nest entrances. After 2 or 3 days, ant activity stopped for up to 16 days but it eventually recovered. More detailed studies are needed in order to have a complete evaluation of the mite as an agent for controlling leaf-cutting ants.

Eucalyptus Species Selection and Host-Plant Resistance

A study has been conducted by Anjos and collaborators (personal communication) at the University of Vicosa with the aim of finding Eucalyptus species less susceptible to leaf-cutting ant attacks. Preliminary results suggest that 2 species out of the 20 most commonly planted in Brazil, Eucalyptus maculata and E. nezophyla, are attacked the least by A. sexdens rubropilosa, the most common leaf-cutting ant species in the state of Minas Gerais. This work is part of the first project to study Eucalyptus resistance to leaf-cutting ants.

Use of "Trap Crop" Food Plants

It is a common practice among growers to attempt to control the ants by planting ginger (Sesamum indicum, Pedaliacea) between the tree rows, as a highly attractive but non-sustainable food source. The plant leaves, which are readily preferred by the ants, do not appear to support the growth of fungi; therefore, it was thought that this practice might result in nest mortality. However, there is no clear indication that the use of ginger or other aromatic plants can successfully control leaf-cutting ants, even in experimental areas.

CONTROL STRATEGIES

In order to overcome the failures of each of the available methods and to maximize the efforts to control leaf-cutting ants, forestry firms in Brazil have adopted the following strategies:

1. apply control measures just before an area is cleared of its natural vegetation, using bait and/or thermal-fogs. Baits are not used when the vegetation is too dense, because of the problem of nest location.

2. apply control measures immediately after seedling trees have been planted using mechanical control, dusts, and methyl bromide.

3. conduct maintenance measures within the area and its surroundings at 6-month intervals. Thermal-fogs and baits are the methods of choice.

A firm with a planted area of 150,000 ha and with 10% new area will need 60 teams of 13 to 15 people each to obtain successful control of the leaf-cutting ants. For each team, 4 machines are required, and for every 4 teams, a trailer is needed.

DISCUSSION AND CONCLUSIONS

From the survey results, we can conclude that leaf-cutting ants of the genera Atta and Acromyrmex are the main insect pests in forest plantations in Brazil. Baits are still the most reliable method of control, and their occasional unsuccessful use on large farms or plantations could be due to the short life span of the bait particles after they have been broadcast in the environment and their concomitant loss of attractiveness to the ants. Application of baits on large farms to coincide with the foraging time of the colonies is impracticable as the bait is exposed to severe tropical weather conditions that cause it to deteriorate very rapidly. New strategies of bait usage or bait preparation during the wet season would minimize this problem.

Small farmers seem to be more successful with the use of baits, as they are able to avoid long bait exposure times that could result in spoilage. However, they practice subsistence agriculture, in which leaf-cutting ant control has a low priority. This lack of interest in leaf-cutting ant control, or lack of resources for bait materials, could provide conditions for a permanent reservoir of leaf-cutting ants.

The limitations and failures of baits have recently led those concerned with forest protection to take particular interest in the thermal-fogging method as an option to control leaf-cutting ants. The great advantage of this method is that large or small colonies are equally susceptible, and it is independent of the season or colony foraging behavior. Moreover, control of nests is far greater.

The exact list of side-effects will vary from one method to another. It must be emphasized that the high toxicity of the particles produced by the thermal-fogging devices presents a hazard to the operators. The tropical climatic conditions are not conducive to the wearing of heavy protective clothing as the temperature is frequently high. Inhalation of the insecticidal fogs is extremely dangerous, and regulations regarding it have been introduced by authorities. Beyond this, we must consider the following inconveniences of fogging in comparison with baits: (1) the need for equipment maintenance because it must be operated under very dusty conditions; (2) the need for transportation in the open field; (3) cost of operations and especially the initial costs (a machine plus burner was priced at about U.S. 200 dollars in January 1985), and one firm with an area of 100,000 ha would need at least 120 machines; (4) the need for fuel that has increased considerably in cost over a very short period of time; and finally, (5) the motorized equipment results in high noise production that could cause permanent hearing impairment of the operators. The costs of the total operation go far beyond the costs of the machines and fuel and include insecticide, labor, transport, and maintenance costs, but these are compensated for by the efficiency of control.

Although bait toxicants are regarded as safe, we have to consider the possibility that bait remaining for a long time in the environment could be dangerous to nontarget organisms. When the baits are very attractive to the ants, they should be picked up more readily, thereby minimizing the danger of them being picked up by other organisms. However, we should bear in mind that the clearance of land to plant exotic Eucalyptus, for instance, far outweighs any environmental consequences of baiting ant nests.

Another problem with baits is related to the active ingredient they contain and their potential toxicity to man and hazards to the environment. The regulations for the use of insecticides in Brazil are still weak, allowing, for instance, the use of mirex bait, which has been banned in some other countries due to its possible carcinogenicity and persistence.

The problem of controlling leaf-cutting ants in Brazil, therefore, tends to increase, rather than decrease, despite the more systematic control undertaken in recent years by the large commercial firms. The agricultural frontiers have been expanding and will continue to do so. Thermal-fogging systems and mirex bait may not remain available. The situation with pests in Brazil is such that the large firms do not have precise estimates of the economic damage that pests cause to their crops. In the case of the leaf-cutting ants, the forestry firms do not have even a rough estimate of damage losses. The damage may be worse than is thought and require more intensive control. Some partial estimates have been published in recent years based on certain well-known situations. For example, Simoes et al. (1981) have pointed out that, in the absence of control measures, (1) a given Eucalyptus plantation with an initial average of 2 nests/ha has 6 nests per hectare after one year; (2) a Eucalyptus plantation with 4 nests/ha suffered a loss in young trees of 14%; and (3) a Eucalyptus area with 200 nests/ha of Acromyrmex spp. had 30%

of the trees killed, probably by repeated defoliation.

However, one must say that the damage caused by leaf-cutting ants can sometimes be exaggerated. To clarify the actual economic importance of leaf-cutting ants in areas like central Brazil, more studies will have to be conducted.

REFERENCES CITED

Clark, E. W. 1972. Status and future needs of forest entomology research in Brazil. Report of the consultant in forest entomology, Project working document FAO No. 7. 33 pp.

Couto, L., J. C. Zanuncio, J. E. M. Alves, E. Campinhos, Jr., L. Soresini, and J. A. Vargas. 1977. Avaliacao da eficiencia e custo do controle de Atta sexdens rubropilosa atraves do sistema termo-nebulizador, na regiao de Aracruz, E. S. Arvore, Vicosa. 1: 9-16.

Cross, J. H., R. C. Byler, U. Ravid, R. M. Silverstein, S. W. Robinson, P. M. Baker, J. S. Oliveira, A. R. Jutsum, and J. M. Cherrett. 1979. The major component of the trail pheromone of the leaf-cutting ant Atta sexdens rubropilosa Forel: 3-Ethyl-2,5-Dimethyl-pyrazine. J. Chem. Ecol. 5: 187-203.

Flechtmann, C. H. W. 1981. Um possivel agente de controle biolog ico da sauva. VII Congresso Brasileiro de Entomologia, Fortaleza. pp. 1.

Gaston, L. K., H. H. Shorey, and C. A. Saario. 1967. Insect population control by using sex pheromones to inhibit orientation between the sexes. Nature 213: 1155.

Gonzales, R. H. 1976. Crop protection in Latin America. FAO Plant Protection Bull. 24: 65-75.

Howse, P. E., and J. W. S. Bradshaw. 1977. Some aspects of the biology and chemistry of the leaf-cutting ants. Out. Agric. 9: 160-166.

Jutsum, A. R., and J. M. Cherrett. 1981. A new matrix for toxic baits for control of the leaf-cutting ant Acromyrmex octospinosus (Reich) (Hym.: Formicidae). Bull. Entomol. Res. 71: 607-616.

Lewis, T. 1972. Aerial baiting to control leaf-cutting ant. PANS. 18: 71-74.

Lewis, T., and F. T. Phillips. 1973. Aerial baiting to control leaf-cutting ants (Formicidae: Attini) in Trinidad. 1. The bait, its production, and the effects of toxicants. Bull. Entomol. Res. 63: 263-273.

Loeck, A. E. 1982. Efeito de novas substancias visando o controle das sauvas Atta spp. (Hym.: Formicidae). Tese M.S. USP/ Piracicaba (SP). 45 pp.

Loeck, A. E., and O. Nakano. 1984. Efeito de novas substancias visando o controle de sauveiros novos de Atta laevigata (Smith, 1858) (Hymenoptera: Formicidae). O Solo, Piracicaba. 76: 25-30.

Mariconi, F. A. M. 1970. As sauvas. Agronomica Ceres, Sao Paulo. 167 pp.

Moser, J. C. 1967. Trails of the leaf-cutters. Nat. Hist. 76: 32-35.

Nogueira, S. B. Avermectin, inseticida eficiente no controle de formigas (Hymenoptera: Formicidae). Rev. Seiva, Vicosa. (In press).

Phillips, F. T., P. Etheridge, and G. C. Scott. 1976. Formulation and field evaluation of experimental baits for the control of leaf-cutting ants (Hymenoptera: Formicidae) in Brazil. Bull. Entomol. Res. 66: 579-585.

Robinson, S. W., and J. M. Cherrett. 1973. Studies on the use of leaf-cutting ant scent trail pheromones as attractants in baits. Proc. VII Congr. IUSSI. pp. 332-338.

Robinson, S. W., and J. M. Cherrett. 1974. Laboratory investigations to evaluate the possible use of brood pheromones of the leaf-cutting ant Atta cephalotes (L.) as a component in an attractive bait. Bull. Entomol. Res. 63: 519-529.

Robinson, S. W., and J. M. Cherrett. 1978. The possible use of methyl-4-methylpyrrole-2-carboxylate, an ant trail pheromone, as a component of an improved bait for leaf-cutting ants. Bull. Entomol. Res. 68: 159-170.

Robinson, S. W., A. R. Jutsum, J. M. Cherrett, and R. J. Quinlan. 1982. Field evaluation of methyl-4-methylpyrrole-2-carboxylate, an ant trail pheromone, as a component of baits for leaf-cutting ant (Hymenoptera: Formicidae) control. Bull. Entomol. Res. 72: 345-356.

Simoes, J. W., R. M. Brandi, N. B. Leite, and E. A. Balloni. 1981. Formacao, manejo e exploracao de florestas com especies de rapido crescimento. Pamphlet, Instituto Brasileiro de Desenvolvimento Florestal. Brasilia.

Vilela, E. F. 1983. Behavior and control of leaf-cutting ants (Hymenoptera: Attini). Ph.D. Thesis, University of Southampton, U.K. 209 pp.

34
Control of *Atta* and *Acromyrmex* spp. in Pine Tree Plantations in the Venezuelan Llanos

K. Jaffe

The practice of integrated pest management is essential to the future of agriculture (Metcalf and Luckmann 1975). It is defined (Geier and Clark 1961) as a consolidation of several available control techniques into a unified program to manage pest populations at a sub-economic level and to minimize adverse effects on the environment (National Academy of Science 1969). Studies in Venezuela described below are demonstrating its value in managing pest ants in pine tree plantations.

Natural wood resources are becoming increasingly scarce; therefore, artificial forest plantations are very important since they provide wood for construction, furniture, and paper industries. In Venezuela, a large forestation project was started in 1969 in the southeast (Corporacion Venezolana de Guayana and Compania Nacional de Reforstacion) to convert dry Trachypogon savannas to Pinus caribaea forests. By 1985, about 150,000 of 500,000 ha had already been planted to pines. The only pest problem reported resulted from attacks by leaf-cutting ants on the trees.

To find a solution to the problem, a 10-stage research project was designed as follows:

1. Identify pest species.
2. For each pest species, evaluate losses in wood production.
3. Study the biology and ecology of the pest species.
4. Design possible control systems.
5. Evaluate the efficiency of these control systems.
6. Introduce selected control systems on a semi-industrial scale.
7. Establish security measures for human and environmental protection.
8. Evaluate the costs of control systems.
9. Introduce the plausible control systems on an industrial scale.

10. Initiate a continuing system to evaluate pest density and control efficiency.

This paper is a progress report of the first five steps.

PEST SPECIES

Three leaf-cutting ant species were found in the study area, Acromyrmex landolti, Atta laevigata, and A. sexdens. Ac. landolti was the most abundant ant species present with densities of 340 ± 130 nests/ha (mean ± standard deviation, n=6 plots of 1 ha each) in natural savanna and 61 ± 39 nests/ha in 9-year-old pine forest. Mature A. laevigata nests were very scarce in natural savanna (0.04 ± 0.02 nests/ha) but were common in the pine forest (8.8 ± 2.4 mature nests/ha). Up to 20 nests/ha were found in heavily infested areas. A. sexdens was very rare in both savanna and pine forest but common in gallery forests in the area.

LOSSES TO WOOD PRODUCTION

Ac. landolti normally attacks grasses (Rubio and Timaure 1977) and not pine trees; however, when pine trees are first planted, the savanna is burned and cleared of vegetation. This forces surviving Ac. landolti to feed on the only vegetation available, pine trees. Thus, we studied the rate of attacks on young pine trees (20 cm in height) by starved ant colonies by isolating their nests with metal walls smeared with oil. It was found after 2 weeks that 60% of the colonies (n=12) had attacked the trees placed in this area; however, only 25% of these colonies had killed the pines after 4 weeks (Navarro 1983). A single mature colony can kill 48% of the young pines in an area of up to 2 ha by defoliation and inhibit the growth of another 20 to 40% of the trees.

Mature pines are attacked by Atta colonies but not by Acromyrmex colonies. Height and diameter measurements of Pinus caribaea trunks were made in areas where different densities of A. laevigata colonies were found but which had equivalent soil types. Mean volume of wood lost in plots with >10 nests/ha was 14% based on measurements from 200 10-year-old trees selected randomly in 5 infested and non-infested plots (p < 0.05; analysis of variance) (Naccarata 1983a, b).

BIOLOGY AND ECOLOGY OF PEST SPECIES

Much data are available on the biology of leaf-cutting ants and their ant-fungus relationship, chemical communication systems, and ecology (see other sections of these proceedings). Our research concentrated on two main subjects: The communication systems involved in agonistic behavior (see Jaffe, Chapter 18) and the population dynamics of the pest species in pine plantations.

Densities of incipient colonies of Ac. landolti and A. laevigata are very high in natural savanna but decline dramatically after it is planted to pine forest (Table 1). In contrast, the number of mature A. laevigata colonies is greater after forestation while mature Ac. landolti colonies level off to an average nest density of 45 to 61/ha.

TABLE 1. Number of nests found in one-ha plots in natural savanna and in pine forest.

Species and type of nest	Savanna before clearing	Mean no. of nests (±SD)		
		Years after forestation		
		1	2	9
A. laevigata				
mature	0.04(±0.02)	0.0	0.16(±0.40)	8.8(±2.4)
incipient	121(±219)	3.0(±2.6)	2.8(±2.9)	2.2(±2.9)
Ac. landolti				
mature	340(±130)	58(±47)	45(±34)	61(±39)
incipient	3980(±2074)	0.0	—	—

The results suggest a strong correlation between savanna habitat and nest founding by new queens. Mating flights of both species always occur at the beginning of the rainy season. Thus, at the time of evaluation in October, new colonization in the pine forest was expected to be comparable to that in the savanna since this is the month that newly founded claustral nests are opened and workers begin to forage. Surprisingly, the number of incipient colonies decreased dramatically in the forested area when compared to natural savanna.

These results cannot be explained by the elimination of sexuals inside mature colonies during the burning and clearing process, since forest plots were chosen at the border of a 100,000 ha natural savanna. In the case of the 1-year-old forest, a possible explanation is available. After burning and ploughing, it takes about 2 years for the savanna to partially recover its natural flora. Thus, many incipient colonies of Ac. landolti may starve as they have no food in 1-year-old plantations. This is not the case for incipient A. laevigata colonies that do feed on pine trees. Increased predation on founding queens in the cleared savanna could also partially explain the results in Table 1. The increase in adult A. laevigata colonies in the forest in spite of a significant decrease in colonization can be

412

explained by a greater survival of the few new nests, probably because of the increase in available biomass provided by the growing pine trees.

The data were analyzed with a Spearman multiple correlation test to determine if the decrease in the number of mature Ac. landolti nests was due only to a change in habitat or if intraspecific competition between the two attine species was important. The results are shown in Table 2. We see that mature A. laevigata colonies are negatively correlated with the incipient nests of this species but are positively correlated with Ac. landolti nests. Incipient Ac. landolti nests correlate positively with mature nests of the same species.

TABLE 2. Correlation coefficient and probability of rejection of the null hypothesis given by the Spearman multiple correlation test. Data from 12 one-ha plots in natural savanna.

	Ac. landolti incipient	A. laevigata mature	A. laevigata incipient
Ac. landolti			
mature	0.80 (p < 0.002)	NS	0.82 (p < 0.002)
incipient		ND	NS
A. laevigata			
mature			−0.41 (p < 0.048)

NS indicates no significant correlations ($\alpha = 0.10$), and ND indicates that not enough data was available.

These results suggest that A. laevigata colonies interact intraspecifically. Mature colony workers probably kill founding queens as reported for A. capiguara (Fowler et al. 1984). This does not appear to be the case with Ac. landolti. The positive correlation between incipient and adult colonies of this species suggests a preference for certain habitats similar to the situation reported by Cherrett (1968) for A. cephalotes. The possibility of execution of incipient colonies by resident Ac. landolti colonies cannot be excluded. The results also suggest an absence of interaction in nest founding and nest survival between the two genera. This may be due to poor overlap in their trophic niches. Although both species consume fresh leaves of Trachypogon spp. (the most common plant in the savanna) and Rinchelitrum roseum, only A. laevigata consumes pine needles and tree leaves.

The absence of aggressive interaction between Ac. landolti colonies could be a primitive social trait (Jaffe 1984). This species only exhibits incipient territorial behaviour (Jaffe and Navarro, in press), whereas other Atta spp. have very well-developed territorial behaviour (Jaffe et al. 1979; Jaffe 1983; Vilela 1983). Thus, intra- and interspecific competition for each of these species may be quite different.

DESIGN OF CONTROL SYSTEMS

Toxic Baits

The use of toxic baits for control of leaf-cutting ants is described by Cherrett and Vilela in Chapters 29 and 33 and will not be discussed further here.

Use of Territorial Pheromones

Territorial pheromones do not seem to be adaptable for control of leaf-cutting ants. Although intercolony conflicts are regulated partially by territorial pheromones (Jaffe et al. 1979; Vilela 1983), we rarely observed fights between colonies in the field, even when trails crossed. The only time Atta colonies were seen fighting was when they competed for an especially rare or palatable food. For example, in the pine forest fights have been observed around artificial baits of citrus pulp but not around pine trees. Ac. landolti colonies are often seen carrying presumed foreign conspecifics away from their nest entrance in a ritualized way (Jaffe and Navarro, in press), but we never observed fights.

Disruption of Nestmate Recognition Systems

This is a promising technique because the introduction into the nest of sufficient quantities of worker alarm pheromones could induce aggression to a degree that the colony would eventually eliminate itself. Intra-colony fights have been observed when excess alarm pheromone is presented to laboratory colonies (personal observation). Thus, it is possible that if components of alarm pheromones of another Atta species are introduced, a higher degree of aggression would be elicited, since interspecific competition would be induced.

Natural Enemies

Very few natural predators are known for leaf-cutting ants. Birds and anteaters eat Atta workers, but the quantities they ingest are small in relation to the work force of millions in mature colonies. Thus, natural enemies do not seem to be promising. Even among other arthropods, no reports of important Attini predators are known. A possible exception could be the driver ants (Ecitoninae); however, they are very rare in the savanna.

Natural Diseases

Few cases of Atta colonies dying of disease have been observed in the field. Of 80 A. laevigata nests examined, 2 showed signs of attack by an illness as indicated by piles of dead nestmates found at nest entrances. Founding queens may be more susceptible to diseases and predation, as a very low percentage of them survive after the nuptial flight.

Planting Techniques

This technique seems to be one of the more promising approaches to control, at least for control of the pest in young pine trees. Atta nests are scarce and are easily detected and controlled with conventional baits after the savanna is cleared of vegetation. This is not true of Ac. landolti nests. Thus, changing the planting technique in such a way that the savanna is not cleared completely, leaving part of the original vegetation, should give the ants sufficient food so as to reduce attacks on the young pine trees.

EFFICIENCY OF SOME CONTROL SYSTEMS

Toxic Baits

Citrus pulp treated with mirex (0.45%) is the most efficient bait for control of A. laevigata, as 100% of the treated colonies were controlled (n=50); i.e., they showed no activity 6 months after treatment. Aldrin on citrus pulp (0.5%) controlled 92% of the colonies (n=120). The same bait sealed into polyethylene plastic bags controlled only 60% of the treated colonies (n=80), although the bait was still active after 6 months of exposure to the climate (Naccarata and Jaffe, in press). Other insecticides and one bacterium tested on the same bait were not effective as they controlled less than 40% of the nests (Acefato®, chlordane, VC-1-13, Permetrina®, dioxathion, dieldrin, Dipel® (=Bacillus thuringiensis)).

Mechanical Control

Burning of the savanna followed by plowing 2 months later killed a mean of 51% of 581 Ac. landolti nests in 16 1-ha plots. A second plowing 2 months after the first one killed 59% of the 288 remaining nests. Most nests probably died of starvation, as practically no vegetation grew during this period between January and June, the period of lowest rainfall. Plowing 50% of the savanna in 3-m-wide fringes killed only 15% of 474 Ac. landolti nests per ha (n=16). Pine survival in these plots will be evaluated at the end of 1985.

ACTUAL SITUATION AND PERSPECTIVES

Research at the moment is focused on the distribution pattern and densities of nest colonies and on the evaluation of the control efficiency of different planting techniques. After planting, control of Atta colonies at critical moments during the productive life of the forest should be enough to maintain pest densities below the economic threshold. A certain number of leaf-cutting ant nests in the forest could even have a positive long-term effect, equivalent to that reported in the succession of pasture to forest in Paraguay (Jonkman 1977). It is also known that leaf-cutting ants have made an important contribution to the addition of nutrients into the soil (Lugo et al. 1973), although opposite effects have been observed in tropical rain forest (Haines 1985). Thus, the population of leaf-cutting ant nests which give an equilibrium between damage and benefit to the pine forest must be determined.

REFERENCES CITED

Cherrett, J. M. 1968. Some aspects of the distribution of pest species of leaf-cutting ants in the Caribbean. Proc. Am. Soc. Hort. Sci. Trop. Reg. 12: 295-310.

Fowler, H. G., S. W. Robinson, and J. Diehl. 1984. Effect of mature colony density on colonization and initial colony survivorship in Atta capiguara, a leaf-cutting ant. Biotropica 16: 51-54.

Geier, P. W., and L. R. Clark. 1961. An ecological approach to pest control. Proc. Tech. Mtg. Int. Union Conserv. Nature Nat. Res. Warsaw. 8: 10-18.

Haines, B. 1983. Leaf-cutting ants bleed mineral elements out of rainforest in southern Venezuela. Trop. Ecol. 24: 85-93.

Jaffe, K. 1983. Chemical communication among workers of leaf-cutting ants, pp. 165-180. In P. Jaisson (ed.), Social insects in the tropics, Vol. 2. Univ. Paris-Nord. 252 pp.

Jaffe, K. 1984. Negentropy and the evolution of chemical recruitment in ants. J. Theor. Biol. 106: 587-604.

Jaffe, K., and J. J. Navarro. Comunicacion quimica entre obreras de la hormiga cortadora de grama Acromyrmex landolti. Rev. Bras. Entomol. (In press).

Jaffe, K., M. Bazire-Benazet, and P. E. Howse. 1979. Territorial marking with a colony specific pheromone from an integumentary gland in leaf-cutting ants. J. Insect Physiol. 25: 833-839.

Jonkman, J. C. M. 1977. Ant nests as accelerators of succession in Paraguayan pastures, pp. 85-87. In W. J. Mattson (ed.), The role of arthropods in forest ecosystems. Springer Verlag, Berlin.

Lugo, A. E., E. G. Farnworth, D. Pool, P. Jerez, and G. Kaufman. 1973. The impact of the leaf-cutting ant Atta colombica on the energy flow of a tropical wet forest. Ecology 54: 1292-1301.

Metcalf, R., and W. Luckmann. 1975. Introduction to insect pest management. Wiley Interscience, New York. 587 pp.

Naccarata, V. 1983a. Biologia y control de Atta laevigata, en las plantaciones de pino caribe en el Estado Monagas. Tesis de Licenciatura, Universidad Simon Bolivar, Caracas, Venezuela.

Naccarata, V. 1983b. Estudio sobre Atta laevigata, plaga de las plantaciones de pino caribe al sur del Estado Monagas. Venezuela Forestal. 2: 16-39.

Naccarata, V., and K. Jaffe. Formulacion y desarrollo de un cebo atractivo toxico para control de bachacos, Atta spp. Bol. Entomol. Venezolana. (In press).

National Academy of Science (USA). 1969. Insect pest management and control. Publ. 1695, Washington, D.C. 508 pp.

Navarro, J. G. 1983. Estudio de algunos aspectos ecologicos y sistemas de comunicacion quimica en Acromyrmex landolti (Forel). Tesis de Licenciatura, Universidad Simon Bolivar, Caracas, Venezuela.

Rubio, E., and A. Timaure. 1977. Caracteristicas de los nidos de Acromyrmex landolti (Forel) en el Oeste de Venezuela. Rev. Fac. Agron. Univ. Zulia 4: 53-62.

Vilela, E. 1983. Behaviour and control of leaf-cutting ants (Hymenoptera: Attini). Ph.D. Thesis, University of Southampton, G.B.

List of Contributors

Dr. Claude T. Adams
U.S. Department of Agriculture
Agricultural Research Service
1600 S.W. 23rd Drive
Gainesville, Florida 32604

Mr. William A. Banks
U.S. Department of Agriculture
Agricultural Research Service
1600 S.W. 23rd Drive
Gainesville, Florida 32604

Dr. J. Malcolm Cherrett
The Shiel, CAER Gelach
Llandegfan, Menai Bridge
Gwynedd LL59 5UF
University College of
North Wales, United Kingdom

Mr. David M. Claborn
Department of Entomology
Texas Tech University
Lubbock, Texas 79409

Mr. James C. Cokendolpher
Department of Biological
   Science
Texas Tech University
Lubbock, Texas 79409

Dr. M. Decharme
Ministere de l'Agriculture
ENSH, Chaire de
   Phytopathologie
F-78000 Versailles, France

Dr. Jacques Delabie
INRA-CNRS
Laboratoire de Neurobiologie
Sensorielle de l'Insecte
La Guyonnerie
91440 Bures sur Yvette, France

Dr. Gerard Febvay
Ministere de l'Agriculture
INRA, Station de Zoologie et
   Lutte Biologique
F-97170 Petit-Bourg
Guadeloupe (FWI)

Dr. David J. C. Fletcher
Department of Entomology
University of Georgia
Athens, Georgia 30602

Dr. Luis Carlos Forti
Instituto Basico de Biologia
   Medica e Agricola
Universidade Estadual Paulista
18600 Botucatu
Sao Paulo, Brazil

418

Dr. Harold G. Fowler
Instituto de Biociencias
Universidade Estadual Paulista
13500 Rio Claro
Sao Paulo, Brazil

Dr. Oscar F. Francke
Department of Biological
    Science
Texas Tech University
Lubbock, Texas 79409

Dr. B. Michael Glancey
U.S. Department of Agriculture
Agricultural Research Service
1600 S.W. 23rd Drive
Gainesville, Florida 32604

Dr. Jerome J. Howard
Department of Biology
University of Iowa
Iowa City, Iowa 52242

Dr. Philip E. Howse
Department of Biology and
    Chemistry
University of Southampton
Southampton, 5095NH
United Kingdom

Dr. Klaus Jaffe
Universidad Simon Bolivar
Apartado 80659
Caracas 1080
Venezuela

Mr. Stanley R. Jones
Department of Entomology
Texas Tech University
Lubbock, Texas 79409

Dr. Donald P. Jouvenaz
U.S. Department of Agriculture
Agricultural Research Service
1600 S.W. 23rd Drive
Gainesville, Florida 32604

Dr. Alain Kermarrec
Ministere de l'Agriculture
INRA, Station de Zoologie et
    Lutte Biologique
F-97170 Petit-Bourg
Guadeloupe (FWI)

Dr. Clifford S. Lofgren
U.S. Department of Agriculture
Agricultural Research Service
1600 S.W. 23rd Drive
Gainesville, Florida 32604

Dr. Claudine Masson
INRA-CNRS
Station de Recherches
sur l'Abeille et les
    Insectes Sociaux
91440 Bures sur Yvette, France

Dr. Jeremy N. McNeil
Department of Biology
Universite Laval
Quebec, P.Q.
Canada G1K 7P4

Dr. Virgilio Pereira-da-Silva
Instituto Basico de Biologia
    Medica e Agricola
Universidade Estadual Paulista
18600 Botucatu
Sao Paulo, Brazil

Dr. Sherman A. Phillips, Jr.
Department of Entomology
Texas Tech University
Lubbock, Texas 79409

Dr. T. E. Reagan
Department of Entomology
Louisiana State University
402 Life Sciences Building
Baton Rouge, Louisiana 70803

419

Dr. Norma Bianca Saes
Instituto de Biociencias
Universidade Estadual Paulista
13500 Rio Claro
Sao Paulo, Brazil

Dr. Pierre Therrien
Institute of Animal Resource
  Ecology
University of British Columbia
Vancouver, B.C. V6T 1W5
Canada

Dr. Walter R. Tschinkel
Department of Biological
  Science
The Florida State University
Tallahassee, Florida 32306

Dr. Robert K. Vander Meer
U.S. Department of Agriculture
Agricultural Research Service
1600 S.W. 23rd Drive
Gainesville, Florida 32604

Dr. Evaldo F. Vilela
Univ. of Vicosa
Department of Entomology
Vicosa Mg 36570, Brazil

Dr. S. Bradleigh Vinson
Department of Entomology
Texas A&M University
College Station, Texas 77843

Dr. Deborah A. Waller
Department of Entomology
Louisiana State University
Baton Rouge, Louisiana 70803

Dr. W. G. Wellington
Institute of Animal Resource
  Ecology
University of British Columbia
Vancouver, B.C. V6T 1W5
Canada

Dr. David F. Wiemer
Department of Chemistry
University of Iowa
Iowa City, Iowa 52242

Dr. David F. Williams
U.S. Department of Agriculture
Agricultural Research Service
1600 S.W. 23rd Drive
Gainesville, Florida 32604

Dr. Edward O. Wilson
Museum of Comparative
  Zoology
Harvard University
Cambridge, Massachusetts
  02138

Dr. Daniel P. Wojcik
U.S. Department of Agriculture
Agricultural Research Service
1600 S.W. 23rd Drive
Gainesville, Florida 32604

Taxonomic Index

Subject Index

To assist you in using this index, we separated the page references for fire ants (FA) and leaf-cutting ants (LCA) unless the ant group to which the reference applies is either obvious or non-specific.

glucans; LCA: 275,282
glycogen; LCA: 274,284
by larvae; FA: 294;
LCA: 275,278-280,
282
lipids; FA: 295,296;
LCA: 274,277
nitrogen metabolism; LCA:
277
plant sap; FA: 50,54;
LCA: 248,275
polyscharrides; LCA: 274,
279-284
protein; FA: 295,296,383;
LCA: 274,275,282
sterols; LCA: 274
triglycerides; LCA: 277
trophallaxis; LCA: 277
Digestive tract anatomy; LCA:
275-277,283
Diseases; FA: 96,97,327,333;
LCA: 414 (see also
natural enemies,
parasites, pathogens)

Ecdysis; FA: 308,309
Ecological techniques; FA:
62,65,94,116; LCA:
147,172,176-180
Ecology, FA; 43,54,58,61,72-
86,88-99; LCA: 18,27,
31,172-181,262,263
compared to weeds; FA: 54,
72-76,86
effect of rainfall on; LCA:
173
Economic damage; FA: 37,43,
53,54,58,399; LCA: 1,
11,14,20,21,23-27,
260,406,407
animal injury; LCA: 28
beans; FA: 48
benefical insects; FA: 62,65
cabbage; FA: 48
cattle production; LCA: 25-
27,29
citrus; FA: 49-51,54

corn; FA: 48,49
damage to roads; FA: 52,
LCA: 29
decreased soil fertility;
LCA: 29
eggplant; FA: 49
eucalyptus; LCA: 401-407
forestry; FA: 52
machinery damage; FA: 49,
52,53; LCA: 28
okra; FA: 49
pastures; LCA: 23-30
peanuts; FA: 48
pine trees; LCA: 409-415
potatoes; FA: 48-50,54
scale and aphid association;
FA: 50,52,59,93
soybean; FA: 49,54
sugarcane; LCA: 25,27
weed increases; LCA: 28
wildlife; FA: 53
Ectothermy; FA: 109
Egg production; FA: 290;
LCA: 126,127
Electroantennagrams; LCA:
303,305,310,340,345,
347,348
Endocrinology; FA: 297
Endothermy; FA: 109
Environmental Protection
Agency; FA: 42,397
Eradication program; FA: 37,
40-42
Evolution, ants; FA: 3-5,8,
319; LCA: 5-8,10,14
ant-fungus; LCA: 256
Execution pheromone; FA: 291

Feeding deterrents; LCA: 248,
260,263-270
chemistry; LCA: 250-254,
263,264,268-270
isolation; LCA: 250-254,
266,267
Feeding
repellents; LCA: 248
stimulants; LCA: 260

9780367158279